MONOGRAPHS AND RESEARCH NOTES IN MATHEMATICS

Iterative Methods without Inversion

Anatoly Galperin

CRC Press
Taylor & Francis Group
Boca Raton London New York

CRC Press is an imprint of the
Taylor & Francis Group, an **informa** business
A CHAPMAN & HALL BOOK

MONOGRAPHS AND RESEARCH NOTES IN MATHEMATICS

Published Titles

Actions and Invariants of Algebraic Groups, Second Edition, Walter Ferrer Santos and Alvaro Rittatore

Analytical Methods for Kolmogorov Equations, Second Edition, Luca Lorenzi

Application of Fuzzy Logic to Social Choice Theory, John N. Mordeson, Davender S. Malik and Terry D. Clark

Blow-up Patterns for Higher-Order: Nonlinear Parabolic, Hyperbolic Dispersion and Schrödinger Equations, Victor A. Galaktionov, Enzo L. Mitidieri, and Stanislav Pohozaev

Complex Analysis: Conformal Inequalities and the Bieberbach Conjecture, Prem K. Kythe

Computational Aspects of Polynomial Identities: Volume I, Kemer's Theorems, 2nd Edition Alexei Kanel-Belov, Yakov Karasik, and Louis Halle Rowen

A Concise Introduction to Geometric Numerical Integration, Fernando Casas and Sergio Blanes

Cremona Groups and Icosahedron, Ivan Cheltsov and Constantin Shramov

Delay Differential Evolutions Subjected to Nonlocal Initial Conditions Monica-Dana Burlică, Mihai Necula, Daniela Roșu, and Ioan I. Vrabie

Diagram Genus, Generators, and Applications, Alexander Stoimenow

Difference Equations: Theory, Applications and Advanced Topics, Third Edition Ronald E. Mickens

Dictionary of Inequalities, Second Edition, Peter Bullen

Finite Element Methods for Eigenvalue Problems, Jiguang Sun and Aihui Zhou

Introduction to Abelian Model Structures and Gorenstein Homological Dimensions Marco A. Pérez

Iterative Methods without Inversion, Anatoly Galperin

Iterative Optimization in Inverse Problems, Charles L. Byrne

Line Integral Methods for Conservative Problems, Luigi Brugnano and Felice Iavernaro

Lineability: The Search for Linearity in Mathematics, Richard M. Aron, Luis Bernal González, Daniel M. Pellegrino, and Juan B. Seoane Sepúlveda

Modeling and Inverse Problems in the Presence of Uncertainty, H. T. Banks, Shuhua Hu, and W. Clayton Thompson

Monomial Algebras, Second Edition, Rafael H. Villarreal

Nonlinear Functional Analysis in Banach Spaces and Banach Algebras: Fixed Point Theory Under Weak Topology for Nonlinear Operators and Block Operator Matrices with Applications, Aref Jeribi and Bilel Krichen

Published Titles Continued

Partial Differential Equations with Variable Exponents: Variational Methods and Qualitative Analysis, Vicenţiu D. Rădulescu and Dušan D. Repovš

A Practical Guide to Geometric Regulation for Distributed Parameter Systems Eugenio Aulisa and David Gilliam

Reconstruction from Integral Data, Victor Palamodov

Signal Processing: A Mathematical Approach, Second Edition, Charles L. Byrne

Sinusoids: Theory and Technological Applications, Prem K. Kythe

Special Integrals of Gradshteyn and Ryzhik: the Proofs – Volume I, Victor H. Moll

Special Integrals of Gradshteyn and Ryzhik: the Proofs – Volume II, Victor H. Moll

Stochastic Cauchy Problems in Infinite Dimensions: Generalized and Regularized Solutions, Irina V. Melnikova

Submanifolds and Holonomy, Second Edition, Jürgen Berndt, Sergio Console, and Carlos Enrique Olmos

The Truth Value Algebra of Type-2 Fuzzy Sets: Order Convolutions of Functions on the Unit Interval, John Harding, Carol Walker, and Elbert Walker

Forthcoming Titles

Geometric Modeling and Mesh Generation from Scanned Images, Yongjie Zhang

Groups, Designs, and Linear Algebra, Donald L. Kreher

Handbook of the Tutte Polynomial, Joanna Anthony Ellis-Monaghan and Iain Moffat

Microlocal Analysis on R^n and on NonCompact Manifolds, Sandro Coriasco

Practical Guide to Geometric Regulation for Distributed Parameter Systems, Eugenio Aulisa and David S. Gilliam

Symmetry and Quantum Mechanics, Scott Corry

CRC Press
Taylor & Francis Group
6000 Broken Sound Parkway NW, Suite 300
Boca Raton, FL 33487-2742

Printed on acid-free paper
Version Date: 20160613

International Standard Book Number-13: 978-1-4987-5892-5 (Hardback)

Library of Congress Cataloging-in-Publication Data

Names: Galperin, Anatoly.
Title: Iterative methods without inversion / Anatoly Galperin.
Description: Boca Raton, FL : CRC Press, [2016] | Series: Monographs and research notes in mathematics | Includes bibliographical references and index.
Identifiers: LCCN 2016025099| ISBN 9781498758925 (hardback : alk. paper) | ISBN 9781498758963 (e-book)
Subjects: LCSH: Iterative methods (Mathematics) | Numerical analysis. | Banach spaces. | Hilbert space.
Classification: LCC QA297.8 .G35 2016 | DDC 518/.26--dc23
LC record available at https://lccn.loc.gov/2016025099

Visit the Taylor & Francis Web site at
http://www.taylorandfrancis.com

and the CRC Press Web site at
http://www.crcpress.com

Printed and bound in the United States of America by Publishers Graphics, LLC on sustainably sourced paper.

Contents

INTRODUCTION

This book is about iterative methods for solving nonlinear operator equations

$$\mathbf{f}(x) = 0 \ , \ \mathbf{f} : \mathbb{X} \supset D \to \mathbb{Y} \ , \tag{0.1}$$

in Banach or Hilbert spaces. The most widely known (and used) examples of such methods are Newton's method

$$x_+ := x - \mathbf{f}'(x)^{-1}\mathbf{f}(x) \tag{0.2}$$

and the secant method

$$x_+ := x - [x, x_- \,|\, \mathbf{f}]^{-1}\mathbf{f}(x) \ , \tag{0.3}$$

where the symbol $[x, x_- \,|\, \mathbf{f}]$ stands for so called divided difference operator (the reader will find the formal definition of this notion in Chapter 3). However, these methods and their numerous variants and derivates are left out of this book because they require inversion of a linear operator at each iteration or at least solution of a linear operator equation. We are exploring only those methods whose implementation does not involve inversions. One of the first such methods was proposed by J. Moser [39, 40] in the 1960s. Given a starting pair $(x_0, \mathbf{A}_0)$, $x_0 \in D$, $\mathbf{A}_0 \in \mathcal{L}(\mathbb{Y}, \mathbb{X})$ (the space of bounded linear operators acting from $\mathbb{Y}$ to $\mathbb{X}$), Moser constructed successive iterations $(x_n, \mathbf{A}_n)$ according to the following rule:

$$x_+ := x - \mathbf{A}\mathbf{f}(x) \ , \ \mathbf{A}_+ := 2\mathbf{A} - \mathbf{A}\mathbf{f}'(x)\mathbf{A} \ .$$

Ulm [58] improved Moser's method replacing $\mathbf{f}'(x)$ with $\mathbf{f}'(x_+)$. Detailed analysis of Ulm's method is the subject of Chapters 2 and 3.

Another example of inversion-free methods is Broyden's method [4]

$$x_+ := x - \mathbf{A}\mathbf{f}(x) \ , \ \mathbf{A}_+ := \mathbf{A} - \frac{\mathbf{A}\mathbf{f}(x_+)}{\langle \mathbf{A}^*\mathbf{A}\mathbf{f}(x) \, , \mathbf{f}(x_+) - \mathbf{f}(x) \rangle} \langle \mathbf{A}^*\mathbf{A}\mathbf{f}(x) \, , \cdot \rangle \ , \tag{0.4}$$

where $\langle a \, , b \rangle$ is the inner product of two vectors of underlying Hilbert space. In addition to being inversion-free, it is also derivative-free, which makes it suitable for solving equations with nondifferentiable operators. This method is studied in Chapter 4.

Broyden's method represents the class of iterative methods sometimes called in literature the secant-update methods. These methods generate the iterations $(x_n, \mathbf{A}_n)$ according to the rule

$$x_+ := x - \mathbf{A}\mathbf{f}(x) \ , \ \mathbf{A}_+ := \mathbf{A} + \mathbf{B} \ ,$$

where the update $\mathbf{B}$ is chosen so that the updated operator $\mathbf{A}_+$ satisfies so-called secant equation

$$\mathbf{A}_+(x_+ - x) = \mathbf{f}(x_+) - \mathbf{f}(x)\,. \tag{0.5}$$

Usually, the update $\mathbf{B}$ is a linear operator of low rank (most often 1 or 2), that has the form $ul(\cdot)$ or $u_1l_1(\cdot) + u_2l_2(\cdot)$, where u, u_1, u_2 are vectors and l, l_1, l_2 are linear functionals. The most widely known example of a secant-update of rank 2 is BFGS update

$$s := x_+ - x\,,\; y := \mathbf{f}(x_+) - \mathbf{f}(x)\,,\; \mathbf{A}_+^{-1} - \mathbf{A}^{-1} = \frac{y}{\langle y\,,s\rangle}\langle y\,,\cdot\rangle - \frac{\mathbf{A}^{-1}s}{\langle \mathbf{A}^{-1}s\,,s\rangle}\langle \mathbf{A}^{-1}s\,,\cdot\rangle\,, \tag{0.6}$$

proposed independently by Broyden [5], Fletcher [10], Goldfarb [25], and Shanno [53] in 1970 for finite-dimensional unconstrained minimization.

The secant equation (0.5) admits a great variety of solutions $\mathbf{A}_+$ for given vectors $x_+ - x \in \mathbb{X}$ and $\mathbf{f}(x_+) - \mathbf{f}(x) \in \mathbb{Y}$. This fact inevitably provokes the question: which one of all solutions (and the corresponding secant-update methods) is more preferable? The answer to this question depends on a criterion enabling one to compare any two given methods and to decide which one is better than the other. As such a criterion, we use the entropy of a solution's position within a set of its guaranteed existence and uniqueness. The existence of such a set is established by a theorem proved in [21] and recalled in Chapter 5. The notion of entropy is basic in the theory of information. It is used to measure the degree of uncertainty of random events in physical systems, given some (usually incomplete) information about the current system's status. The goal of any iterative method designed to solve some problem is to reduce the uncertainty in a solution's whereabouts using the information obtained at one iteration. So, it is quite natural to borrow the notion of entropy to measure the efficiency of iterative methods. A particular representative of a class of methods for solving operator equations is optimal if its iteration reduces the uncertainty of a solution's position (measured by the entropy) as much as possible for the methods of this class. This optimality criterion was introduced in [21] and used in [22] to determine the most efficient secant-type methods. In Chapter 5, the entropy criterion is applied to characterize optimal secant-updates of rank 2. As it turns out, there are many such updates. So, one can try to optimize his choice for $\mathbf{A}_+$ in (0.5) further. For example, he may wish to get $\mathbf{A}_+$ with the least condition number possible. We show in Chapter 5 that, in the case of secant-updates of rank 1, this problem has a nice analytical solution, which leads to a new iterative method.

The next logical step is to analyze the more general class of secant-type iterative methods,

$$x_+ := x - F\big(x\,,\mathbf{f}(x)\big)\,,$$

which (like the generic secant method (0.3)) require only one evaluation of the operator $\mathbf{f}$ (no derivatives) per iteration. Ulm's and Broyden's methods can be viewed as representatives of the methods of this class. Its members differ from each other by the mapping F used to generate the next approximation x_+ from the current iteration $\big(x\,,\mathbf{f}(x)\big)$. Hence the inevitable question: which F is the best? In Chapter 6, we try to shed some light on this question in a one-dimensional setting, using the same entropy optimality criterion as in Chapter 5.

Prerequisites for reading the book are very modest. It is readable for anyone with minimal exposure to nonlinear functional analysis. In fact, one feeling comfortable with the notions of continuity and differentiability of nonlinear operators acting between Banach and Hilbert spaces should have no difficulty in understanding developments in the book.

The book is addressed first of all to graduate students and young researchers beginning their career in the field of computational mathematics. In their interest I suggest several research projects which, in my opinion, are important for further improvement of the methods discussed in the book. However, a practitioner also may find something of interest in parts of the book dealing with examples of the application of those methods to the numerical solution of various infinite-dimensional problems. Having in mind the needs of this group of potential readers, I make space for details of computer implementation. My hope is that even an experienced reader will find some fresh ideas not found elsewhere. I would mention among them regular continuity, the use of invariants of difference equations in convergence analyses, and the entropy optimality criterion for iterative methods.

Most (but not all) of the material included in the book has appeared in journal articles [12]–[24]. However, the process of writing the book has involved revision of ideas, refinement of some proofs, adding new applications, putting forward conjectures, and suggesting research directions. The result is that the exposition in some places deviates considerably from what can be found in those articles.

Chapter 1

Some useful tools of the trade

This introductory chapter gives some known facts that will be needed later. The reader can skip it on the first reading and return to it only after being prompted by a reference.

1.1 Banach's lemma on perturbations

Lemma 1.1. 1° *A linear bounded operator* $\mathbf{A}$ *acting from one Banach space* $\mathbb{X}$ *into another* $\mathbb{Y}$ *is boundedly invertible if and only if*

$$l(\mathbf{A}) := \min_{\|x\|=1} \|\mathbf{A}x\| > 0\,. \tag{1.1}$$

In this case,

$$\left\|\mathbf{A}^{-1}\right\| = \frac{1}{l(\mathbf{A})}\,. \tag{1.2}$$

2° *For every two linear operators* $\mathbf{A}$ *and* $\mathbf{B}$,

$$\left|l(\mathbf{A}) - l(\mathbf{B})\right| \leq \|\mathbf{A} - \mathbf{B}\|\,.$$

Proof. 1° Let $\mathbf{A}$ be boundedly invertible. If $l(\mathbf{A}) = 0$, then there exists a sequence x_n with $\|x_n\| = 1$ & $\|\mathbf{A}x_n\| < 1/n$, so that

$$\left\|\mathbf{A}^{-1}\right\| := \sup_{\|y\|=1} \left\|\mathbf{A}^{-1}y\right\| \geq \left\|\mathbf{A}^{-1}\frac{\mathbf{A}x_n}{\|\mathbf{A}x_n\|}\right\| = \frac{1}{\|\mathbf{A}x_n\|} > n\,.$$

It follows that $\left\|\mathbf{A}^{-1}\right\| = \infty$, contrary to the hypothesis. Hence, invertibility of $\mathbf{A}$ implies (1.1). Conversely, (1.1) ensures bounded invertibility and (1.2). Indeed, if $\mathbf{A}$ is not invertible, then there can be found a nonzero $x_0 \in \mathbb{X}$ with $\mathbf{A}x_0 = 0$, so that $0 \leq l(\mathbf{A}) \leq \left\|\mathbf{A}\frac{x_0}{\|x_0\|}\right\| = 0$, which contradicts the hypothesis $l(\mathbf{A}) > 0$. Besides, by the definition of $l(\mathbf{A})$, there exists a sequence x_n with $\|x_n\| = 1$ & $\lim \|\mathbf{A}x_n\| = l(\mathbf{A})$. For this sequence, $1 = \left\|\mathbf{A}^{-1}\mathbf{A}x_n\right\| \leq \left\|\mathbf{A}^{-1}\right\| \cdot \|\mathbf{A}x_n\| \to l(\mathbf{A}) \left\|\mathbf{A}^{-1}\right\|$, whence $l(\mathbf{A}) \geq \left\|\mathbf{A}^{-1}\right\|^{-1}$. On the other hand, by the definition of $\|\mathbf{A}^{-1}\|$, $\exists\{y_n\}\subset\mathbb{Y}$ with $\|y_n\|=1$ & $\lim\|\mathbf{A}^{-1}y_n\|=\|\mathbf{A}^{-1}\|$, so that

$$1=\|\mathbf{A}\mathbf{A}^{-1}y_n\|=\left\|\mathbf{A}\frac{\mathbf{A}^{-1}y_n}{\|\mathbf{A}^{-1}y_n\|}\right\|\cdot\|\mathbf{A}^{-1}y_n\|\geq l(\mathbf{A})\,\|\mathbf{A}^{-1}y_n\|\to l(\mathbf{A})\,\|\mathbf{A}^{-1}\|$$

and $l(\mathbf{A})\leq\|\mathbf{A}^{-1}\|^{-1}$.

2° $\forall x\in\mathbb{X}$ with $\|x\|=1$,

$$\|\mathbf{B}x\|=\|\mathbf{A}x+(\mathbf{B}-\mathbf{A})x\|\geq\|\mathbf{A}x\|-\|(\mathbf{B}-\mathbf{A})x\|\geq l(\mathbf{A})-\|\mathbf{B}-\mathbf{A}\|.$$

So, $l(\mathbf{B})\geq l(\mathbf{A})-\|\mathbf{B}-\mathbf{A}\|$. By the same reason, $l(\mathbf{A})\geq l(\mathbf{B})-\|\mathbf{A}-\mathbf{B}\|$. Therefore, $|l(\mathbf{A})-l(\mathbf{B})|\leq\|\mathbf{A}-\mathbf{B}\|$. □

Corollary 1.2. *If* $\mathbf{A}$ *is boundedly invertible and* $\|\mathbf{B}-\mathbf{A}\|\leq\|\mathbf{A}^{-1}\|^{-1}$, *then* $\mathbf{B}$ *is boundedly invertible too,*

$$\|\mathbf{B}^{-1}\|\leq\frac{\|\mathbf{A}^{-1}\|}{1-\|\mathbf{A}^{-1}\|\cdot\|\mathbf{B}-\mathbf{A}\|},$$

and

$$\|\mathbf{B}^{-1}-\mathbf{A}^{-1}\|\leq\frac{\|\mathbf{A}^{-1}\|^2\,\|\mathbf{B}-\mathbf{A}\|}{1-\|\mathbf{A}^{-1}\|\cdot\|\mathbf{B}-\mathbf{A}\|}.$$

Proof. By the lemma, $\|\mathbf{B}^{-1}\|^{-1}\geq\|\mathbf{A}^{-1}\|^{-1}-\|\mathbf{B}-\mathbf{A}\|$, and, consequently,

$$\|\mathbf{B}^{-1}\|\leq\frac{1}{\|\mathbf{A}^{-1}\|^{-1}-\|\mathbf{B}-\mathbf{A}\|}=\frac{\|\mathbf{A}^{-1}\|}{1-\|\mathbf{A}^{-1}\|\cdot\|\mathbf{B}-\mathbf{A}\|}.$$

As $\|\mathbf{B}^{-1}-\mathbf{A}^{-1}\|=\|\mathbf{A}^{-1}(\mathbf{A}-\mathbf{B})\mathbf{B}^{-1}\|\leq\|\mathbf{A}^{-1}\|\cdot\|\mathbf{B}^{-1}\|\cdot\|\mathbf{B}-\mathbf{A}\|$, it follows that

$$\|\mathbf{B}^{-1}-\mathbf{A}^{-1}\|\leq\frac{\|\mathbf{A}^{-1}\|^2\,\|\mathbf{B}-\mathbf{A}\|}{1-\|\mathbf{A}^{-1}\|\cdot\|\mathbf{B}-\mathbf{A}\|}.$$

□

1.2 Sherman–Morrison formula

Lemma 1.3. *Let* $\mathbb{H}$ *and* $\mathbf{A}$ *be a Hilbert space with the inner product* $\langle\cdot\,,\cdot\rangle$ *and a linear invertible operator on* $\mathbb{H}$. *For any two vectors* *u and v of $\mathbb{H}$, the operator $\mathbf{A}+u\langle v\,,\cdot\rangle$ is invertible if and only if*

$$\langle \mathbf{A}^{-1}u\,,v\rangle \neq -1\,,$$

in which case

$$(\mathbf{A}+u\langle v\,,\cdot\rangle)^{-1} = \mathbf{A}^{-1} - \frac{\mathbf{A}^{-1}u}{1+\langle \mathbf{A}^{-1}u\,,v\rangle}\langle (\mathbf{A}^{-1})^{*}\,v\,,\cdot\rangle\,.$$

($\mathbf{T}^*$ denotes the adjoint of $\mathbf{T}$.)

Proof. The proof is by direct verification of the equalities

$$(\mathbf{A}+u\langle v\,,\cdot\rangle)\left(\mathbf{A}^{-1} - \frac{\mathbf{A}^{-1}u}{1+\langle \mathbf{A}^{-1}u\,,v\rangle}\langle (\mathbf{A}^{-1})^{*}\,v\,,\cdot\rangle\right) = \mathbf{I}$$

and

$$\left(\mathbf{A}^{-1} - \frac{\mathbf{A}^{-1}u}{1+\langle \mathbf{A}^{-1}u\,,v\rangle}\langle (\mathbf{A}^{-1})^{*}\,v\,,\cdot\rangle\right)(\mathbf{A}+u\langle v\,,\cdot\rangle) = \mathbf{I}\,.$$

□

1.3 Lemma on sections

Lemma 1.4. *Let Z and f be a set of pairs (x,y) of arbitrary kind and a function defined on Z. Denote*

$$Y(x) := \{y \mid (x,y)\in Z\}\,,\ X := \{x \mid Y(x)\neq\emptyset\}\,.$$

Then 1° $Z\neq\emptyset \iff X\neq\emptyset$.

2° $\inf\limits_{Z} f(x,y) = \inf\limits_{x\in X}\ \inf\limits_{y\in Y(x)} f(x,y)$.

3° $\sup\limits_{Z} f(x,y) = \sup\limits_{x\in X}\ \sup\limits_{y\in Y(x)} f(x,y)$.

Proof. 1°

$$\begin{aligned} Z\neq\emptyset &\Longrightarrow \exists\,(x_0,y_0)\in Z \Longrightarrow y_0\in Y(x_0) \Longrightarrow Y(x_0)\neq\emptyset \Longrightarrow x_0\in X\\ &\Longrightarrow X\neq\emptyset \Longrightarrow \exists\,x_* \text{ with } Y(x_*)\neq\emptyset \Longrightarrow \exists\,y_*\in Y(x_*)\\ &\Longrightarrow (x_*,y_*)\in Z \Longrightarrow Z\neq\emptyset\,. \end{aligned}$$

2° By definition of $\inf\limits_{Z} f(x,y)$, $\forall\varepsilon>0\ \exists\,(x_\varepsilon,y_\varepsilon)\in Z$ with $f(x_\varepsilon,y_\varepsilon) < \inf\limits_{Z} f(x,y)+\varepsilon$. Since $y_\varepsilon \in Y(x_\varepsilon)$, $f(x_\varepsilon,y_\varepsilon) \geq \inf\limits_{y\in Y(x_\varepsilon)} f(x_\varepsilon,y)$. So,

$\inf_{y\in Y(x_\varepsilon)} f(x_\varepsilon, y) < \inf_Z f(x,y)+\varepsilon$. Forcing ε to zero results in $\inf_{y\in Y(x_\varepsilon)} f(x_\varepsilon, y) \le \inf_Z f(x,y)$. As $x_\varepsilon \in X$,

$$\inf_{x\in X} \inf_{y\in Y(x)} f(x,y) \le \inf_{y\in Y(x_\varepsilon)} f(x_\varepsilon, y) \le \inf_Z f(x,y). \tag{1.3}$$

On the other hand, $\forall \varepsilon > 0 \; \exists x_\varepsilon \in X$ with $\inf_{y\in Y(x_\varepsilon)} f(x_\varepsilon, y) < \inf_{x\in X} \inf_{y\in Y(x)} f(x,y) + \varepsilon$. For all $y \in Y(x_\varepsilon)$ $(x_\varepsilon\, , y) \in Z$ and so $f(x_\varepsilon\, , y) \ge \inf_Z f(x,y)$. Then $\inf_{y\in Y(x_\varepsilon)} f(x_\varepsilon\, , y) \ge \inf_Z f(x,y)$ too. It follows that $\inf_Z f(x,y) < \inf_{x\in X} \inf_{y\in Y(x)} f(x,y) + \varepsilon$ and, consequently, $\inf_Z f(x,y) \le \inf_{x\in X} \inf_{y\in Y(x)} f(x,y)$. Together with (1.3) this gives the claim.

3° is proved similarly. □

If, in particular, f does not depend on y, then the objective in the interior extremum is constant, so that $\inf_{x\in X} \inf_{y\in Y(x)} f(x) = \inf_{x\in X} f(x)$ and $\sup_{x\in X} \sup_{y\in Y(x)} f(x) = \sup_{x\in X} f(x)$.

1.4 Entropy

Entropy is a basic notion of information theory [59] used for measuring of uncertainty of random events in physical systems. It was introduced by C. Shennon in his seminal paper [54]. If a system X can be in a finite number of states $x_1, \dots, x_n$ with probabilities (respectively) $p_1, \dots, p_n$, then the uncertainty of the current status of the system is measured by its entropy

$$H(X) := -\sum_{i=1}^{n} p_i \log p_i \ , \tag{1.4}$$

where the logarithm can be taken on any base (most often 2,e, or 10): a change of a base results in multiplying the entropy by a positive constant. If all states are equiprobable: $p_1 = \dots = p_n = 1/n$, then $H(X) = \log n$. For systems whose possible states are continuously distributed with probability density $f(x)$, the entropy is defined analogously to (1.4) as

$$H(X) := -\int_{-\infty}^{\infty} f(x) \log f(x)\, dx \ .$$

In particular, if the states of the system are distributed uniformly on a segment $[a, b]$ (so that $f(x) = 1/(b-a)$ for $x \in [a, b]$ and zero for others), then

$$H(X) = -\int_a^b \frac{1}{b-a} \log \frac{1}{b-a}\, dx = \log(b-a) \ ,$$

the logarithm of the size of the segment. Because the logarithm is an increasing function, the size itself can be used as another measure of uncertainty.

1.5 Generalized inversions in Hilbert spaces

It was shown by Penrose [44] that for every (real or complex) matrix A there exists a unique matrix $A^\dagger$ (Moore–Penrose pseudoinverse of A) such that

$$AA^\dagger A = A \ \ \& \ \ A^\dagger AA^\dagger = A^\dagger \ \ \& \ \ \left(AA^\dagger\right)^* = AA^\dagger \ \ \& \ \ \left(A^\dagger A\right)^* = A^\dagger A\,, \quad (1.5)$$

where the asterisk denotes the Hermite conjugation. Penrose [45] has shown also that

(a) for any matrix B (of compatible dimensions), the matrix $A^\dagger B$ is the unique minimizer of the Frobenius norm $\|AX - B\|_F$ of minimal Frobenius norm $\|X\|_F$ and

(b) $A^\dagger$ is the unique minimizer of $\min\limits_X\{\|X\|_F \mid AXA = A\}$.

The following properties of $A^\dagger$ are easily verifiable consequences of (1.5):

(i) if $rank(A)$ is equal to the number of rows of A (the number of columns), then

$$A^\dagger = A^*\left(A^*A\right)^{-1} \ \ \left(A^\dagger = \left(A^*A\right)^{-1}A^*\right),$$

(ii) $(A^\dagger)^\dagger = A$,

(iii) $(A^*)^\dagger = (A^\dagger)^*$,

(iv) $\left(AA^*\right)^\dagger = \left(A^\dagger\right)^* A^\dagger, \ \ \left(A^*A\right)^\dagger = A^\dagger\left(A^*\right)^\dagger$.

Penrose's result does not hold in general for linear operators between Hilbert spaces. We can have its analog only for special classes of operators. Namely, according to one of the corollaries of Theorem 5.1 in [41], the system of operator equations

$$\mathbf{AXA} = \mathbf{A} \ \ \& \ \ \mathbf{XAX} = \mathbf{X} \ \ \& \ \ (\mathbf{AX})^* = \mathbf{AX} \ \ \& \ \ (\mathbf{XA})^* = \mathbf{XA}\,. \quad (1.6)$$

is uniquely solvable for $\mathbf{X}$, if $\mathbf{A}$ is bounded and has closed range. Here $\mathbf{A}^*$ denotes the adjoint of $\mathbf{A}$. The solution is called in citeNashed the *orthogonal generalized inverse* of $\mathbf{A}$ and denoted $\mathbf{A}^\dagger$. For example, if $\mathbf{A}$ is an orthoprojector: $\mathbf{A}^2 = \mathbf{A}^* = \mathbf{A}$, then $\mathbf{A}^\dagger = \mathbf{A}$. As another example, consider the (rank n) operator

$$\mathbf{T}_n : \mathbb{H} \to \mathbb{E}^n \ , \ \ \mathbf{T}_n x := [\langle e_1\,, x\rangle\,, \ldots\,, \langle e_n\,, x\rangle]\ ,$$

which acts from a Hilbert space $\mathbb{H}$ into the Euclidean space $\mathbb{E}^n$. Here $e_1\,, \ldots\,, e_n$ are orthonormal vectors in $\mathbb{H}$.

Lemma 1.5. 1° $\|\mathbf{T}_n\| = 1$.

2° *The operator*

$$\mathbf{T}_n^\dagger : \mathbb{E}^n \to \mathbb{H} \ , \ \mathbf{T}_n^\dagger y := \sum_{i=1}^n y_i e_i \ , \ \forall\, y = [y_1, \ldots, y_n] \in \mathbb{E}^n,$$

is the orthogonal generalized inverse of $\mathbf{T}_n$.

3° $\|\mathbf{T}_n^\dagger\| = 1$.

4° $\overset{n}{\underset{i=1}{\&}} \langle e_i\,, x\rangle = \alpha_i \Longleftrightarrow x \in \sum_{i=1}^n \alpha_i e_i + \left(\mathbf{I} - \mathbf{T}_n^\dagger \mathbf{T}_n\right)\mathbb{H}$ *and* $\sum_{i=1}^n \alpha_i e_i$ *is the minimum norm solution of the system* $\overset{n}{\underset{i=1}{\&}} \langle e_i, x\rangle = \alpha_i$.

5° *The system*

$$\overset{n}{\underset{i=1}{\&}} \langle e_i, x\rangle = \alpha_i \ \ \& \ \ \|x\| = \alpha_0 \tag{1.7}$$

is solvable if and only if $\alpha_0 \geq 0 \ \& \ \sum_{i=1}^n \alpha_i^2 \leq \alpha_0^2$. *In this case, it is equivalent to*

$$x = \sum_{i=1}^n \alpha_i e_i + z \ \ \& \ \ z \in \left(\mathbf{I} - \mathbf{T}_n^\dagger \mathbf{T}_n\right)\mathbb{H} \ \ \& \ \ \|z\|^2 = \alpha_0^2 - \sum_{i=1}^n \alpha_i^2 .$$

Proof. 1° As $x = \sum_{i=1}^n \langle e_i, x\rangle e_i + x'$ and $x' = x - \sum_{i=1}^n \langle e_i, x\rangle e_i$ is orthogonal to all e_i, it is clear that $\|x\| \leq 1 \Longrightarrow \sum_{i=1}^n \langle e_i, x\rangle^2 \leq 1$. So,

$$\|\mathbf{T}_n\|^2 := \max_{\|x\|\leq 1} \|\mathbf{T}_n x\|^2 = \max_{\|x\|\leq 1} \|[\langle e_1, x\rangle, \ldots, \langle e_n, x\rangle]\|^2 = \max_{\|x\|\leq 1} \sum_{i=1}^n \langle e_i, x\rangle^2 \leq 1 .$$

On the other hand, $\max_{\|x\|\leq 1} \sum_{i=1}^n \langle e_i, x\rangle^2 \geq \sum_{i=1}^n \langle e_i, e_1\rangle^2 = 1$.

2° By Theorem 5.1 in [41], it is enough to verify that $\mathbf{T}_n^\dagger$ satisfies the system

$$\mathbf{T}_n \mathbf{X} = \mathbf{Q} \ \ \& \ \ \mathbf{X}\mathbf{T}_n = \mathbf{I} - \mathbf{P} \ \ \& \ \ \mathbf{X}\mathbf{T}_n\mathbf{X} = \mathbf{X}\ ,$$

where $\mathbf{P}$ and $\mathbf{Q}$ are the orthogonal projectors onto the null space $\mathcal{N}(\mathbf{T}_n)$ and the range $\mathcal{R}(\mathbf{T}_n)$ of $\mathbf{T}_n$, respectively:

$$\mathbf{P} : \mathbb{H} \to \mathbb{H}, \ \ \mathbf{P}x = x - \sum_{i=1}^n \langle e_i, x\rangle e_i, \quad \mathbf{Q} : \mathbb{E}^n \to \mathbb{E}^n, \ \ \mathbf{Q}y = y,$$

(i.e. $\mathbf{Q} = \mathbf{I}$, the identity operator on $\mathbb{E}^n$). Indeed, $\forall\, y \in \mathbb{E}^n$,

$$\mathbf{T}_n \mathbf{T}_n^\dagger y = \mathbf{T}_n \left(\sum_1^n y_i e_i\right) = \sum_1^n y_i \mathbf{T}_n e_i = \sum_1^n y_i [\langle e_1, e_i\rangle, \ldots, \langle e_n, e_i\rangle]$$

$$= \sum_1^n y_i [0, \ldots, 0, 1_i, 0, \ldots, 0] = \sum_1^n [0, \ldots, 0, y_i, 0, \ldots, 0] = [y_1, \ldots, y_n] = y,$$

i.e. $\mathbf{T}_n\mathbf{T}_n^\dagger = \mathbf{I} = \mathbf{Q}$. Then $\mathbf{T}_n^\dagger\mathbf{T}_n\mathbf{T}_n^\dagger = \mathbf{T}_n^\dagger\mathbf{I} = \mathbf{T}_n^\dagger$. Besides, $\forall\, x \in \mathbb{H}$

$$\mathbf{T}_n^\dagger\mathbf{T}_n x = \mathbf{T}_n^\dagger[\langle e_1, x\rangle, \ldots, \langle e_n, x\rangle] = \sum_1^n \langle e_i, x\rangle e_i = (\mathbf{I} - \mathbf{P})x\ ,$$

i.e. $\mathbf{T}_n^\dagger\mathbf{T}_n = \mathbf{I} - \mathbf{P}$.

3° $\forall\, y \in \mathbb{E}^n$, $\left\|\mathbf{T}_n^\dagger y\right\|^2 = \|\sum_1^n y_i e_i\|^2 = \sum_1^n y_i^2 = \|y\|^2$. So, $\left\|\mathbf{T}_n^\dagger\right\| = 1$.

4°

$$\mathop{\&}_{i=1}^{n} \langle e_i\,, x\rangle = \alpha_i \iff \mathop{\&}_{i=1}^{n} \left\langle e_i, x - \textstyle\sum_{j=1}^n \alpha_j e_j\right\rangle = 0$$

$$\iff x - \sum_{j=1}^n \alpha_j e_j \in \mathcal{N}(\mathbf{T}_n) = \mathcal{R}(\mathbf{P}) = \mathcal{R}(\mathbf{I} - \mathbf{T}_n^\dagger\mathbf{T}_n)$$

$$\iff x \in \textstyle\sum_{j=1}^n \alpha_j e_j + \left(\mathbf{I} - \mathbf{T}_n^\dagger\mathbf{T}_n\right)\mathbb{H}\ .$$

Since $\sum_{j=1}^n \alpha_j e_j \perp \mathcal{N}(\mathbf{T}_n)$, for each $x = \sum_{j=1}^n \alpha_j e_j + z$ with $z \in \mathcal{N}(\mathbf{T}_n)$, we have $\left\|\sum_{j=1}^n \alpha_i e_j + z\right\|^2 = \left\|\sum_{j=1}^n \alpha_j e_j\right\|^2 + \|z\|^2 \geq \left\|\sum_{j=1}^n \alpha_j e_j\right\|^2$.

5° By 4°, the inequality $\|x\|^2 \geq \|\sum_{i=1}^n \alpha_i e_i\|^2 = \sum_{i=1}^n \alpha_i^2$ holds for any solution x of the system $\mathop{\&}_{i=1}^{n} \langle e_i, x\rangle = \alpha_i$. So, if x solves (1.7), then $\alpha_0 \geq 0\ \ \&\ \ \alpha_0^2 = \|x\|^2 \geq \sum_{i=1}^n \alpha_i^2$. Conversely, if this is true, then all vectors $x = \sum_{i=1}^n \alpha_i e_i + z$, where $z \in (\mathbf{I} - \mathbf{T}^\dagger\mathbf{T})\mathbb{H}\ \ \&\ \ \|z\|^2 = \alpha_0^2 - \sum_{i=1}^n \alpha_i^2$, and only these solve (1.7):

$$\mathop{\&}_{i=1}^{n} \langle e_i, x\rangle = \left\langle e_i, \textstyle\sum_{j=1}^n \alpha_j e_j + z\right\rangle = \alpha_i\ \ \&\ \ \|x\|^2 = \left\|\textstyle\sum_{j=1}^n \alpha_j e_j + z\right\|^2$$

$$= \textstyle\sum_{i=1}^n \alpha_i^2 + \|z\|^2 = \alpha_0^2.$$

□

Corollary 1.6. *The system* $\langle a\,, x\rangle = \alpha\ \ \&\ \ \|x\| = \beta$ *is solvable for* x *if and only if* $|\alpha| \leq \beta\|a\|$. *In this case, it is equivalent to*

$$x = \frac{\alpha}{\|a\|^2}a + z - \frac{\langle a\,, z\rangle}{\|a\|^2}a\ \ \&\ \ \|z\|^2 - \frac{\langle a\,, z\rangle^2}{\|a\|^2} = \beta^2 - \frac{\alpha^2}{\|a\|^2}.$$

Corollary 1.7. *The system*

$$\langle a\,, x\rangle = \alpha\ \ \&\ \ \langle b\,, x\rangle = \beta\ \ \&\ \ \|x\| = \gamma \qquad (1.8)$$

is solvable for x *if and only if*

$$\gamma \geq 0\ \ \&\ \ \|\alpha b - \beta a\|^2 \leq \gamma^2\left(\|a\|^2\|b\|^2 - \langle a\,, b\rangle^2\right).$$

In this case, if $\langle a\,,b\rangle^2 = \|a\|^2\|b\|^2$ *, the system is equivalent to*

$$x = \frac{\alpha}{\|a\|^2}a + z - \frac{\langle a\,,z\rangle}{\|a\|^2}a \ \ \&\ \ \|z\|^2 - \frac{\langle a\,,z\rangle^2}{\|a\|^2} = \gamma^2 - \frac{\alpha^2}{\|a\|^2}\ .$$

If $\langle a\,,b\rangle^2 < \|a\|^2\|b\|^2$ *, then it is equivalent to*

$$x = \frac{\alpha\|b\|^2 - \beta\langle a\,,b\rangle}{\|a\|^2\|b\|^2 - \langle a\,,b\rangle^2}a + \frac{\beta\|a\|^2 - \alpha\langle a\,,b\rangle}{\|a\|^2\|b\|^2 - \langle a,b\rangle^2}b + z\ ,$$

where z *is any vector satisfying*

$$\langle a\,,z\rangle = \langle b\,,z\rangle = 0 \ \ \&\ \ \|z\|^2 \le \gamma^2 - \frac{\|\alpha b - \beta a\|^2}{\|a\|^2\|b\|^2 - \langle a,b\rangle^2}\ .$$

Proof. If $\langle a\,,b\rangle^2 = \|a\|^2\|b\|^2$, then $b = \langle a\,,b\rangle\|a\|^{-2}a$ and so

$$\langle a\,,x\rangle = \alpha \ \ \&\ \ \langle b\,,x\rangle = \beta \Longleftrightarrow \langle a\,,x\rangle = \alpha \ \ \&\ \ \frac{\langle a\,,b\rangle}{\|a\|^2}\langle a\,,x\rangle = \beta$$

$$\Longrightarrow \frac{\langle a\,,b\rangle}{\|a\|^2} = \frac{\beta}{\alpha} \Longrightarrow b = \frac{\beta}{\alpha}a\ .$$

It follows that $\langle a\,,x\rangle = \alpha \Longrightarrow \langle b\,,x\rangle = \beta$, i.e. the system (1.8) is equivalent to $\langle a\,,x\rangle = \alpha$ & $\|x\| = \gamma$, which, by Corollary 1.6, is equivalent to

$$x = \frac{\alpha}{\|a\|^2}a + z - \frac{\langle a\,,z\rangle}{\|a\|^2}a \ \ \&\ \ \|z\|^2 - \frac{\langle a\,,z\rangle^2}{\|a\|^2} = \gamma^2 - \frac{\alpha^2}{\|a\|^2}\ .$$

If $\langle a\,,b\rangle^2 < \|a\|^2\|b\|^2$, then the system $\langle a\,,x\rangle = \alpha$ & $\langle b\,,x\rangle = \beta$ is equivalent to

$$\langle a',x\rangle = \frac{\alpha}{\|a\|} \ \ \&\ \ \langle b',x\rangle = \frac{\beta\|a\| - \alpha\langle a',b\rangle}{\sqrt{\|a\|^2\|b\|^2 - \langle a\,,b\rangle^2}}\ ,$$

where

$$a' := \frac{a}{\|a\|}\ ,\quad b' := \frac{b - \langle b\,,a'\rangle a'}{\|b - \langle b\,,a'\rangle a'\|} \perp a'.$$

So, the system (1.8) is solvable simultaneously with the system

$$\langle a',x\rangle = \frac{\alpha}{\|a\|} \ \ \&\ \ \langle b',x\rangle = \frac{\beta\|a\| - \alpha\langle a',b\rangle}{\sqrt{\|a\|^2\|b\|^2 - \langle a\,,b\rangle^2}} \ \ \&\ \ \|x\| = \gamma\ ,$$

which is solvable, by Lemma 1.5,

$$\Longleftrightarrow \frac{\alpha^2}{\|a\|^2} + \frac{(\beta\|a\| - \alpha\langle a',b\rangle)^2}{\|a\|^2\|b\|^2 - \langle a\,,b\rangle^2} \le \gamma^2$$

$$\Longleftrightarrow \gamma^2(\|a\|^2\|b\|^2 - \langle a\,,b\rangle^2) \ge \alpha^2(\|b\|^2 - \langle a',b\rangle^2) + \beta^2\|a\|^2 - 2\alpha\beta\langle a\,,b\rangle + \alpha^2\langle a',b\rangle^2$$

$$= \alpha^2\|b\|^2 - 2\alpha\beta\langle a\,,b\rangle + \beta^2\|a\|^2 = \|\alpha b - \beta a\|^2.$$

□

In terms of generalized inverses, the question of solvability of the general linear equation

$$\mathbf{A}x = y \tag{1.9}$$

for the linear bounded operator $\mathbf{A} : \mathbb{H}_1 \to \mathbb{H}_2$ acting between two Hilbert spaces gets complete resolution. Namely, according to Theorem 5.1 in [41], the equation (1.9) is solvable for x if and only if $\mathbf{A}\mathbf{A}^\dagger y = y$, in which case

$$\mathbf{A}x = y \Longleftrightarrow x \in \mathbf{A}^\dagger y + (\mathbf{I} - \mathbf{A}^\dagger \mathbf{A})\, \mathbb{H}_1$$

and $\mathbf{A}^\dagger y$ is the unique solution of (1.9) of minimal norm.

1.6 Difference equations

Carrying out convergence analyses of various iterative methods for solving operator equations, we will have more than one opportunity to consider systems of difference equations of the form

$$x_{n+1} = \mathbf{g}(x_n) \ , \ \ \mathbf{g} : \mathbb{R}^k \supset D \to D \ , \tag{1.10}$$

where $\mathbf{g}$ is a given function (the *generator* of the sequence x_n) defined on a subset D of $\mathbb{R}^k$ and mapping it into itself. Difference equation of k-th order

$$y_{n+1} = G(y_n, y_{n-1}, \ldots, y_{n+1-k})$$

is a particular case of (1.10), where

$$x_{n,i} := y_{n+1-i} \ , \ \ i = 1, \ldots, k \, , g_1(x_{n,1}, \ldots, x_{n,k}) := G(x_{n,1}, \ldots, x_{n,k}) \, ,$$

and $g_i(x_{n,1}, \ldots, x_{n,k}) := x_{n,i-1} \ , \ \ i = 2, \ldots, k$.

When the sequence x_n generated by the generator $\mathbf{g}$ from a starter x_0 converges, it does necessarily to a fixed point of $\mathbf{g} : x_\infty = \mathbf{g}(x_\infty)$. It follows directly from (1.10) by forcing n to infinity. A fixed point x of $\mathbf{g}$ is called attracting if it has a *basin of attraction* that is a subset $B(x)$ of D such that $x \neq x_0 \in B(x) \Longrightarrow x_n \to x$. In other words, a basin of attraction of x is the set of all starters x_0 different from x that cause convergence of x_n to x. The union

$$Q(\mathbf{g}) := \bigcup_{x \in Fix(\mathbf{g})} B(x)$$

of basins of attraction of all fixed points of $\mathbf{g}$ can be called the *convergence domain* of $\mathbf{g}$. It comprises all starters $x_0 \in D$, except for fixed points, that result in convergence of x_n to a fixed point.

A nonconstant continuous function $I : D \to \mathbb{R}$, which is constant on the sequence x_n: $\underset{n}{\&}\, I(x_{n+1}) = I(x_n)$, is called an invariant of the generator $\mathbf{g}$ [35, Ch. 4]. The classical example is the complete elliptic integral of the first kind

$$I(x_1, x_2) := \int_0^{\pi/2} \frac{dt}{\sqrt{(x_1 \cos t)^2 + (x_2 \sin t)^2}}\,,$$

which is an invariant of the generator

$$x_{n+1,1} := \frac{x_{n,1} + x_{n,2}}{2}\,, \quad x_{n+1,2} := \sqrt{x_{n,1} x_{n,2}}\,.$$

The problem of finding an invariant for a given generator $\mathbf{g}$ of the kind (1.10) is closely related to solution of a functional equation. This relation is conveniently explained in the special case $k = 2$. If $I(x_1, x_2)$ is an invariant of $\mathbf{g}$, then $\underset{n}{\&}\, I(x_{n,1}, x_{n,2}) = c$, where c is some constant. Solving this equation for $x_{n,2}$ produces a function f such that $\underset{n}{\&}\, x_{n,2} = f(x_{n,1}, c)$. In particular (and with (1.10) in mind),

$$x_{n+1,2} = f(x_{n+1,1}, c) = f\big(g_1(x_{n,1}\,, x_{n,2}, c)\big) = f\big(g_1\big(x_{n,1}\,, f(x_{n,1}, c)\big), c\big)\,.$$

At the same time, $x_{n+1,2} = g_2(x_{n,1}, x_{n,2}) = g_2\big(x_{n,1}\,, f(x_{n,1}, c)\big)$. So,

$$f\big(g_1(t\,, f(t, c)), c\big) = g_2\big(t\,, f(t, c)\big)\,. \tag{1.11}$$

This is a functional equation for f. Conversely, if f is its continuous solution and $f(x_{0,1}, c) = x_{0,2}$, then $I(x_1\,, x_2) := f(x_1, c) - x_2$ is an invariant. Indeed, by (1.10),

$$I(x_{1,1}\,, x_{1,2}) = f(x_{1,1}, c) - x_{1,2} = f\big(g_1(x_{0,1}\,, x_{0,2}), c\big) - g_2(x_{0,1}\,, x_{0,2})\,,$$

so that $x_{0,2} = f(x_{0,1}, c)$ implies

$$\begin{aligned} I(x_{1,1}\,, x_{1,2}) &= f\big(g_1\big(x_{0,1}\,, f(x_{0,1}, c)\big)\big) - g_2(x_{0,1}\,, f(x_{0,1}, c)) = 0 \\ &= f(x_{0,1}, c) - x_{0,2} = I(x_{0,1}\,, x_{0,2}) \end{aligned}$$

by (1.11). It follows (by induction) that

$$I(x_{0,1}, x_{0,2}) = c \Longrightarrow \underset{n}{\&}\, I(x_{n,1}, x_{n,2}) = c\,.$$

Note that the operator $\mathbf{f}(I)(x) := I\big(\mathbf{g}(x)\big) - I(x)$ induced by this equation is linear.

Availability of an invariant $I(x)$ greatly facilitates determining the convergence domain of a generator. Namely, an x_0 belongs to the convergence domain if and only if it is not a fixed point and $I(x_0) = I(x_\infty)$.

1.7 Minimax and maximin

Let X and Y be two sets of arbitrary kind and f be a function defined on $X \times Y$. The simple fact is that always

$$\inf_{x\in X} \sup_{y\in Y} f(x,y) \geq \sup_{y\in Y} \inf_{x\in X} f(x,y) . \tag{1.12}$$

Indeed, obviously for all $y \in Y$ $f(x,y) \geq \inf_{x\in X} f(x,y)$ and so $\sup_{y\in Y} f(x,y) \geq \sup_{y\in Y} \inf_{x\in X} f(x,y)$. As this is true for all $x \in X$, it is also for $\inf_{x\in X} \sup_{y\in Y} f(x,y)$.

The inequality (1.12) provokes the question: for which triples (X, Y, f) the inequality becomes equality. The answer is not trivial. Theorems that give it are referred to as minimax theorems. Such a theorem is the following one due to Kneser [34]:

Theorem 1.8. *Let the sets $X \subset \mathbb{R}^m$ and $Y \subset \mathbb{R}^n$ be convex and the function f be convex on X and concave on Y. If one of the sets X, Y is compact (that is closed and bounded) and f is continuous in the corresponding variable, then*

$$\inf_{x\in X} \sup_{y\in Y} f(x,y) = \sup_{y\in Y} \inf_{x\in X} f(x,y) .$$

Its generalizations for infinite-dimensional spaces are found in [55], [3].

1.8 Diagonal operators

The linear matrix operators of the form

$$\mathbf{A}x := ax \ , \ \forall\, x \in \mathbb{R}^{n\times n}, \tag{1.13}$$

for some $a \in \mathbb{R}^{n\times n}$, are invariant with respect to summation and composition:

$$\mathbf{A}x = ax \ \ \& \ \ \mathbf{B}x = bx \Longrightarrow \begin{cases} (\mathbf{A}+\mathbf{B})x = \mathbf{A}x + \mathbf{B}x = ax + bx = (a+b)x \\ \mathbf{AB}x = \mathbf{A}(bx) = a(bx) = (ab)x . \end{cases}$$

The same is true for operators of the form

$$(\mathbf{A}x)(t) := a(t)x(t) \tag{1.14}$$

acting on $\mathbb{C}[0\,,1]$. This invariance is of interest in connection with Ulm's methods studied in the next two chapters. Therefore, the operators possessing this property deserve a name. We will call them *diagonal*. The formal definition follows.

Definition 1.9. *Let $\mathcal{S}$ be a subset of the space $\mathcal{L}(\mathbb{X})$ of linear bounded operators acting on a Banach space $\mathbb{X}$, which is invariant with respect to summation and composition:*

$$\mathbf{A} \in \mathcal{S} \;\; \& \;\; \mathbf{B} \in \mathcal{S} \Longrightarrow \mathbf{A} + \mathbf{B} \in \mathcal{S} \;\; \& \;\; \mathbf{AB} \in \mathcal{S} \,. \tag{1.15}$$

The operators of $\mathcal{S}$ are called diagonal.

Obvious examples are operators (1.13) and (1.14). Another example is the integral operator

$$(\mathbf{A}x)(t) := a(t)x(t) + \int_0^1 K(s,t)x(s)\,ds \;,\; \forall\, x \in \mathbb{C}[0\,,1]\,. \tag{1.16}$$

Indeed, if $\mathbf{A}$ acts as in (1.16) and $(\mathbf{B}x)(t) = b(t)x(t) + \int_0^1 L(s,t)x(s)\,ds$, then

$$\begin{aligned} \big((\mathbf{A}+\mathbf{B})x\big)(t) &= a(t)x(t) + \int_0^1 K(s,t)x(s)\,ds + b(t)x(t) + \int_0^1 L(s,t)x(s)\,ds \\ &= \big(a(t)+b(t)\big)x(t) + \int_0^1 \big(K(s,t)+L(s,t)\big)x(t)\,ds \end{aligned}$$

and

$$\begin{aligned} (\mathbf{AB})x(t) &= \mathbf{A}\left(b(t)x(t) + \int_0^1 L(s,t)x(s)\,ds\right) \\ &= a(t)\left(b(t)x(t) + \int_0^1 L(s,t)x(s)\,ds\right) + \\ &\qquad \int_0^1 K(s,t)\left(b(s)x(s) + \int_0^1 L(\sigma,s)x(\sigma)\,d\sigma\right)ds \\ &= a(t)b(t)x(t) + \int_0^1 \big(a(t)L(s,t) + b(s)K(s,t)\big)x(s)\,ds + \\ &\qquad \int_0^1 K(s,t)\int_0^1 L(\sigma,s)x(\sigma)\,d\sigma\,ds\,. \end{aligned}$$

The last double integral

$$= \int_0^1\int_0^1 K(s,t)L(\sigma,s)ds\,x(\sigma)d\sigma = \int_0^1\int_0^1 K(\sigma,t)L(s,\sigma)d\sigma\,x(s)ds\,.$$

Therefore,

$$\begin{aligned} (\mathbf{AB})x(t) &= a(t)b(t)x(t) + \\ &\qquad \int_0^1 \left(a(t)L(s,t) + b(s)K(s,t) + \int_0^1 K(\sigma,t)L(s,\sigma)\,d\sigma\right)x(s)ds\,. \end{aligned}$$

Thus, both $\mathbf{A}+\mathbf{B}$ and $\mathbf{AB}$ retain the form (1.16).

Particularly interesting (in our context) are those diagonal operators which are also *parametric.* This is the case when $\mathcal{S}$ is a family diagonally parametrized by a parameter, that is, a function $f(a,x)$, which is diagonal with respect to the first argument:

$$f(a,x)+f(b,x)=f(\varphi(a,b),x) \quad \& \quad f\big(a,f(b,x)\big)=f\big(\psi(a,b),x\big)$$

and linear with respect to the second:

$$f(a,\alpha x+\beta y)=\alpha f(a,x)+\beta f(a,y)\,,\ \forall x,y\in\mathbb{X}\,,\ \forall\,\alpha,\beta\in\mathbb{R}\,,$$

is known such that

$$\mathcal{S}=\big\{\mathbf{A}\in\mathcal{L}(\mathbb{X})\ \big|\ \mathbf{A}x=f(a,x)\big\}\,. \tag{1.17}$$

The operators of the family $\mathcal{S}$ will be referred to as *parametrically diagonal* (briefly pd-operators). Such are the operators (1.13), (1.14), (1.16). Also a pd-operator is any scalar multiple of a linear operator $\mathbf{A}$ on a Banach space $\mathbb{X}$:

$$\forall\,x\in\mathbb{X}\,,\ \mathbf{f}(\alpha,x):=\alpha\mathbf{A}x$$

and this $\mathbf{f}$ is diagonal with respect to the first argument:

$$\mathbf{f}(\alpha,x)+\mathbf{f}(\beta,x)=\alpha\mathbf{A}x+\beta\mathbf{A}x=(\alpha+\beta)\mathbf{A}x=\mathbf{f}(\alpha+\beta,x)$$

and

$$\mathbf{f}\big(\alpha,\mathbf{f}(\beta,x)\big)=\alpha\mathbf{A}\mathbf{f}(\beta,x)=\alpha\beta\mathbf{A}x=\mathbf{f}(\alpha\beta,x)\,.$$

The diagonality of f in (1.17) guarantees invariance (1.15) of the family of pd-operators with respect to summation and composition.

Chapter 2

Ulm's method

Ulm's method [58],

$$x_+ := x - \mathbf{A}\mathbf{f}(x) \ , \ \mathbf{A}_+ := 2\mathbf{A} - \mathbf{A}\mathbf{f}'(x_+)\mathbf{A} \ , \tag{2.1}$$

has several attractive properties. In addition to being inversion-free, it (like Newton's method) is self-correcting. Besides, if the iterations $(x_n, \mathbf{A}_n)$ converge, they do so to a solution of the system

$$\mathbf{f}(x) = 0 \ \ \& \ \ \mathbf{X}\mathbf{f}'(x) = \mathbf{I} \tag{2.2}$$

for the pair $(x, \mathbf{X})$, $x \in \mathbb{X}$, $\mathbf{X} \in \mathcal{L}(\mathbb{Y}, \mathbb{X})$. This property is very helpful when one is interested in the solution's sensitivity to small perturbations in the data. Moreover, as we are going to see, under natural assumptions the convergence is quadratic.

2.1 Motivation

Ulm's method is motivated by the following

Proposition 2.1. *Let* $\mathbf{A} \in \mathcal{L}(\mathbb{X}, \mathbb{Y})$ *and* $\mathbf{B} \in \mathcal{L}(\mathbb{Y}, \mathbb{X})$. *The following statements are equivalent:*

(i)
$$\|\mathbf{I} - \mathbf{B}\mathbf{A}\| < 1 \ ; \tag{2.3}$$

(ii) $\mathbf{C} := (\mathbf{B}\mathbf{A})^{-1}\mathbf{B}$ *is a left-inverse of* $\mathbf{A}$ *and*

$$\|\mathbf{C}\| \le \frac{\|\mathbf{B}\|}{1 - \|\mathbf{I} - \mathbf{B}\mathbf{A}\|} \ ;$$

(iii) $\mathbf{D} := \mathbf{A}(\mathbf{B}\mathbf{A})^{-1}$ *is a right-inverse of* $\mathbf{B}$ *and*

$$\|\mathbf{D}\| \le \frac{\|\mathbf{A}\|}{1 - \|\mathbf{I} - \mathbf{B}\mathbf{A}\|} \ ;$$

(iv) the null space of $\mathbf{A}$ *is zero:* $\mathcal{N}(\mathbf{A}) = \{0\}$ *;*

(v) the range of $\mathbf{B}$ *is* $\mathbb{X}$: $\mathcal{R}(\mathbf{B}) = \mathbb{X}$.

If (i) is true, then

1° $\mathbf{A}$ *is also right-invertible if and only if its range coincide with* $\mathbb{Y}$: $\mathcal{R}(\mathbf{A}) = \mathbb{Y}$.

In this case, both $\mathbf{A}$ *and* $\mathbf{B}$ *are invertible,* $\mathbf{A}^{-1} = \mathbf{C}$, $\mathbf{B}^{-1} = \mathbf{D}$, *and*

$$\|\mathbf{I} - \mathbf{AB}\| \le \kappa(\mathbf{A})\|\mathbf{I} - \mathbf{BA}\| ,$$

where $\kappa(\mathbf{A}) := \|\mathbf{A}\| \cdot \|\mathbf{A}^{-1}\|$ *(the condition number of* $\mathbf{A}$*).*

2° $\mathbf{B}$ *is also left-invertible* $\iff \mathcal{N}(\mathbf{B}) = \{0\}$, *in which case both* $\mathbf{A}$ *and* $\mathbf{B}$ *are invertible,* $\mathbf{A}^{-1} = \mathbf{C}$, $\mathbf{B}^{-1} = \mathbf{D}$, *and* $\|\mathbf{I} - \mathbf{AB}\| \le \kappa(\mathbf{B})\|\mathbf{I} - \mathbf{BA}\|$.

Proof. (i) $\iff$ (ii). If $\mathbf{B}$ satisfies (2.3), then the operator $\mathbf{BA}$ is invertible: $(\mathbf{BA})^{-1} = \sum_0^\infty (\mathbf{I} - \mathbf{BA})^n$, and $\mathbf{C}$ is a left inverse of $\mathbf{A}$. Moreover,

$$\|\mathbf{C}\| \le \|\mathbf{B}\| \cdot \|(\mathbf{BA})^{-1}\| \le \frac{\|\mathbf{B}\|}{1 - \|\mathbf{I} - \mathbf{BA}\|} .$$

Conversely, if $\mathbf{C}$ is a left inverse of $\mathbf{A}$, then $\|\mathbf{I} - \mathbf{CA}\| = 0 < 1$.

(i) $\iff$ (iii) is proved similarly.

(ii) $\iff$ (iv). If $\mathbf{C}$ is a left inverse of $\mathbf{A}$, then $x \in \mathcal{N}(\mathbf{A}) \implies \mathbf{A}x = 0 \implies x = \mathbf{CA}x = 0$, i.e. $\mathcal{N}(\mathbf{A}) = \{0\}$. Conversely, if $\mathcal{N}(\mathbf{A}) = \{0\}$, then $\forall y \in \mathcal{R}(\mathbf{A})$ the equation $\mathbf{A}x = y$ is uniquely solvable:

$$\mathbf{A}x_1 = \mathbf{A}x_2 = y \;\; \& \;\; x_1 \ne x_2 \implies \mathbf{A}(x_1 - x_2) = 0$$

contrary to $\mathcal{N}(\mathbf{A}) = \{0\}$. Let $\mathbf{C}$ be the operator which takes each $y \in \mathcal{R}(\mathbf{A})$ to the only solution of $\mathbf{A}x = y$, so that $\mathbf{AC}y = y$, $\forall y \in \mathcal{R}(\mathbf{A})$. $\mathbf{C}$ is linear operator. Indeed, if $y_1, y_2 \in \mathcal{R}(\mathbf{A})$ and $x_i = \mathbf{C}y_i$, then

$$\begin{aligned}\mathbf{AC}(\alpha_1 y_1 + \alpha_2 y_2) &= \alpha_1 y_1 + \alpha_2 y_2 \\ &= \alpha_1 \mathbf{A}x_1 + \alpha_2 \mathbf{A}x_2 = \mathbf{A}(\alpha_1 x_1 + \alpha_2 x_2) \\ &= \mathbf{A}(\alpha_1 \mathbf{C}y_1 + \alpha_2 \mathbf{C}y_2) ,\end{aligned}$$

that is, both $\mathbf{C}(\alpha_1 y_1 + \alpha_2 y_2)$ and $\alpha_1 \mathbf{C}y_1 + \alpha_2 \mathbf{C}y_2$ solve the equation $\mathbf{A}x = \alpha_1 y_1 + \alpha_2 y_2$ and so have to be the same: $\mathbf{C}(\alpha_1 y_1 + \alpha_2 y_2) = \alpha_1 \mathbf{C}y_1 + \alpha_2 \mathbf{C}y_2$. It remains to show that $\mathbf{CA} = \mathbf{I}$.

$$x \in \mathbb{X} \;\; \& \;\; y := \mathbf{A}x \implies \mathbf{CA}x = \mathbf{CAC}y = \mathbf{C}y = x,$$

i.e., $\mathbf{CA}x = x$, $\forall x \in \mathbb{X}$.

(iii) $\iff$ (v). If $\mathbf{D}$ is a right-inverse of $\mathbf{B}$, then the equation $\mathbf{B}y = x$ is solvable $\forall x \in \mathbb{X}$: $y = \mathbf{D}x$ is a solution. So, $\mathcal{R}(\mathbf{B}) = \mathbb{X}$. Conversely, if $\mathcal{R}(\mathbf{B}) = \mathbb{X}$, then the equation $\mathbf{B}y = x$ has a solution $\forall x \in \mathbb{X}$. Let $\mathbf{D}$ be an operator (the generalized inverse $\mathbf{B}^\dagger$ of $\mathbf{B}$, for example) that takes each $x \in \mathbb{X}$ to a certain solution of this equation, so that $\mathbf{BD}x = x$. $\mathbf{D}$ is a linear operator. Indeed, if $x_i \in \mathbb{X}$ and $y_i := \mathbf{D}x_i$, $i = 1, 2$, then

$$\begin{aligned}\mathbf{BD}(\alpha_1 x_1 + \alpha_2 x_2) &= \alpha_1 x_1 + \alpha_2 x_2 \\ &= \alpha_1 \mathbf{B}y_1 + \alpha_2 \mathbf{B}y_2 = \mathbf{B}(\alpha_1 y_1 + \alpha_2 y_2) \\ &= \mathbf{B}(\alpha_1 \mathbf{D}x_1 + \alpha_2 \mathbf{D}x_2)\ ,\end{aligned}$$

which shows that $\alpha_1 \mathbf{D}x_1 + \alpha_2 \mathbf{D}x_2$ solves the equation $\mathbf{B}y = \alpha_1 x_1 + \alpha_2 x_2$, so that $\alpha_1 \mathbf{D}x_1 + \alpha_2 \mathbf{D}x_2 = \mathbf{D}(\alpha_1 x_1 + \alpha_2 x_2)$. Thus, $\mathbf{D}$ is a linear operator from $\mathbb{X}$ into $\mathbb{Y}$ satisfying $\mathbf{BD}x = x$, $\forall x \in \mathbb{X}$, that is a right-inverse of $\mathbf{B}$.

1° Suppose now that (2.3) is true. If $\mathbf{AX} = \mathbf{I}$ for some $\mathbf{X} \in \mathcal{L}(\mathbb{Y}, \mathbb{X})$, then $\forall y \in \mathbb{Y}\ y = \mathbf{AX}y \in \mathcal{R}(\mathbf{A})$, that is, $\mathcal{R}(\mathbf{A}) = \mathbb{Y}$. Conversely, let $\mathcal{R}(\mathbf{A}) = \mathbb{Y}$. As $\mathcal{N}(\mathbf{A}) = \{0\}$, by (iv), it follows that $\mathbf{A}$ maps $\mathbb{X}$ onto $\mathbb{Y}$ one-to-one and so is invertible, by the inverse mapping theorem. Besides, $\mathbf{CA} = \mathbf{I} = \mathbf{BD}$, by (ii) and (iii), and so $\mathbf{DBA} = \mathbf{A} \Longrightarrow \mathbf{DB} = \mathbf{DBAA}^{-1} = \mathbf{AA}^{-1} = \mathbf{I}$, i.e. $\mathbf{B}$ is left invertible. As it is also right invertible, it is invertible and $\mathbf{B}^{-1} = \mathbf{D}$. Now, $\mathbf{BAC} = \mathbf{B} \Longrightarrow \mathbf{AC} = \mathbf{I} \Longrightarrow \mathbf{C} = \mathbf{A}^{-1}$. Finally, $\mathbf{I} - \mathbf{AB} = \mathbf{A}(\mathbf{I} - \mathbf{BA})\mathbf{A}^{-1}$, and $\|\mathbf{I} - \mathbf{AB}\| \le \|\mathbf{A}\| \cdot \|\mathbf{A}^{-1}\| \cdot \|\mathbf{I} - \mathbf{BA}\|$.

2° is proved similarly to 1°. □

If $\mathbf{B}$ is such that $\|\mathbf{I} - \mathbf{Bf}\,'(x)\| < 1$, each partial sum of the operator series $\sum_{n=0}^{\infty}(\mathbf{I} - \mathbf{Bf}\,'(x))^n\mathbf{B}$ provides better approximation to $(\mathbf{Bf}\,'(x))^{-1}\mathbf{B}$ than the preceding one. In particular,

$$\begin{aligned}\big\|\mathbf{I} - (2\mathbf{B} - \mathbf{Bf}\,'(x)\mathbf{B})\mathbf{f}\,'(x)\big\| = \big\|(\mathbf{I} - \mathbf{Bf}\,'(x))^2\big\| &\le \|\mathbf{I} - \mathbf{Bf}\,'(x)\|^2 \\ &< \|\mathbf{I} - \mathbf{Bf}\,'(x)\|\ .\end{aligned}$$

The same can be expected when x_+ is close to x (and so $\mathbf{f}\,'(x_+)$ is close to $\mathbf{f}\,'(x)$ if $\mathbf{f}$ is smooth in the neighborhood of x).

2.2 Regular smoothness

The Kantorovich-type convergence analysis of the method (2.1) that is carried out in this section is based on the more general and more flexible smoothness assumption than the usual Lipschitz and similar assumptions. We call it *regular smoothness*. Its formal definition follows.

Let Ω denote the class of nondecreasing functions $\omega : [0, \infty) \mapsto [0, \infty)$ that are concave (i.e., have convex subgraphs $\{(s, t) \mid s \ge 0\ \ \&\ \ t \le \omega(s)\}$, cf. [51]) and vanishing at zero. Being monotone, a function $\omega \in \Omega$ has left- and right-hand derivatives $\omega^{\backprime}, \omega'$ at each $s > 0$ (they coincide everywhere except, perhaps, for a countable number of points). The typical representatives of this class are the functions $s \mapsto as^p$, $a > 0$, $0 < p \le 1$. We also need the notation

$$\underline{h}(\mathbf{f}\,') := \inf_{x \in D} \|\mathbf{f}\,'(x)\|\ .$$

Definition 2.2. [13] *Given an $\omega \in \Omega$, we say that* $\mathbf{f}$ *is ω-regularly smooth on D or, equivalently, that ω is a regular smoothness modulus of* $\mathbf{f}$ *on D, if the inequality*

$$\omega^{-1}\Big(\min\Big\{\|\mathbf{f}'(x)\|, \|\mathbf{f}'(x')\|\Big\} - \underline{h} + \|\mathbf{f}'(x') - \mathbf{f}'(x)\|\Big) \\ - \omega^{-1}\Big(\min\Big\{\|\mathbf{f}'(x)\|, \|\mathbf{f}'(x')\|\Big\} - \underline{h}\Big) \le \|x' - x\| \tag{2.4}$$

holds for some $\underline{h} \in [0, \underline{h}(\mathbf{f}'))]$ and all $x, x' \in D$. The operator $\mathbf{f}$ *is regularly smooth on D if it has a regular smoothness modulus there.*

This definition admits zero as a possible choice for $\underline{h}$. However, the closer $\underline{h}$ is to $\underline{h}(\mathbf{f}')$, the less the condition (2.4) requires of a potential candidate for a regularity modulus, the greater their stock to choose from is, and so the better the chances are to find a good modulus.

Here and in the rest of the book the symbol ω^{-1} denotes the function whose closed epigraph $cl\{(s,t) | s \ge 0 \;\&\; t \ge \omega^{-1}(s)\}$ is symmetrical to closure of the subgraph of ω with respect to the axis $t = s$. Clearly, ω^{-1} is a convex function on $[0, \infty)$ vanishing at zero, non-decreasing in $[0, \omega(\infty))$, and equal to ∞ for all $s > \omega(\infty)$ (if any). Because of the convexity of ω^{-1}, each ω-regularly smooth operator $\mathbf{f}$ is also ω-smooth in the sense that

$$\|\mathbf{f}'(x') - \mathbf{f}'(x)\| \le \omega(\|x' - x\|) \,, \quad \forall x, x' \in D \,.$$

To see that the converse is not true, let us show that an ω-regularly smooth on D operator $\mathbf{f}$ must be Lipschitz smooth on the set

$$D_h := \big\{x \in D \;\big|\; \|\mathbf{f}'(x)\| - \underline{h} \ge h\big\} \tag{2.5}$$

for each $h > 0$. Indeed,

$$(2.4) \Longrightarrow \sup_{x,x' \in D_h} \frac{\|\mathbf{f}'(x') - \mathbf{f}'(x)\|}{\|x' - x\|} \le \sup_{x,x' \in D_h} \frac{\omega\left(\omega^{-1}(h) + \|x' - x\|\right) - h}{\|x' - x\|} \\ \le \lim_{t \searrow 0} \frac{\omega\left(\omega^{-1}(h) + t\right) - \omega\left(\omega^{-1}(h)\right)}{t} = \omega'\big(\omega^{-1}(h)\big) < \infty$$

if $h > 0$. Obviously, any function $f : \mathbb{R} \to \mathbb{R}$ with $f'(t) := 1 + \operatorname{sign}(t)\sqrt{|t|}$ does not meet this requirement while it has smoothness moduli ($t \mapsto \sqrt{2t}$, for instance). On the other hand, each Lipschitz smooth (with the modulus $t \mapsto ct$) operator on D is also regularly smooth there with the same modulus.

It is easy to deduce from (2.4) that, if ω_1 is a regular smoothness modulus of an operator $\mathbf{f}$, $\omega_2 \in \Omega$, and $\omega_1' \le \omega_2'$ on $[0, \infty)$, then ω_2 is a modulus too. This remark allows to assume, without loss of generality, that a modulus ω under discussion is increasing in $[0, \infty)$, since one can replace it with $\omega_\varepsilon(t) := \omega(t) + \varepsilon t$ when necessary and then force ε to zero in the result.

Likewise, if ω_1 and ω_2 are two regular smoothness moduli of an operator, then their pointwise minimum $\omega(t) := \min\{\omega_1(t), \omega_2(t)\}$ is too. It follows that among all such moduli there is (pointwise) the least one.

Another immediate consequence of regular smoothness is the following

Lemma 2.3. 1° *If an operator* $\mathbf{f}: D \subset \mathbb{X} \to \mathcal{L}(\mathbb{X}, \mathbb{Y})$ *is* ω*-regularly smooth on* D *relative to an* $x_0 \in D$*, then for all* $x \in D$*,*

$$|\omega^{-1}(\|\mathbf{f}'(x)\| - \underline{h}) - \omega^{-1}(\|\mathbf{f}'(x_0)\| - \underline{h})| \le \|x - x_0\|.$$

2° *If* $\mathbf{f}$ *is* ω*-regularly smooth on* D*, then this inequality holds for all* $x, x_0 \in D$*.*

Proof. If $\|\mathbf{f}'(x)\| \le \|\mathbf{f}'(x_0)\|$, then (2.4) becomes

$$\omega^{-1}\Big(\|\mathbf{f}'(x)\| - \underline{h} + \|\mathbf{f}'(x) - \mathbf{f}'(x_0)\|\Big) \le \omega^{-1}\Big(\|\mathbf{f}'(x)\| - \underline{h}\Big) + \|x - x_0\| \ .$$

By monotonicity of ω^{-1}, $\omega^{-1}\Big(\|\mathbf{f}'(x_0)\| - \underline{h}\Big) \le \omega^{-1}\Big(\|\mathbf{f}'(x_0) - \mathbf{f}'(x)\| + \|\mathbf{f}'(x)\|) - \underline{h}\Big)$. Hence, $\omega^{-1}\Big(\|\mathbf{f}'(x_0)\| - \underline{h}\Big) \le \omega^{-1}\Big(\|\mathbf{f}'(x)\| - \underline{h}\Big) + \|x - x_0\|$. The more so, if $\|\mathbf{f}'(x)\| > \|\mathbf{f}'(x_0)\|$. The argument remains intact when the roles of x_0 and x are exchanged. □

In particular,

$$\omega^{-1}(\|\mathbf{f}'(x)\| - \underline{h}) \ge \big(\omega^{-1}(\|\mathbf{f}'(x_0)\| - \underline{h}) - \|x - x_0\|\big)^+ . \tag{2.6}$$

Here and in the rest of the book the superscripts $^+$ and $^-$ denote the positive and the negative parts of a real number:

$$r^+ := \max\{r, 0\} \ , \ r^- := \max\{-r, 0\} \ ,$$

so that $r^+ - r^- = r$, $r^+ + r^- = |r|$, and $r^+ r^- = 0$.

The ω-regular smoothness of an operator $\mathbf{f}$ implies a bound on the remainder of the Taylor formula, parallel to the classical bound

$$r(x, x_0) := \|\mathbf{f}(x) - \mathbf{f}(x_0) - \mathbf{f}'(x_0)(x - x_0)\| \le 0.5c\|x - x_0\|^2 \tag{2.7}$$

known for Lipschitz smooth operators. To state it, adopt suitable notations. Let $\omega \in \Omega$ and denote

$$w(t) := \int_0^t \omega(\tau)d\tau \ , \ e(\alpha, t) := \omega\big((\alpha - t)^+ + t\big) - \omega\big((\alpha - t)^+\big) \ , \tag{2.8}$$

and

$$\Psi(\alpha, t) := \int_0^t e(\alpha, \tau)d\tau = \begin{cases} t\omega(\alpha) - w(\alpha) + w(\alpha - t) \ , & \text{if } 0 \le t \le \alpha \ , \\ \alpha\omega(\alpha) - 2w(\alpha) + w(t) \ , & \text{if } t \ge \alpha \ge 0 \ . \end{cases} \tag{2.9}$$

As the derivatives of the functions w and $t \mapsto \Psi(\alpha, t)$ are increasing, it is clear that both are convex.

Lemma 2.4. *If the operator* $\mathbf{f}$ *is* ω*-regularly smooth on* D, *then*
1° *for every pair of points* $x\,,x'$ *in* D,

$$r(x'\,,x) \le \Psi(\bar{\alpha}\,,\bar{\delta})\,, \tag{2.10}$$

where $\bar{\alpha} := \omega^{-1}(\|\mathbf{f}'(x)\| - \underline{h})$ *and* $\bar{\delta} := \|x' - x\|$;
2° *the function* Ψ *is not increasing in the first argument and is increasing in the second.*

Proof. 1° By the Newton–Leibnitz theorem,

$$\mathbf{f}(x') - \mathbf{f}(x) - \mathbf{f}'(x)(x'-x) = \int_0^1 \big[\mathbf{f}'(x+s(x'-x)) - \mathbf{f}'(x)\big](x'-x)ds\ ,$$

whence

$$r(x'\,,x) \le \int_0^1 \|\mathbf{f}'(x+s(x'-x)) - \mathbf{f}'(x)\| \cdot \|x'-x\|ds\ .$$

As seen from (2.4), $\|\mathbf{f}'(x+s(x'-x)) - \mathbf{f}'(x)\|$

$$\begin{aligned}
&\le \omega\left(\omega^{-1}(\min\{\|\mathbf{f}'(x+s(x'-x))\|\,,\|\mathbf{f}'(x)\|\} - \underline{h}) + s\|x'-x\|\right) - \\
&\qquad\qquad \min\{\|\mathbf{f}'(x+s(x'-x))\|\,,\|\mathbf{f}'(x)\|\} + \underline{h} \\
&= \omega\left(\min\left\{\omega^{-1}\left(\|\mathbf{f}'(x+s(x'-x))\| - \underline{h}\right),\omega^{-1}\left(\|\mathbf{f}'(x)\| - \underline{h}\right)\right\} + s\|x'-x\|\right) - \\
&\qquad \omega\left(\min\left\{\omega^{-1}\left(\|\mathbf{f}'(x+s(x'-x))\| - \underline{h}\right),\omega^{-1}\left(\|\mathbf{f}'(x)\| - \underline{h}\right)\right\}\right).
\end{aligned} \tag{2.11}$$

By (2.6),

$$\omega^{-1}(\|\mathbf{f}'(x+s(x'-x))\| - \underline{h}) \ge \left(\omega^{-1}(\|\mathbf{f}'(x)\| - \underline{h}) - s\|x'-x\|\right)^+ = (\bar{\alpha} - s\bar{\delta})^+$$

and so (due to concavity of ω) the difference (2.11)

$$\begin{aligned}
&\le \omega\Big(\min\Big\{(\bar{\alpha}-s\bar{\delta})^+\,,\,\bar{\alpha}\Big\} + s\bar{\delta}\Big) - \omega\Big(\min\Big\{(\bar{\alpha}-s\bar{\delta})^+\,,\bar{\alpha}\Big\}\Big) \\
&= \omega\Big((\bar{\alpha}-s\bar{\delta})^+ + s\bar{\delta}\Big) - \omega\Big((\bar{\alpha}-s\bar{\delta})^+\Big).
\end{aligned}$$

Therefore,

$$\begin{aligned}
r(x'\,,x) &\le \int_0^1 \Big[\omega\Big((\bar{\alpha}-s\bar{\delta})^+ + s\bar{\delta}\Big) - \omega\Big((\bar{\alpha}-s\bar{\delta})^+\Big)\Big]\bar{\delta}ds \\
&= \int_0^{\bar{\delta}} \Big[\omega\Big((\bar{\alpha}-t)^+ + t\Big) - \omega\Big((\bar{\alpha}-t)^+\Big)\Big]dt = \Psi(\bar{\alpha}\,,\bar{\delta})\ .
\end{aligned}$$

2°

$$\frac{\partial\Psi}{\partial\alpha} = \begin{cases} t\left(\omega'(\alpha) - \dfrac{\omega(\alpha)-\omega(\alpha-t)}{\delta}\right)\,, & \text{if } 0 \le t < \alpha\,, \\[2ex] \alpha\left(\omega'(\alpha) - \dfrac{\omega(\alpha)}{\alpha}\right)\,, & \text{if } 0 \le \alpha < t\,. \end{cases}$$

Because of the concavity of ω, $\omega'(\alpha) \le \min\{(\omega(\alpha)-\omega(\alpha-t))/t, \omega(\alpha)/\alpha\}$. So, $\partial\Psi/\partial\alpha \le 0$, while

$$\frac{\partial\Psi}{\partial t} = \begin{cases} \omega(\alpha)-\omega(\alpha-t)\ , & \text{if } 0 \le t < \alpha\ , \\ \omega(t)\ , & \text{if } 0 \le \alpha < t \end{cases} > 0$$

owing to monotonicity of ω. □

In the Lipschitz case ($\omega(t) = ct$) the bound (2.10) reduces (independently of $\underline{h}$) to (2.7). In contrast, for a nonlinear ω, the bound (2.10) is sharper than $w(\bar{\delta})$, the bound implied by ω-smoothness. In particular, for $\omega(t) = ct^p$, $0 < p \le 1$,

$$\Psi(\bar{\alpha}, \bar{\delta}) < \frac{c\bar{\delta}^{p+1}}{p+1}\ , \ \forall\, \bar{\alpha} > 0\ ,\ \bar{\delta} \in [0\,, \bar{\alpha})\,.$$

We apply Definition 2.2 to prove that the best (i.e., pointwise the least) regular smoothness modulus of any quadratic operator is linear. The operator $\mathbf{Q}$ is quadratic if

$$\mathbf{Q}(x) := \frac{1}{2}\mathbf{B}(x,x) + \mathbf{A}x + y\ ,\ \forall\, x \in \mathbb{X}\ , \tag{2.12}$$

where $\mathbf{B}$ is a bilinear operator acting from $\mathbb{X}^2$ into $\mathbb{Y}$ (see, for example, [32, Ch. XVII] for the definition), $\mathbf{A} \in \mathcal{L}(\mathbb{X}, \mathbb{Y})$, and $y \in \mathbb{Y}$. We can assume without loss of generality that $\mathbf{B}$ is symmetric: $\mathbf{B}(x,y) = \mathbf{B}(y,x)$, for one can always replace $\mathbf{B}$ with $0.5(\mathbf{B}(x,y) + \mathbf{B}(y,x))$ without affecting $\mathbf{Q}$. As

$$\begin{aligned}\mathbf{Q}(x+h) - \mathbf{Q}(x) &= \frac{1}{2}\big(\mathbf{B}(x+h\,, x+h) - \mathbf{B}(x+h\,, x) + \mathbf{B}(x+h\,, x) - \\ &\qquad\qquad \mathbf{B}(x\,, x)\big) + \mathbf{A}h \\ &= \frac{1}{2}\big(\mathbf{B}(x\,, h) + \mathbf{B}(h\,, h) + \mathbf{B}(x\,, h)\big) + \mathbf{A}h = \big(\mathbf{B}(x\,,\cdot) + \mathbf{A}\big)h,\end{aligned}$$

it is clear that

$$\mathbf{Q}'(x) = \mathbf{B}(x\,,\cdot) + \mathbf{A} \tag{2.13}$$

and

$$\big\|\mathbf{Q}'(x+u) - \mathbf{Q}'(x)\big\| = \big\|\mathbf{B}(u\,,\cdot)\big\|\ . \tag{2.14}$$

According to definition, regular smoothness modulus $\omega \in \Omega$ must be such that $\forall\ x\,, x+u$

$$\begin{aligned}\omega^{-1}\big(\min\{\|\mathbf{Q}'(x+u)\|, \|\mathbf{Q}'(x)\|\} - \underline{h} + \|\mathbf{Q}'(x+u) - \mathbf{Q}'(x)\|\big)& \\ - \ \omega^{-1}\big(\min\{\|\mathbf{Q}'(x+u)\|, \|\mathbf{Q}'(x)\|\} - \underline{h}\big) \le \|u\|\ .&\end{aligned} \tag{2.15}$$

Because of the symmetry $x \leftrightarrow x+u$, we can restict ourselves to the case $\|\mathbf{Q}'(x+u)\| \ge \|\mathbf{Q}'(x)\|$. In this case, (2.15) becomes

$$\omega^{-1}\big(\|\mathbf{Q}'(x)\| - \underline{h} + \|\mathbf{Q}'(x+u) - \mathbf{Q}'(x)\|\big) - \ \omega^{-1}\big(\|\mathbf{Q}'(x)\| - \underline{h}\big) \le \|u\|\ ,$$

or, in view of (2.13) and (2.14),

$$\omega^{-1}\big(\,\|\mathbf{B}(x\,,\cdot)+\mathbf{A}\|-\underline{h}+\|\mathbf{B}(u\,,\cdot)\|\big)-\omega^{-1}\big(\,\|\mathbf{B}(x\,,\cdot)+\mathbf{A}\|-\underline{h}\big)\le\|u\|\;,$$

$\forall\,x\,,x+u$ with $\|\mathbf{B}(x+u\,,\cdot)+\mathbf{A}\|\ge\|\mathbf{B}(x\,,\cdot)+\mathbf{A})\|=:s$ (for brevity). This condition is restated as

$$\sup_u\Big\{\omega^{-1}\big(\,s-\underline{h}+\|\mathbf{B}(u\,,\cdot)\|\big)-\|u\|\;\Big|\;\|\mathbf{B}(x+u\,,\cdot)+\mathbf{A}\|\ge s\Big\}\le\omega^{-1}(s-\underline{h})\;. \tag{2.16}$$

Here $\omega^{-1}\Big(s-\underline{h}+\|\mathbf{B}(u\,,\cdot)\|\Big)\ge\omega^{-1}(s-\underline{h})+\omega^{-1}\Big(\|\mathbf{B}(u\,,\cdot)\|\Big)$, since convexity of ω^{-1} together with the equality $\omega^{-1}(0)=0$ implies its superadditivity. It follows that the supremum in (2.16)

$$\begin{aligned}&\ge\omega^{-1}(s-\underline{h})+\sup_u\Big\{\omega^{-1}\big(\|\mathbf{B}(u\,,\cdot)\|\big)-\|u\|\;\Big|\;\|\mathbf{B}(x+u\,,\cdot)+\mathbf{A}\|\ge s\Big\}\\&\ge\omega^{-1}(s-\underline{h})\;,\;\forall\,x\;.\end{aligned}$$

Thus, (2.16) implies

$$\begin{aligned}\omega^{-1}(s-\underline{h})\le\omega^{-1}(s-\underline{h})+\sup_u\Big\{\omega^{-1}\big(\|\mathbf{B}(u\,,\cdot)\|\big)-\|u\|\;\Big|\;\|\mathbf{B}(x+u\,,\cdot)+\mathbf{A}\|\ge s\Big\}\\\le\omega^{-1}(s-\underline{h})\;,\end{aligned}$$

that is,

$$\sup_u\Big\{\omega^{-1}\big(\|\mathbf{B}(u\,,\cdot)\|\big)-\|u\|\;\Big|\;\|\mathbf{B}(x+u\,,\cdot)+\mathbf{A}\|\ge s\Big\}=0\;.$$

We see that, for any regular smoothness modulus ω of $\mathbf{Q}$, this supremum does not depend on x, so that the subjection can be dropped: $\sup_u\Big(\omega^{-1}\big(\|\mathbf{B}(u\,,\cdot)\|\big)-\|u\|\Big)=0$, that is $\|\mathbf{B}(u\,,\cdot)\|\le\omega(\|u\|)$, $\forall\,u$. In particular, this must be true for any u with $\|\mathbf{B}(u\,,\cdot)\|=\|\mathbf{B}\|\cdot\|u\|$. We have proved

Proposition 2.5. *An* $\omega\in\Omega$ *is a regular smoothness modulus of* $\mathbf{Q}$ *if and only if* $\omega(t)\ge\|\mathbf{B}\|t\;,\;\forall\,t\ge0$.

Consider, for example, Chandrasekhar's integral operator [6, p. 357]

$$\mathbf{f}(x)(t):=x(t)-0.5tx(t)\int_0^1\frac{x(s)}{s+t}ds-1\;,\;0\le t\le1\;, \tag{2.17}$$

acting on the space $\mathbb{C}$ of continuous functions on $[0,1]$. For it, the symmetric

$$\mathbf{B}(x_1\,,x_2)(t)=-0.5t\left[x_1(t)\int_0^1\frac{x_2(s)}{s+t}ds+x_2(t)\int_0^1\frac{x_1(s)}{s+t}ds\right],$$

$(\mathbf{A}x)(t) = x(t)$ (i.e., $\mathbf{A} = \mathbf{I}$), and $y(t) = -1$, $\forall t \in [0,1]$. The $\mathbb{C}$-norm $\|\mathbf{B}(x_1, x_2)\|$ of the function $\mathbf{B}(x_1, x_2)(t)$

$$
\begin{aligned}
&= \max_{0\le t\le 1}\left|0.5t\left[x_1(t)\int_0^1 \frac{x_2(s)}{s+t}ds + x_2(t)\int_0^1 \frac{x_1(s)}{s+t}ds\right]\right| \\
&\le 0.5\left[\max_{0\le t\le 1}\left|tx_1(t)\int_0^1 \frac{x_2(s)}{s+t}ds\right| + \max_{0\le t\le 1}\left|tx_2(t)\int_0^1 \frac{x_1(s)}{s+t}ds\right|\right] \\
&\le \|x_1\|\cdot\|x_2\| \max_{0\le t\le 1}\int_0^1 \frac{ds}{s+t},
\end{aligned}
$$

so that $\|\mathbf{B}\| := \sup_{x_1,x_2}\left\{\|\mathbf{B}(x_1, x_2)\| \;\middle|\; \|x_1\| \le 1 \;\;\&\;\; \|x_2\| \le 1\right\}$

$$
\le \max_{0\le t\le 1}\int_0^1 \frac{ds}{s+t} = \max_{0\le t\le 1} t\ln\left(1+\frac{1}{t}\right) = \ln 2.
$$

On the other hand, $\left|\mathbf{B}(1,1)(t)\right| = t\ln(1+1/t)$, $\|\mathbf{B}(1,1)\| = \ln 2$, and $\|\mathbf{B}\| \ge \|\mathbf{B}(1,1)\| = \ln 2$. So, $\|\mathbf{B}\| = \ln 2$ and, by the proposition, the best regular smoothness modulus of the operator (2.17) is $(\ln 2)t$.

Another example is the Riccati matrix operator

$$
\mathbf{f}(x) := \frac{1}{2}xax + bx + xc + d \tag{2.18}
$$

acting on the space $\mathbb{R}^{n\times n}$ of $n\times n$ real matrices. For it, $\mathbf{B}(x,y) = xay$ and

$$
\begin{aligned}
\|\mathbf{B}\| &= \sup_{x,y}\{\|xay\| \mid \|x\| = \|y\| = 1\} \\
&\le \sup_{x,y}\{\|x\|\,\|a\|\,\|y\| \mid \|x\| = \|y\| = 1\} = \|a\|\,.
\end{aligned}
$$

On the other hand, taking for x and y the unit matrix e, we obtain $\|\mathbf{B}\| \ge \|eae\| = \|a\|$. Thus, $\|\mathbf{B}\| = \|a\|$ and, by the proposition, the operator (2.18) is Lipschitz smooth with the Lipschitz constant $\|a\|$.

2.3 Majorant generator and convergence lemma

It is more convenient to investigate the convergence properties of Ulm's method (2.1) in its equivalent formulation

$$
\mathbf{A}_+ := 2\mathbf{A} - \mathbf{A}\mathbf{f}'(x)\mathbf{A}\ ,\ x_+ := x - \mathbf{A}_+\mathbf{f}(x)\ . \tag{2.19}
$$

Each pair $(\mathbf{A}, x) \in \mathcal{L}(\mathbb{Y},\mathbb{X})\times D$ induces the quadruple $\bar{q} = \left(\bar{t},\bar{\beta},\bar{\gamma},\bar{\delta}\right)$ of non-negative real numbers

$$
\bar{t} := \|x - x_0\|\ ,\ \bar{\beta} := \|\mathbf{A}\|\ ,\ \bar{\gamma} := \|\mathbf{I} - \mathbf{A}\mathbf{f}'(x)\|\ ,\ \bar{\delta} := \|\mathbf{A}_+\mathbf{f}(x)\|\ . \tag{2.20}
$$

The following lemma relates the next quadruple $\bar{q}_+ = (\bar{t}_+, \bar{\beta}_+, \bar{\gamma}_+, \bar{\delta}_+)$ to $\bar{q}$.

Lemma 2.6. 1° $\bar{t}_+ := \|x_+ - x_0\| \le \bar{t} + \bar{\delta}$.

2° $\bar{\beta}_+ := \|\mathbf{A}_+\| \le \bar{\beta}\,(1 + \bar{\gamma})$.

If the operator $\mathbf{f}$ *is* ω*-regularly smooth on* D *and* $\bar{\alpha}_0 := \omega^{-1}\big(\|\mathbf{f}'(x_0)\| - \underline{h}\big)$, *then*

3° $\bar{\gamma}_+ := \|\mathbf{I} - \mathbf{A}_+\mathbf{f}'(x_+)\| \le \bar{\gamma}^2 + \bar{\beta}_+ e\big(\bar{\alpha}_0 - \bar{t}, \bar{\delta}\big)$.

4° $\bar{\delta}_+ := \|\mathbf{A}_{++}\mathbf{f}(x_+)\| \le (1 + \bar{\gamma}_+)\big[\bar{\beta}_+\Psi\big(\bar{\alpha}_0 - \bar{t}, \bar{\delta}\big) + \bar{\gamma}^2\bar{\delta}\,\big]$.

5° *All these upper bounds are exact: they are attained for the scalar* ω*-regularly smooth function*

$$f_a : [0,\infty) \to \mathbb{R}\ ,\ \ f_a(x) := w(x) - a\,,$$

where a *is a positive constant.*

Proof. 1° is a direct consequence of the triangle inequality.

2° $\bar{\beta}_+ = \big\|\big[\mathbf{I} + (\mathbf{I} - \mathbf{A}\mathbf{f}'(x))\big]\mathbf{A}\big\| \le \big(1 + \|\mathbf{I} - \mathbf{A}\mathbf{f}'(x)\|\big)\|\mathbf{A}\| = \bar{\beta}(1 + \bar{\gamma})$.

3°

$$\begin{aligned}
\mathbf{I} - \mathbf{A}_+\mathbf{f}'(x_+) &= \mathbf{I} - \big(2\mathbf{I} - \mathbf{A}\mathbf{f}'(x)\big)\mathbf{A}\big(\mathbf{f}'(x_+) - \mathbf{f}'(x) + \mathbf{f}'(x)\big)\\
&= \mathbf{I} - \big(2\mathbf{I} - \mathbf{A}\mathbf{f}'(x)\big)\mathbf{A}\mathbf{f}'(x) - \mathbf{A}_+\big(\mathbf{f}'(x_+) - \mathbf{f}'(x)\big)\\
&= \big(\mathbf{I} - \mathbf{A}\mathbf{f}'(x)\big)^2 - \mathbf{A}_+\big(\mathbf{f}'(x_+) - \mathbf{f}'(x)\big)\ .
\end{aligned}$$

So, $\bar{\gamma}_+ \le \Big\|\big(\mathbf{I} - \mathbf{A}\mathbf{f}'(x)\big)^2\Big\| + \|\mathbf{A}_+\| \cdot \big\|\mathbf{f}'(x_+) - \mathbf{f}'(x)\big\|$

$$\le \big\|\mathbf{I} - \mathbf{A}\mathbf{f}'(x)\big\|^2 + \|\mathbf{A}_+\| \cdot \big\|\mathbf{f}'(x_+) - \mathbf{f}'(x)\big\| = \bar{\gamma}^2 + \bar{\beta}_+\|\mathbf{f}'(x_+) - \mathbf{f}'(x)\|\ .$$

By (2.4), $\|\mathbf{f}'(x_+) - \mathbf{f}'(x)\|$

$$\begin{aligned}
&\le \omega\left(\omega^{-1}\big(\min\{\|\mathbf{f}'(x)\|, \|\mathbf{f}'(x_+)\|\} - \underline{h}\big) + \bar{\delta}\right) - \min\{\|\mathbf{f}'(x)\|, \|\mathbf{f}'(x_+)\|\} + \underline{h}\\
&= \omega\left(\min\left\{\omega^{-1}\big(\|\mathbf{f}'(x)\| - \underline{h}\big), \omega^{-1}\big(\|\mathbf{f}'(x_+)\| - \underline{h}\big)\right\} + \bar{\delta}\right) -\\
&\qquad\qquad \omega\left(\min\left\{\omega^{-1}\big(\|\mathbf{f}'(x)\| - \underline{h}\big), \omega^{-1}\big(\|\mathbf{f}'(x_+)\| - \underline{h}\big)\right\}\right),
\end{aligned}$$

where by (2.6) $\omega^{-1}\big(\|\mathbf{f}'(x)\| - \underline{h}\big) \ge \big(\omega^{-1}\big(\|\mathbf{f}'(x_0)\| - \underline{h}\big) - \|x - x_0\|\big)^+ = (\bar{\alpha}_0 - \bar{t})^+$ and $\omega^{-1}\big(\|\mathbf{f}'(x_+)\| - \underline{h}\big) \ge (\bar{\alpha}_0 - \bar{t}_+)^+ \ge (\bar{\alpha}_0 - \bar{t} - \bar{\delta})^+$. So, owing to concavity of ω,

$$\begin{aligned}
\big\|\mathbf{f}'(x_+) - \mathbf{f}'(x)\big\| &\le \omega\left(\min\left\{(\bar{\alpha}_0 - \bar{t})^+, (\bar{\alpha}_0 - \bar{t} - \bar{\delta})^+\right\} + \bar{\delta}\right)\\
&\qquad - \omega\left(\min\left\{(\bar{\alpha}_0 - \bar{t})^+, (\bar{\alpha}_0 - \bar{t} - \bar{\delta})^+\right\}\right)\\
&= \omega\left((\bar{\alpha}_0 - \bar{t} - \bar{\delta})^+ + \bar{\delta}\right) - \omega\left((\bar{\alpha}_0 - \bar{t} - \bar{\delta})^+\right) = e\big(\bar{\alpha}_0 - \bar{t}, \bar{\delta}\big)
\end{aligned}$$

and $\bar{\gamma}_+ \le \bar{\gamma}^2 + \bar{\beta}_+ e\big(\bar{\alpha}_0 - \bar{t}, \bar{\delta}\big)$.

4° As $\mathbf{f}(x_+) = \mathbf{f}(x_+) - \mathbf{f}(x) - \mathbf{f}'(x)(x_+ - x) + (\mathbf{I} - \mathbf{f}'(x)\mathbf{A}_+)\mathbf{f}(x)$, we have

$$\mathbf{A}_{++}\mathbf{f}(x_+) = \big(2\mathbf{I} - \mathbf{A}_+\mathbf{f}'(x_+)\big)\Big[\mathbf{A}_+\big(\mathbf{f}(x_+) - \mathbf{f}(x) - \mathbf{f}'(x)(x_+ - x)\big) + \big(\mathbf{I} - \mathbf{A}_+\mathbf{f}'(x)\big)\mathbf{A}_+\mathbf{f}(x)\Big]$$

and so

$$\bar{\delta}_+ \le \big\|2\mathbf{I} - \mathbf{A}_+\mathbf{f}'(x_+)\big\|\Big[\|\mathbf{A}_+\|\big\|\mathbf{f}(x_+) - \mathbf{f}(x) - \mathbf{f}'(x)(x_+ - x)\big\| + \|\mathbf{I} - \mathbf{A}_+\mathbf{f}'(x)\|\|\mathbf{A}_+\mathbf{f}(x)\|\Big].$$

The first norm $\le 1 + \bar{\gamma}_+$,

$$\begin{aligned}\big\|\mathbf{f}(x_+) - \mathbf{f}(x) - \mathbf{f}'(x)(x_+ - x)\big\| &\le \Psi\big(\omega^{-1}\big(\|\mathbf{f}'(x)\| - \underline{h}\big), \|x_+ - x\|\big)\\ &\le \Psi\Big(\big(\omega^{-1}\big(\|\mathbf{f}'(x_0)\| - \underline{h}\big) - \|x - x_0\|\big)^+, \bar{\delta}\Big)\\ &= \Psi\big((\bar{\alpha}_0 - \bar{t})^+, \bar{\delta}\big),\end{aligned}$$

by Lemma 2.4, $\|\mathbf{I} - \mathbf{A}_+\mathbf{f}'(x)\| = \Big\|\big(\mathbf{I} - \mathbf{A}\mathbf{f}'(x)\big)^2\Big\| \le \|\mathbf{I} - \mathbf{A}\mathbf{f}'(x)\|^2 = \bar{\gamma}^2$ and $\|\mathbf{A}_+\mathbf{f}(x)\| = \bar{\delta}$. Therefore, $\bar{\delta}_+ \le (1 + \bar{\gamma}_+)\,\big[\bar{\beta}_+\Psi\big((\bar{\alpha}_0 - \bar{t})^+, \bar{\delta}\big) + \bar{\gamma}^2\bar{\delta}\,\big]$.

5° First, we show that the mapping (Ulm's iteration)

$$\mathbf{U}_a : \mathbb{R}^2 \to \mathbb{R}^2\,,\ \mathbf{U}_a(A, x) := (A_+, x - A_+ f_a(x))\,,\ A_+ := 2A - A^2\omega(x),$$

maps the set

$$M := \Big\{(A, x)\ \Big|\ A > 0\ \ \&\ \ w^{-1}(a) < x < x_0\ \ \&\ \ A\omega(x) < 1\Big\}$$

to itself: $\mathbf{U}_a(M) \subset M$. For $(A, x) \in M$, $A_+ = A(2 - Af_a'(x)) = A(1 + 1 - A\omega(x)) > A > 0$, $f_a(x) > 0$, and so $x_+ - x = -A_+ f_a(x) < 0$. Hence, $x_+ - x_0 = x - x_0 - A_+ f_a(x) < x - x_0 < 0$. Besides, as f_a is a convex function,

$$0 = f_a\big(w^{-1}(a)\big) \ge f_a(x) + f_a'(x)\big(w^{-1}(a) - x\big) = f_a(x) + \omega(x)\big(w^{-1}(a) - x\big),$$

where $\omega(x) < A_+^{-1}$ because $1 - A_+\omega(x) = \big(1 - A\omega(x)\big)^2 > 0$. So, $0 > f_a(x) + A_+^{-1}\big(w^{-1}(a) - x\big)$, $0 > A_+ f_a(x) + w^{-1}(a) - x = w^{-1}(a) - x_+$, and $x_+ > w^{-1}(a)$. Finally,

$$\begin{aligned}0 < x_+ < x &\Longrightarrow 0 < \omega(x_+) \le \omega(x)\\ &\Longrightarrow 1 - A_+\omega(x_+) = 1 - \big(2A - A^2\omega(x)\big)\omega(x) + A_+\big(\omega(x) - \omega(x_+)\big)\\ &\qquad = \big(1 - A\omega(x)\big)^2 + A_+\big(\omega(x) - \omega(x_+)\big) \ge \big(1 - A\omega(x)\big)^2 > 0\,.\end{aligned}$$

Thus, $(A, x) \in M \Longrightarrow (A_+\,, x_+) \in M$.

Now define the mappings

$$\mathbf{T}_a : \mathbb{R}^2 \to \mathbb{R}^4\,,\ \mathbf{T}_a(A, x) := \Big(|x - x_0|\,, |A|\,, |1 - A\omega(x)|\,, |A_+ f_a(x)|\Big)$$

and $\mathbf{g}:\mathbb{R}^4\to\mathbb{R}^4$, $\mathbf{g}(t,\beta,\gamma,\delta):=(t_+\,,\beta_+\,,\gamma_+\,,\delta_+)$, where

$$t_+ := t+\delta\ ,\ \beta_+ := \beta(1+\gamma)\ ,\ \gamma_+ := \gamma^2+\beta_+ e(\bar{\alpha}_0-t,\delta)\ ,$$

$$\delta_+ := (1+\gamma_+)\big[\beta_+\Psi(\bar{\alpha}_0-t,\delta)+\gamma^2\delta\big]. \tag{2.21}$$

We have to prove that $\mathbf{g}\mathbf{T}_a=\mathbf{T}_a\mathbf{U}_a$ on M. Since $\underline{h}(f_a'(x))=\inf\limits_{x\geq 0}\omega(x)=0$, the parameter $\underline{h}$ is necessarily zero, so that $\bar{\alpha}_0:=\omega^{-1}(|f_a'(x_0)|-\underline{h})=\omega^{-1}(\omega(x_0))=x_0$ and for $(A,x)\in M$, $\bar{t}=|x-x_0|=x_0-x$, $\bar{\beta}=|A|=A$, $\bar{\gamma}=|1-Af_a'(x)|=1-A\omega(x)$, and $\bar{\delta}=|A_+f_a(x)|=A_+(w(x)-a)$. In other words,

$$(A,x)\in M\Longrightarrow \mathbf{T}_a(A,x)=\big(x_0-x\ ,A\ ,1-A\omega(x)\ ,A_+(w(x)-a)\big).$$

Then $\mathbf{g}\mathbf{T}_a(A,x)=(t_+\,,\beta_+\,,\gamma_+\,,\delta_+)$, where, according to (2.21), $t_+=x_0-x+A_+(w(x)-a)=x_0-x_+=\bar{t}_+$, $\beta_+=A(1+1-A\omega(x))=2A-A^2\omega(x)=A_+=\bar{\beta}_+$,

$$\begin{aligned}\gamma_+ &= \big(1-A\omega(x)\big)^2+A_+e\big((\bar{\alpha}_0-x_0+x)^+,A_+f_a(x)\big)\\ &=1-\big(2A-A^2\omega(x)\big)\omega(x)+A_+e\big(x\,,A_+f_a(x)\big)\\ &=1-A_+\omega(x)+A_+\Big(\omega\big((x-A_+f_a(x))^++A_+f_a(x)\big)-\omega\big((x-A_+f_a(x))^+\big)\Big)\\ &=1-A_+\omega(x)+A_+\big(\omega(x)-\omega(x_+)\big)=1-A_+\omega(x_+)=\bar{\gamma}_+\ .\end{aligned}$$

Besides, $\delta=|x_+-x|=x-x_+<x=x_0-(x_0-x)=\bar{\alpha}_0-\bar{t}$ and so

$$\begin{aligned}\Psi\big((\bar{\alpha}_0-\bar{t})^+,\bar{\delta}\big)&=\bar{\delta}\omega\big((\bar{\alpha}_0-\bar{t})^+,\bar{\delta}\big)-w\big((\bar{\alpha}_0-\bar{t})^+\big)+w\big((\bar{\alpha}_0-\bar{t})^+-\bar{\delta}\big)\\ &=(x-x_+)\omega(x)-w(x)+w(x_+)\ .\end{aligned}$$

It follows that $\beta_+\Psi\big((\bar{\alpha}_0-\bar{t})^+,\delta\big)+\gamma^2\delta$

$$\begin{aligned}&=A_+\big(\omega(x)(x-x_+)-w(x)+w(x_+)\big)+(x-x_+)\big(1-A\omega(x)\big)^2\\ &=A_+\big(w(x_+)-w(x)\big)+(x-x_+)\big(A_+\omega(x)+\big(1-A\omega(x)\big)^2\\ &=A_+\big(w(x_+)-w(x)\big)+(x-x_+)\big(\omega(x)\big(2A-A^2\omega(x)\big)\\ &\qquad\qquad+1-2A\omega(x)+\big(A\omega(x)\big)^2\big)\\ &=A_+\big(w(x_+)-w(x)\big)+x-x_+=A_+\big(w(x_+)-w(x)\big)+A_+\big(w(x)-a\big)\\ &=A_+\big(w(x_+)-a\big).\end{aligned}$$

Therefore,

$$\begin{aligned}\delta_+ &:= (1+\gamma_+)\big(\beta_+\Psi(\alpha,\delta)+\gamma^2\delta\big)=\big(2-A_+\omega(x_+)\big)A_+\mathbf{f}_a(x_+)=A_{++}\mathbf{f}_a(x_+)\\ &=x_+-x_{++}=\bar{\delta}_+\ .\end{aligned}$$

Thus, $\mathbf{g}\mathbf{T}_a(A,x)=\mathbf{T}_a(A_+,x_+)=\mathbf{T}_a\mathbf{U}_a(A,x)$, $\forall\,(A,x)\in M$. □

We say that a quadruple $q' = (t', \beta', \gamma', \delta')$ majorizes $q = (t, \beta, \gamma, \delta)$ (symbolically $q \prec q'$), if

$$t \le t' \ \& \ \beta \le \beta' \ \& \ \gamma \le \gamma' \ \& \ \delta \le \delta' .$$

The lemma asserts that

$$\bar{q}_+ = (\bar{t}_+, \bar{\beta}_+, \bar{\gamma}_+, \bar{\delta}_+) \prec \mathbf{g}(\bar{q}) . \tag{2.22}$$

Given an initial quadruple q_0, the mapping $\mathbf{g}$ iterates

$$q_{n+1} := \mathbf{g}(q_n),$$

producing the infinite sequence q_n. It is called a majorant sequence if each q_n majorizes its prototype quadruple $\bar{q}_n = (\bar{t}_n, \bar{\beta}_n, \bar{\gamma}_n, \bar{\delta}_n)$ induced by the n-th approximation $(\mathbf{A}_n, x_n)$ as in (2.20). Correspondingly, we refer to $\mathbf{g}$ as the *majorant generator* or just *generator*. Using concavity and monotonicity of ω, one can easily prove the following momotonicity property of the generator $\mathbf{g}$.

Lemma 2.7. $0 \le q \prec q' \Longrightarrow 0 \le \mathbf{g}(q) \prec \mathbf{g}(q')$.

As the next lemma testifies, the generator $\mathbf{g}$ plays an important role in the investigation of convergence properties of Ulm's method.

Lemma 2.8. $\bar{q}_0 \prec q_0 \Longrightarrow \underset{n}{\&}\, \bar{q}_n \prec q_n$. *If, in addition,* q_0 *causes the sequences* β_n *and* t_n *to converge:*

$$\beta_\infty := \lim \beta_n < \infty \ \& \ t_\infty := \lim t_n < \infty , \tag{2.23}$$

then

1° *the successive iterations* $(\mathbf{A}_n, x_n)$ *remain in the ball*

$$B\big((\mathbf{A}_0, x_0), (\beta_\infty - \beta_0, t_\infty)\big) := \Big\{(\mathbf{A}, x) \,\Big|\, \|\mathbf{A} - \mathbf{A}_0\| \le \beta_\infty - \beta_0 \ \& \ \|x - x_0\| \le t_\infty\Big\}$$

and converge to a limit $(\mathbf{A}_\infty, x_\infty)$*;*

2° *this limit solves the system*

$$\mathbf{X}\mathbf{f}'(x) = \mathbf{I} \ \& \ \mathbf{A}_0\mathbf{f}(x) = 0 \tag{2.24}$$

for $(\mathbf{X}, x) \in \mathcal{L}(\mathbb{Y}, \mathbb{X}) \times \mathbb{X}$ *;*

3° x_∞ *is the only solution of the equation* $\mathbf{A}_0\mathbf{f}(x) = 0$ *in the ball* $B(x_0, R)$*, where* R *is the unique solution for* t *of the equation*

$$\frac{\Psi(\bar{\alpha}_0, t) - \Psi(\bar{\alpha}_0, t_\infty)}{t - t_\infty} = \frac{1 - \gamma_0}{\beta_0} ;$$

4° *for all* $n = 0, 1, ...$

$$\begin{aligned} &\Delta_n := \|x_\infty - x_n\| \le t_\infty - t_n \,, \\ &\|\mathbf{A}_\infty - \mathbf{A}_n\| \le \beta_\infty - \beta_n \,, \\ &\frac{\Delta_{n+1}}{\Delta_n} \le \beta_{n+1} \frac{w(\Delta_n)}{\Delta_n} + \gamma_n^2 ; \end{aligned} \tag{2.25}$$

5° *the convergence condition* $t_\infty < \infty \;\&\; \beta_\infty < \infty$ *is exact: if it is violated, then there can be found an* ω*-regularly smooth operator* $\mathbf{f}$, *for which Ulm's method starting from the starter* $(\mathbf{A}_0, x_0)$ *generates divergent iterations* $(\mathbf{A}_n, x_n)$.

Proof. Suppose that $\bar{q}_n \prec q_n$ for some $n \ge 0$. Then Lemma 2.7 yields $\mathbf{g}(\bar{q}_n) \prec \mathbf{g}(q_n) = q_{n+1}$. Combining this with (2.22) results in $\bar{q}_{n+1} \prec q_{n+1}$. By induction, $\underset{n}{\&} \bar{q}_n \prec q_n$.

1° In particular, $\underset{n}{\&}\big(\bar{\beta}_n \le \beta_n \;\&\; \bar{\gamma}_n \le \gamma_n\big)$ and so

$$\begin{aligned} \|\mathbf{A}_{m+n} - \mathbf{A}_n\| &\le \sum_{k=n}^{m+n-1} \|\mathbf{A}_{k+1} - \mathbf{A}_k\| \le \sum_{k=n}^{m+n-1} \|\mathbf{A}_k\| \cdot \|\mathbf{I} - \mathbf{A}_k \mathbf{f}'(x_k)\| \\ &\le \sum_{k=n}^{m+n-1} \beta_k \gamma_k = \sum_{k=n}^{m+n-1} (\beta_{k+1} - \beta_k) = \beta_{m+n} - \beta_n \,. \end{aligned} \tag{2.26}$$

Similarly, $\underset{n}{\&} \bar{\delta}_n \le \delta_n$ implies

$$\begin{aligned} \|x_{m+n} - x_n\| &\le \sum_{k=n}^{m+n-1} \|x_{k+1} - x_k\| \le \sum_{k=n}^{m+n-1} \bar{\delta}_k \le \sum_{k=n}^{n+m-1} \delta_k \le \sum_{k=n}^{n+m-1} (t_{k+1} - t_k) \\ &= t_{m+n} - t_n \,. \end{aligned} \tag{2.27}$$

It follows that $\mathbf{A}_n$ and x_n are Cauchy sequences in respective Banach spaces and so converge to limits $\mathbf{A}_\infty$ and x_∞. Setting $n = 0$ in (2.26) and (2.27) shows that $\underset{m}{\&}(\mathbf{A}_m, x_m) \in B\big((\mathbf{A}_0, x_0), (\beta_\infty - \beta_0, t_\infty)\big)$, while forcing m to ∞ yields the first two inequalities in (2.25).

2° The assumption $\beta_\infty < \infty$ implies, by (2.21), $\gamma_n = \beta_{n+1}/\beta_n - 1 \to 0$, and so $\underset{n}{\&} \gamma_n < 1$, for $\gamma_n \ge 1 \Longrightarrow \gamma_{n+1} \ge 1 \Longrightarrow \underset{k \ge n}{\&} \gamma_k \ge 1$. It follows that all $\mathbf{A}_n$ have the same null space:

$$\underset{n}{\&} \mathcal{N}(\mathbf{A}_n) = \mathcal{N}(\mathbf{A}_0) \,. \tag{2.28}$$

Indeed, as seen from (2.1),

$$\begin{aligned} \mathbf{A}_{n+1} y = 0 &\Longrightarrow -\mathbf{A}_n y = (\mathbf{I} - \mathbf{A}_n \mathbf{f}'(x_n)) \mathbf{A}_n y \\ &\Longrightarrow \|\mathbf{A}_n y\| \le \|\mathbf{I} - \mathbf{A}_n \mathbf{f}'(x_n)\| \cdot \|\mathbf{A}_n y\| = \bar{\gamma}_n \|\mathbf{A}_n y\| \le \gamma_n \|\mathbf{A}_n y\| \\ &\Longrightarrow \mathbf{A}_n y = 0, \end{aligned}$$

if $\gamma_n < 1$. Thus, $\mathcal{N}(\mathbf{A}_{n+1}) \subset \mathcal{N}(\mathbf{A}_n)$. The opposite inclusion is a trivial consequence of (2.1). Now, the distance of $\mathbf{f}(x_\infty)$ from $\mathcal{N}(\mathbf{A}_0)$

$$\begin{aligned} dist(\mathbf{f}(x_\infty), \mathcal{N}(\mathbf{A}_0)) &:= \inf_y \{ \|\mathbf{f}(x_\infty) - y\| \mid y \in \mathcal{N}(\mathbf{A}_0) \} \\ &= \inf_y \{ \|\mathbf{f}(x_\infty) - y\| \mid y \in \mathcal{N}(\mathbf{A}_n) \}. \end{aligned}$$

Inasmuch as $\bar{\gamma}_n := \|\mathbf{I} - \mathbf{A}_n \mathbf{f}\,'(x_n)\| \le \gamma_n < 1$, the operator $\mathbf{A}_n \mathbf{f}\,'(x_n)$ is boundedly invertible, $\|(\mathbf{A}_n \mathbf{f}\,'(x_n))^{-1}\| \le \big(1 - \|\mathbf{I} - \mathbf{A}_n \mathbf{f}\,'(x_n)\|\big)^{-1}$, and $\mathbf{f}\,'(x_n)(\mathbf{A}_n \mathbf{f}\,'(x_n))^{-1}$ is a right-inverse of $\mathbf{A}_n$ (Lemma 2.1, 1°). Therefore,

$$\begin{aligned} y \in \mathcal{N}(\mathbf{A}_n) \iff \mathbf{A}_n y = 0 &\implies y = [\mathbf{I} - \mathbf{f}\,'(x_n)(\mathbf{A}_n \mathbf{f}\,'(x_n))^{-1} \mathbf{A}_n] y \\ &\implies y \in [\mathbf{I} - \mathbf{f}\,'(x_n)(\mathbf{A}_n \mathbf{f}\,'(x_n))^{-1} \mathbf{A}_n] \mathbb{Y}. \end{aligned}$$

Conversely, $y \in [\mathbf{I} - \mathbf{f}\,'(x_n)(\mathbf{A}_n \mathbf{f}\,'(x_n))^{-1} \mathbf{A}_n] \mathbb{Y}$

$$\implies y = [\mathbf{I} - \mathbf{f}\,'(x_n)(\mathbf{A}_n \mathbf{f}\,'(x_n))^{-1} \mathbf{A}_n] z\,, \ z \in \mathbb{Y} \implies \mathbf{A}_n y = 0 \implies y \in \mathcal{N}(\mathbf{A}_n)\,.$$

Hence, $\mathcal{N}(\mathbf{A}_n) = [\mathbf{I} - \mathbf{f}\,'(x_n)(\mathbf{A}_n \mathbf{f}\,'(x_n))^{-1} \mathbf{A}_n] \mathbb{Y}$. Then

$$\begin{aligned} dist(\mathbf{f}(x_\infty), \mathcal{N}(\mathbf{A}_0)) &= \inf_{y \in Y} \big\| \mathbf{f}(x_\infty) - [\mathbf{I} - \mathbf{f}\,'(x_n)(\mathbf{A}_n \mathbf{f}\,'(x_n))^{-1} \mathbf{A}_n] y \big\| \\ &\le \|\mathbf{f}(x_\infty) - [\mathbf{I} - \mathbf{f}\,'(x_n)(\mathbf{A}_n \mathbf{f}\,'(x_n))^{-1} \mathbf{A}_n] \mathbf{f}(x_\infty)\| \\ &\le \frac{\|\mathbf{f}\,'(x_n)\|}{1 - \|\mathbf{I} - \mathbf{A}_n \mathbf{f}\,'(x_n)\|} \|\mathbf{A}_n \mathbf{f}(x_\infty)\|. \end{aligned}$$

The quotient on the right converges to $\|\mathbf{f}\,'(x_\infty)\|$, whereas $\|\mathbf{A}_n \mathbf{f}(x_\infty)\|$

$$\le \|\mathbf{A}_n \mathbf{f}(x_n)\| + \|\mathbf{A}_n [\mathbf{f}(x_\infty) - \mathbf{f}(x_n)]\| \le \delta_n + \beta_n \|\mathbf{f}(x_\infty) - \mathbf{f}(x_n)\| \to 0\,,$$

since $\beta_n \le \beta_\infty < \infty$ by assumption. Hence, $dist(\mathbf{f}(x_\infty), \mathcal{N}(\mathbf{A}_0)) = 0$, which means (because $\mathcal{N}(\mathbf{A}_0)$ is closed) that $\mathbf{f}(x_\infty) \in \mathcal{N}(\mathbf{A}_0)$, i.e., $\mathbf{A}_0 \mathbf{f}(x_\infty) = 0$. The equality $\mathbf{A}_\infty \mathbf{f}\,'(x_\infty) = \mathbf{I}$ is obtained by taking limits in the inequality $\|\mathbf{I} - \mathbf{A}_n \mathbf{f}\,'(x_n)\| \le \gamma_n$, for $\beta_\infty < \infty \implies \gamma_n = \beta_{n+1}/\beta_n - 1 \to 0$.

3° Let x^* be another solution of the equation $\mathbf{A}_0 \mathbf{f}(x) = 0$ and $\bar{R} := \|x^* - x_0\|$. Then

$$0 = \mathbf{A}_0 \big(\mathbf{f}(x^*) - \mathbf{f}(x_\infty)\big) = \mathbf{A}_0 \int_0^1 \mathbf{f}\,'\big(x_\infty + s(x^* - x_\infty)\big)(x^* - x_\infty) ds\,,$$

which shows that the operator $\mathbf{A}_0 \int_0^1 \mathbf{f}\,'\big(x_\infty + s(x^* - x_\infty)\big) ds$ is not invertible. It follows that

$$\left\| \mathbf{I} - \mathbf{A}_0 \int_0^1 \mathbf{f}\,'(x_\infty + s(x^* - x_\infty)) ds \right\| \ge 1\,.$$

On the other hand, $\left\|\mathbf{I}-\mathbf{A}_0\int_0^1\mathbf{f}\,'(x_\infty+s(x^*-x_\infty))ds\right\|$

$$\begin{aligned}
&\le \|\mathbf{I}-\mathbf{A}_0\mathbf{f}\,'(x_0)\|+\left\|\mathbf{A}_0\int_0^1\Big[\mathbf{f}\,'(x_\infty+s(x^*-x_\infty))-\mathbf{f}\,'(x_0)\Big]ds\right\|\\
&\le \bar\gamma_0+\bar\beta_0\int_0^1\Big\|\mathbf{f}\,'(x_\infty+s(x^*-x_\infty))-\mathbf{f}\,'(x_0)\Big\|ds\,.\\
&\le \gamma_0+\beta_0\int_0^1\Big\|\mathbf{f}\,'(x_\infty+s(x^*-x_\infty))-\mathbf{f}\,'(x_0)\Big\|ds\,. \qquad (2.29)
\end{aligned}$$

Since $\mathbf{f}$ is ω-regularly smooth by assumption, the integrand

$$\begin{aligned}
&\le \omega\Big(\omega^{-1}(\min\{\|\mathbf{f}\,'(x_\infty+s(x^*-x_\infty))\|\,,\|\mathbf{f}\,'(x_0)\|\}-\underline{h})+\\
&\qquad \|x_\infty+s(x^*-x_\infty)-x_0\|\Big)-\min\{\|\mathbf{f}\,'(x_\infty+s(x^*-x_\infty))\|\,,\|\mathbf{f}\,'(x_0)\|\}+\underline{h}\\
&=\omega\Big(\min\{\omega^{-1}(\|\mathbf{f}\,'(x_\infty+s(x^*-x_\infty))\|-\underline{h})\,,\omega^{-1}(\|\mathbf{f}\,'(x_0)\|-\underline{h})\}+\\
&\qquad\qquad \|x_\infty+s(x^*-x_\infty)-x_0\|\Big)-\\
&\qquad \omega\Big(\min\{\omega^{-1}(\|\mathbf{f}\,'(x_\infty+s(x^*-x_\infty))\|-\underline{h})\,,\omega^{-1}(\|\mathbf{f}\,'(x_0)\|-\underline{h})\}\Big).
\end{aligned}$$

Here $\|x_\infty+s(x^*-x_\infty)-x_0\|$

$$\begin{aligned}
&=\|s(x^*-x_0)+(1-s)(x_\infty-x_0)\|\le s\|x^*-x_0\|+(1-s)\|x_\infty-x_0\|\\
&\qquad\le s\bar R+(1-s)\bar t_\infty\,.
\end{aligned}$$

Besides, by (2.6), $\omega^{-1}(\|\mathbf{f}\,'(x_\infty+s(x^*-x_\infty))\|-\underline{h})$

$$\begin{aligned}
&\ge\Big(\omega^{-1}(\|\mathbf{f}\,'(x_0)\|-\underline{h})-\|x_\infty+s(x^*-x_\infty)-x_0\|\Big)^+\\
&\ge\Big(\bar\alpha_0-s\|x^*-x_0\|-(1-s)\|x_\infty-x_0\|\Big)^+=(\bar\alpha_0-s\bar R-(1-s)\bar t_\infty)^+.
\end{aligned}$$

Hence, due to concavity and monotonicity of ω, the integrand in eqref191

$$\begin{aligned}
&\le\omega\Big(\min\{(\bar\alpha_0-s\bar R-(1-s)\bar t_\infty)^+\,,\bar\alpha_0\}+s\bar R+(1-s)\bar t_\infty\Big)-\\
&\qquad\qquad \omega\Big(\min\{(\bar\alpha_0-s\bar R-(1-s)\bar t_\infty)^+\,,\bar\alpha_\infty\}\Big)\\
&=\omega\Big((\bar\alpha_0-s\bar R-(1-s)\bar t_\infty)^++s\bar R+(1-s)\bar t_\infty\Big)-\\
&\qquad\qquad \omega\Big((\bar\alpha_0-s\bar R-(1-s)\bar t_\infty)^+\Big)
\end{aligned}$$

and the integral

$$\leq \int_0^1 \Big[\omega\Big(\big(\bar{\alpha}_0 - s\bar{R} - (1-s)\bar{t}_\infty\big)^+ + s\bar{R} + (1-s)\bar{t}_\infty\Big) - \omega\Big(\big(\bar{\alpha}_0 - s\bar{R} - (1-s)\bar{t}_\infty\big)^+\Big)\Big] ds$$

$$= \int_0^1 e\big(\bar{\alpha}_0 \,, s\bar{R} + (1-s)\bar{t}_\infty\big) ds = \int_{t_\infty}^{\bar{R}} e(\bar{\alpha}_0 \,, \tau) \frac{d\tau}{\bar{R} - \bar{t}_\infty}$$

$$= \frac{\Psi(\bar{\alpha}_0 \,, \bar{R}) - \Psi(\bar{\alpha}_0 \,, \bar{t}_\infty)}{\bar{R} - \bar{t}_\infty} \,.$$

It follows that

$$\frac{1-\gamma_0}{\beta_0} \leq \frac{\Psi(\bar{\alpha}_0 \,, \bar{R}) - \Psi(\bar{\alpha}_0 \,, \bar{t}_\infty)}{\bar{R} - \bar{t}_\infty} =: f(\bar{R}) \,.$$

As the function $t \mapsto \Psi(\bar{\alpha}_0 \,, t)$ is convex, f is increasing and so

$$\frac{1-\gamma_0}{\beta_0} \leq f(\bar{R}) \iff \bar{R} \geq f^{-1}\Big(\frac{1-\gamma_0}{\beta_0}\Big) = R.$$

4° To obtain the third inequality in (2.25), use (2.19) to get the identity

$$x_{n+1} - x_\infty = \mathbf{A}_{n+1}\Big(\mathbf{f}(x_\infty) - \mathbf{f}(x_n) - \mathbf{f}'(x_n)(x_\infty - x_n)\Big) - \big(\mathbf{I} - \mathbf{A}_{n+1}\mathbf{f}'(x_n)\big)(x_\infty - x_n).$$

Taking norms gives

$$\Delta_{n+1} \leq \bar{\beta}_{n+1}\big\|\mathbf{f}(x_\infty) - \mathbf{f}(x_n) - \mathbf{f}'(x_n)(x_\infty - x_n)\big\| + \Delta_n\big\|\mathbf{I} - \mathbf{A}_{n+1}\mathbf{f}'(x_n)\big\| \,.$$

By (2.19),

$$\mathbf{I} - \mathbf{A}_{n+1}\mathbf{f}'(x_n) = \mathbf{I} - \big(2\mathbf{A}_n - \mathbf{A}_n\mathbf{f}'(x_n)\mathbf{A}_n\big)\mathbf{f}'(x_n) = \big(\mathbf{I} - \mathbf{A}_n\mathbf{f}'(x_n)\big)^2,$$

so that $\big\|\mathbf{I} - \mathbf{A}_{n+1}\mathbf{f}'(x_n)\big\| \leq \big\|\mathbf{I} - \mathbf{A}_n\mathbf{f}'(x_n)\big\|^2 = \bar{\gamma}_n^2$. Besides, by Lemma 2.4,

$$\big\|\mathbf{f}(x_\infty) - \mathbf{f}(x_n) - \mathbf{f}'(x_n)(x_\infty - x_n)\big\| \leq \Psi(\bar{\alpha}_n \,, \Delta_n) \,,$$

where $\bar{\alpha}_n := \omega^{-1}\big(\|\mathbf{f}(x_n)\| - \underline{h}\big)$. Hence, $\Delta_{n+1} \leq \bar{\beta}_{n+1}\Psi(\bar{\alpha}_n \,, \Delta_n) + \bar{\gamma}^2\Delta_n$. Since the function Ψ is not increasing in the first argument, by Lemma 2.4, $\bar{\alpha}_n \geq 0 \Longrightarrow \Psi(\bar{\alpha}_n \,, \Delta_n) \leq \Psi(0 \,, \Delta_n) = w(\Delta_n)$. Consequently,

$$\Delta_{n+1} \leq \Delta_n\left(\bar{\beta}_{n+1}\frac{w(\Delta_n)}{\Delta_n} + \bar{\gamma}_n^2\right) \leq \Delta_n\left(\beta_{n+1}\frac{w(\Delta_n)}{\Delta_n} + \gamma_n^2\right). \tag{2.30}$$

Because w is convex, $w(t)/t \leq w'(t) = \omega(t)$, whence $\Delta_n \to 0 \Longrightarrow w(\Delta_n)/\Delta_n \to 0$.

5° While proving claim 5° of Lemma 2.6, we show that, for the scalar function $f_a(x) := w(x) - a$, Ulm's method generates from the starter $(A_0 \,, x_0)$ the same sequence $(A_n \,, x_n)$ as the generator (2.21) from $\big(0 \,, A_0 \,, 1 - A_0\omega(x_0) \,, A_1 f_a(x_0)\big)$. So, if the sequence $(t_n \,, \beta_n \,, \gamma_n \,, \delta_n)$ diverges, then $(A_n \,, x_n)$ diverges too. □

2.4 Convergence theorem

Lemma 2.8 raises the question: which q_0 cause the sequences β_n and t_n to converge? They surely converge when $\beta_0 = 0 \leq \gamma_0 < 1$. This possibility, however, should be left out of our analysis, since it implies $\mathbf{A}_0 = \mathbf{0}$ (the iterations stall at $(\mathbf{A}_0, x_0)$). On the other hand, convergence is impossible if $\gamma_n \geq 1$ for some n: by (2.21),

$$\gamma \geq 1 \Longrightarrow \gamma_+ \geq 1 \Longrightarrow \delta_+ = (1+\gamma_+)\Big(\beta_+ \Psi\big((\bar{\alpha}_0 - t)^+, \delta\big) + \gamma^2 \delta\Big) \geq 2\delta \Longrightarrow \delta_n \to \infty .$$

Thus, the condition

$$\beta_0 > 0 \ \ \& \ \ \underset{n}{\&}\, \gamma_n < 1 \tag{2.31}$$

is necessary for convergence. In fact, it is also sufficient. While proving this, we use the abbreviation

$$e_n := \omega\big((\bar{\alpha}_0 - t_{n+1})^+ + \delta_n\big) - \omega\big((\bar{\alpha}_0 - t_{n+1})^+\big) .$$

Lemma 2.9. *Suppose that* (2.31) *is true. Then*

$$t_m < \bar{\alpha}_0 \Longrightarrow \underset{n=0}{\overset{m-1}{\&}} t_n = \bar{\alpha}_0 - \omega^{-1}\left(\omega(\bar{\alpha}_0) - \frac{1-\gamma_0}{\beta_0} + \frac{1-\gamma_n}{\beta_n}\right)$$

$$t_m < \bar{\alpha}_0 \leq t_{m+1} \Longrightarrow \underset{n=m}{\overset{\infty}{\&}} t_n \leq t_m + \omega^{-1}\left(\frac{1-\gamma_0}{\beta_0} - \frac{1-\gamma_n}{\beta_n} - \omega(\bar{\alpha}_0) + \omega(\bar{\alpha}_0 - t_m)\right) .$$

Proof. By (2.21), $\underset{n}{\&} \dfrac{1-\gamma_{n+1}}{\beta_{n+1}} = \dfrac{1-\gamma_n}{\beta_n} - e_n$ and so

$$\sum_0^{n-1} e_k = \frac{1-\gamma_0}{\beta_0} - \frac{1-\gamma_n}{\beta_n} . \tag{2.32}$$

If $t_m < \bar{\alpha}_0$, then $\underset{k=0}{\overset{m-1}{\&}} e_k = \omega(\bar{\alpha}_0 - t_{k+1} + \delta_k) - \omega(\bar{\alpha}_0 - t_{k+1}) = \omega(\bar{\alpha}_0 - t_k) - \omega(\bar{\alpha}_0 - t_{k+1})$ and so $\sum_0^{m-1} e_k = \omega(\bar{\alpha}_0) - \omega(\bar{\alpha}_0 - t_m)$. It follows that

$$\underset{n=0}{\overset{m-1}{\&}} \omega(\bar{\alpha}_0 - t_n) - \frac{1-\gamma_n}{\beta_n} = \omega(\bar{\alpha}_0) - \frac{1-\gamma_0}{\beta_0} \tag{2.33}$$

and

$$\underset{n=0}{\overset{m-1}{\&}} t_n = \bar{\alpha}_0 - \omega^{-1}\left(\omega(\bar{\alpha}_0) - \frac{1-\gamma_0}{\beta_0} + \frac{1-\gamma_n}{\beta_n}\right) .$$

If $\exists m$ with $t_m < \bar{\alpha}_0 \leq t_{m+1}$, then

$$e_n = \begin{cases} \omega(\bar{\alpha}_0 - t_n) - \omega(\bar{\alpha}_0 - t_{n+1}) & \text{for } n < m , \\ \omega(\delta_n) & \text{for } n \geq m , \end{cases}$$

and

$$n \geq m \Longrightarrow \sum_0^{n-1} e_k = \sum_0^{m-1} \left(\omega(\bar{\alpha}_0 - t_n) - \omega(\bar{\alpha}_0 - t_{k+1})\right) + \sum_m^{n-1} \omega(\delta_k)$$
$$= \omega(\bar{\alpha}_0) - \omega(\bar{\alpha}_0 - t_m) + \sum_m^{n-1} \omega(\delta_n) \,.$$

Invoking subadditivity of ω (an obvious consequence of its concavity) and using (2.32), we get

$$0 < \omega\left(\sum_m^{n-1} \delta_k\right) \leq \sum_m^{n-1} \omega(\delta_k) = \sum_0^{n-1} e_k - \omega(\bar{\alpha}_0) + \omega(\bar{\alpha}_0 - t_m)$$
$$= \frac{1-\gamma_0}{\beta_0} - \frac{1-\gamma_n}{\beta_n} - \omega(\bar{\alpha}_0) + \omega(\bar{\alpha}_0 - t_m) \,,$$

$$t_n - t_m = \textstyle\sum_m^{n-1} \delta_k \leq \omega^{-1}\left(\dfrac{1-\gamma_0}{\beta_0} - \dfrac{1-\gamma_n}{\beta_n} - \omega(\bar{\alpha}_0) + \omega(\bar{\alpha}_0 - t_m)\right) \,, \text{ and}$$

$$t_n \leq t_m + \omega^{-1}\left(\frac{1-\gamma_0}{\beta_0} - \frac{1-\gamma_n}{\beta_n} - \omega(\bar{\alpha}_0) + \omega(\bar{\alpha}_0 - t_m)\right) .$$

□

Corollary 2.10.

$$\mathop{\&}_n t_n < \bar{\alpha}_0 \Longrightarrow t_\infty = \bar{\alpha}_0 - \omega^{-1}\left(\omega(\bar{\alpha}_0) - \frac{1-\gamma_0}{\beta_0} + \frac{1}{\beta_\infty}\right)$$
$$\exists\, m \textit{ with } t_m < \bar{\alpha}_0 \leq t_{m+1} \Longrightarrow t_\infty \leq \omega^{-1}\left(\frac{1-\gamma_0}{\beta_0} - \frac{1}{\beta_\infty} - \omega(\bar{\alpha}_0) + 2\omega(\bar{\alpha}_0/2)\right) .$$

Proof. By the lemma,

$$\mathop{\&}_n t_n < \bar{\alpha}_0 \Longrightarrow t_\infty = \bar{\alpha}_0 - \omega^{-1}\left(\omega(\bar{\alpha}_0) - \frac{1-\gamma_0}{\beta_0} + \lim \frac{1-\gamma_n}{\beta_n}\right) .$$

If $\beta_\infty < \infty$, then $\gamma_n = \beta_{n+1}/\beta_n - 1 \to 0$ and $\lim(1-\gamma_n)/\beta_n = 1/\beta_\infty$. Otherwise, $1-\gamma_n \in (0,1]$ by (2.31) and $\lim(1-\gamma_n)/\beta_n = 0$. In both cases, this limit $= 1/\beta_\infty$. If $t_m < \bar{\alpha}_0 \leq t_{m+1}$, then (with $a := (1-\gamma_0)/\beta_0 - 1/\beta_\infty - \omega(\bar{\alpha}_0)$ for short and superadditivity of ω^{-1} in mind) the lemma implies

$$t_\infty \leq \omega^{-1}\left(\omega(t_m)\right) + \omega^{-1}\left(a + \omega(\bar{\alpha}_0 - t_m)\right) \leq \omega^{-1}\left(a + \omega(t_m) + \omega(\bar{\alpha}_0 - t_m)\right)$$
$$\leq \omega^{-1}\left(a + \sup_{0 \leq t \leq \bar{\alpha}_0} \left(\omega(t) + \omega(\bar{\alpha}_0 - t)\right)\right) = \omega^{-1}\left(a + 2\omega(\bar{\alpha}_0/2)\right) .$$

□

The corollary shows that, if $\beta_0 > 0$ & $\beta_\infty < \infty$, then the sequence t_n converges if and only if the starter q_0 satisfies $\underset{n}{\&}\gamma_n < 1$. The following lemma describes all such q_0 (the *convergence domain* of the generator (2.21)). Proving it, we use the abbreviations

$$\alpha := (\bar{\alpha}_0 - t)^+ \ , \ \alpha_n := (\bar{\alpha}_0 - t_n)^+ \ .$$

Proposition 2.11. 1° *If* $\beta_0 > 0$, *then* $\underset{n}{\&}\gamma_n < 1 \Longleftrightarrow \beta_0 \le f_\infty(\alpha_0, \gamma_0, \delta_0)$, *where* $f_0(\alpha, \gamma, \delta) := (1-\gamma)/e(\alpha, \delta)$ *and* $f_{n+1}(\alpha, \gamma, \delta)$ *is the solution for* β *of the equation*

$$f_n\Big((\alpha-\delta)^+,\ \gamma^2 + \beta_+ e(\alpha,\delta)\,,\,(1+\gamma^2+\beta_+ e(\alpha,\delta))\,(\beta_+\Psi(\alpha,\delta)+\gamma^2\delta)\Big) = \beta_+$$

where (for brevity's sake) $\beta_+ := \beta(1+\gamma)$ *and* $e(\alpha,\delta) := \omega\big((\alpha-\delta)^+ + \delta\big) - \omega\big((\alpha-\delta)^+\big)$.

2° *The function* f_∞ *is a solution of the system (a functional equation with an end condition)*

$$x\Big((\alpha-\delta)^+, \gamma^2 + x(\alpha,\gamma,\delta)(1+\gamma)e(\alpha,\delta)\ ,\ (1+\gamma^2+x(\alpha,\gamma,\delta)(1+\gamma)e(\alpha,\delta))$$
$$\big(x(\alpha,\gamma,\delta)(1+\gamma)\Psi(\alpha,\delta)+\gamma^2\delta\big)\Big) = x(\alpha,\gamma,\delta)(1+\gamma)$$
$$x(\alpha,0,0) = \beta_\infty. \tag{2.34}$$

Proof. By (2.21), $\gamma_{n+1} < 1 \Longleftrightarrow \gamma_n^2 + \beta_{n+1}e_n < 1 \Longleftrightarrow \beta_n < f_0(\alpha_n\,,\gamma_n\,,\beta_n)$. As $\gamma_{n+1} < 1 \Longrightarrow \gamma_n < 1$, f_0 is positive, not decreasing in the first argument and decreasing in the second and in the third. Suppose that, for some $k \ge 0$, $\beta_{n-k} < f_k(\alpha_{n-k}, \gamma_{n-k}, \delta_{n-k})$, where f_k is positive, not decreasing in the first argument and decreasing in the second and the third. Using (2.21), we rewrite this inequality as

$$F_k(q_{n-k-1}) > 0\ ,\ q_n := (\alpha_n, \beta_n, \gamma_n, \delta_n),$$

where

$$F_k(q) := f_k\Big((\alpha-\delta)^+,\ \gamma^2 + \beta_+\,,\,(1+\gamma^2+\beta_+ e(\alpha\,,\delta))\ (\beta_+\Psi(\alpha,\delta)+\gamma^2\delta)\Big) - \beta_+ \tag{2.35}$$

and $\beta_+ := \beta(1+\gamma)$ for short. Since f_k is decreasing in the second and the third arguments, the function $\beta \mapsto F_k(q)$ is decreasing in the interval $(0\,,\infty)$ from $f_k((\alpha-\delta)^+\,,\gamma^2)\,,(1+\gamma^2)\,\gamma^2\delta) > 0$ to

$$\lim_{\beta\to\infty}\Big(f_k((\alpha-\delta)^+,\ \gamma^2+\beta_+ e(\alpha\,,\delta)\,,\,(1+\gamma^2+$$
$$\beta_+ e(\alpha\,,\delta))(\beta_+\Psi(\alpha,\delta)+\gamma^2\delta) - \beta_+\Big)$$
$$\le \lim_{\beta\to\infty}\Big(f_k((\alpha-\delta)^+,\ \gamma^2\,,\gamma^2\delta(1+\gamma^2)) - \beta\Big) = -\infty\ .$$

Therefore, the equation $F_k(q) = 0$ for β has a unique positive solution, which we denote $f_{k+1}(\alpha, \gamma, \delta)$:

$$F_k\big(\alpha, f_{k+1}(\alpha,\gamma,\delta),\gamma,\delta\big) = 0 \ . \tag{2.36}$$

Moreover, f_{k+1} is not decreasing in α and decreasing in γ and δ . Indeed, by the induction hypothesis, f_k is not decreasing in the first argument, while the functions e and Ψ are not increasing. Hence, F_k is not decreasing in α and decreasing in β. It follows that

$$\begin{aligned}\alpha < \alpha' \Longrightarrow F_k\big(\alpha, f_{k+1}(\alpha,\gamma,\delta),\gamma,\delta\big) = 0 = F_k\big(\alpha', f_{k+1}(\alpha',\gamma,\delta),\gamma,\delta\big) \\ \geq F_k\big(\alpha, f_{k+1}(\alpha',\gamma,\delta),\gamma,\delta\big) \\ \Longrightarrow f_{k+1}(\alpha',\gamma,\delta) \geq f_{k+1}(\alpha,\gamma,\delta) \ .\end{aligned}$$

Similarly,

$$\gamma < \gamma' \Longrightarrow f_{k+1}(\alpha,\gamma',\delta) < f_{k+1}(\alpha,\gamma,\delta)$$

and

$$\delta < \delta' \Longrightarrow f_{k+1}(\alpha,\gamma,\delta') < f_{k+1}(\alpha,\gamma,\delta).$$

Thus,

$$\beta_{n-k} < f_k(\alpha_{n-k},\gamma_{n-k},\delta_{n-k}) \Longrightarrow \beta_{n-k-1} < f_{k+1}(\alpha_{n-k-1},\gamma_{n-k-1},\delta_{n-k-1}).$$

By induction, $\gamma_{n+1} < 1 \Longleftrightarrow \beta_0 < f_n(\alpha_0,\gamma_0,\delta_0)$. It follows that

$$\mathop{\&}_n \gamma_n < 1 \Longleftrightarrow \beta_0 \leq \inf_n f_n(\alpha_0,\gamma_0),\delta_0).$$

The sequence f_n is pointwise decreasing:

$$\mathop{\&}_n f_{n+1}(\alpha,\beta,\gamma) < f_n(\alpha,\beta,\gamma) \ . \tag{2.37}$$

This is proved inductively. First, we have to verify that $f_1(\alpha,\beta,\gamma) < f_0(\alpha,\beta,\gamma)$. Since F_k is decreasing with respect to its second argument, it is enough to show that $F_0\big(\alpha, f_0(\alpha,\gamma,\delta),\gamma,\delta\big) < F_0\big(\alpha, f_1(\alpha,\gamma,\delta),\gamma,\delta\big) = 0$ (see (2.36). By definition,

$$\begin{aligned}F_0\big(\alpha, f_0(\alpha,\gamma,\delta),\gamma,\delta\big) = f_0\big((\alpha-\delta)^+,\ \gamma^2 + f_0(\alpha,\gamma,\delta)(1+\gamma)e(\alpha,\delta)\ ,\ \big(1+\gamma^2+ \\ f_0(\alpha,\gamma,\delta)(1+\gamma)e(\alpha,\delta)\big)\big(f_0(\alpha,\gamma,\delta)(1+\gamma)\Psi(\alpha,\delta)+\gamma^2\delta\big)\big) - f_0(\alpha,\gamma,\delta)(1+\gamma),\end{aligned} \tag{2.38}$$

where $\gamma^2 + f_0(\alpha,\gamma,\delta)(1+\gamma)e(\alpha,\delta) = 1$. So, $F_0\big(\alpha, f_0(\alpha,\gamma,\delta),\gamma,\delta\big)$

$$\begin{aligned}&= f_0\left((\alpha-\delta)^+, 1\,, 2\big(f_0(\alpha,\gamma,\delta)(1+\gamma)\Psi(\alpha,\delta) + \gamma^2\delta\big)\right) - \frac{1-\gamma}{e(\alpha,\delta)}(1+\gamma) \\ &= -\frac{1-\gamma^2}{e(\alpha,\delta)} < 0.\end{aligned}$$

Suppose now that $f_n(\alpha,\gamma,\delta) < f_{n-1}(\alpha,\gamma,\delta)$ for some $n \ge 1$. Then

$$\begin{aligned}
&F_{n-1}\big(\alpha, f_n(\alpha,\gamma,\delta),\gamma,\delta\big) = 0 = F_n\big(\alpha, f_{n+1}(\alpha,\gamma,\delta),\gamma,\delta\big) = \\
&\qquad f_n\big((\alpha-\delta)^+,\, \gamma^2 + f_{n+1}(\alpha,\gamma,\delta)(1+\gamma)e(\alpha,\delta) \\
&\qquad\qquad (1+\gamma^2 + f_{n+1}(\alpha,\gamma,\delta)(1+\gamma)e(\alpha,\delta))\cdot \\
&\qquad\qquad\qquad \big(f_{n+1}(\alpha,\gamma,\delta)(1+\gamma)\Psi(\alpha,\delta)+\gamma^2\delta\big)\big) - f_{n+1}(\alpha,\gamma,\delta)(1+\gamma) \\
&< f_{n-1}\big((\alpha-\delta)^+,\, \gamma^2 + f_{n+1}(\alpha,\gamma,\delta)(1+\gamma)e(\alpha,\delta) \\
&\qquad (1+\gamma^2 + f_{n+1}(\alpha,\gamma,\delta)(1+\gamma)e(\alpha,\delta))\cdot \\
&\qquad\qquad \big(f_{n+1}(\alpha,\gamma,\delta)(1+\gamma)\Psi(\alpha,\delta)+\gamma^2\delta\big)\big) - f_{n+1}(\alpha,\gamma,\delta)(1+\gamma) \\
&\qquad = F_{n-1}\big(\alpha, f_{n+1}(\alpha,\gamma,\delta),\gamma,\delta\big) \Longrightarrow f_{n+1}(\alpha,\gamma,\delta) < f_n(\alpha,\gamma,\delta)\ ,
\end{aligned}$$

because F_{n-1} is decreasing with respect to the second argument. By induction, (2.37) is proved. Now $\inf_n f_n = f_\infty$. Taking limits in (2.36) yields $F_\infty\big(\alpha, f_\infty(\alpha,\gamma,\delta),\gamma,\delta\big) = 0$, i.e.,

$$\begin{aligned}
&f_\infty\Big((\alpha-\delta)^+,\gamma^2 + f_\infty(\alpha,\gamma,\delta)(1+\gamma)e(\alpha,\delta)\ , \\
&\qquad (1+\gamma^2+f_\infty(\alpha,\gamma,\delta)(1+\gamma)e(\alpha,\delta))\big(f_\infty(\alpha,\gamma,\delta)(1+\gamma)\Psi(\alpha,\delta)+\gamma^2\delta\big)\Big) \\
&= f_\infty(\alpha,\gamma,\delta)(1+\gamma)\ .
\end{aligned}$$

Besides, (2.21) implies $f_\infty(\alpha,0,0) = \beta_\infty$. Thus, the function $f_\infty(\alpha,\gamma,\delta)$ is a solution of the system (2.34). □

In the special case when ω is linear, $\omega(t) = ct$ (this means that the operator $\mathbf{f}$ is Lipschitz smooth), the generator (2.21) simplifies into

$$\beta_+ := \beta(1+\gamma)\ ,\ \gamma_+ := \gamma^2 + c\beta_+\delta\ ,\ \delta_+ := \delta(1+\gamma_+)\big(0.5c\beta_+\delta+\gamma^2\big)\ . \tag{2.39}$$

This generator can be simplified even further by the change of variables

$$r := \gamma^2\ ,\ s := c\beta_+\delta\ .$$

After the change, it becomes

$$\beta_+ := \beta\big(1+\sqrt{r}\big)\ ,\ r_+ := (r+s)^2\ ,\ s_+ := s(1+r+s)^2(0.5s+r). \tag{2.40}$$

For this generator, $\underset{n}{\&}\, r_n < 1 \Longleftrightarrow \beta_0 \le f_\infty(r_0,s_0)$, where f_∞ is a solution of the system (an analog of (2.34))

$$x\Big((r+s)^2\ ,\ s(1+r+s)^2(0.5s+r)\Big) = x(r,s)\left(1+\sqrt{r}\right)\ \&\ x(0,0) = \beta_\infty\ . \tag{2.41}$$

Unlike the system (2.34), this one admits a solution expressed through elementary functions, which in turn leads to an explicit invariant of the generator (2.40) (see Section 1.6).

Proposition 2.12. 1° *The function*

$$f_\infty(r,s) := \frac{\beta_\infty}{1+\sqrt{r}}\sqrt{(1-r)^2-2s}\ ,\ r\geq 0\leq s\leq 0.5(1-r)^2,$$

is the only solution of the system (2.41).

2° *The function*

$$I(\beta,r,s) := \frac{(1-r)^2-2s}{\beta^2\left(1+\sqrt{r}\right)^2}$$

is an invariant of the generator (2.40).

3° *The sequence* $(\beta_n\,,r_n\,,s_n)$, *generated by the generator* (2.40) *from a starter* $(\beta_0\,,r_0\,,s_0)$, *converges* $\big($*to* $(\beta_\infty\,,0\,,0)\big)$ *if and only if*

$$\beta_0\geq 0\ \ \&\ \ r_0\geq 0\ \ \&\ \ 0\leq s_0\leq 0.5(1-r_0)^2.$$

Proof. 1° Obviously, $f_\infty(0,0)=\beta_\infty$. Besides $f_\infty\Big(r+s\,,s(1+r+s)^2(0.5s+r)\Big)$

$$\begin{aligned}
&= \frac{\beta_\infty}{1+r+s}\sqrt{\left(1-(r+s)^2\right)^2-2s(1+r+s)^2(0.5s+r)}\\
&= \frac{\beta_\infty}{1+r+s}\sqrt{(1-r-s)^2(1+r+s)^2-2s(1+r+s)^2(0.5s+r)}\\
&= \beta_\infty\sqrt{(1-r-s)^2-2s(0.5s+r)}\\
&= \beta_\infty\sqrt{(1-r)^2-2s(1-r)+s^2-s^2-2sr}\\
&= \beta_\infty\sqrt{(1-r)^2-2s} = f_\infty(r,s)\left(1+\sqrt{r}\right)\ ,
\end{aligned}$$

i.e., f_∞ also satisfies the functional equation (2.41). To see that this solution is unique, note that the generator (2.40) is invertible. Indeed, (2.40) implies

$$r+s=\sqrt{r_+}\ \ \&\ \ \frac{s_+}{\left(1+\sqrt{r_+}\right)^2}=s(0.5s+r)$$

and so

$$\begin{aligned}
\frac{s_+}{\left(1+\sqrt{r_+}\right)^2} &= \left(\sqrt{r_+}-r\right)\left(0.5\left(\sqrt{r_+}-r\right)+r\right)=0.5\left(\sqrt{r_+}-r\right)\left(\sqrt{r_+}+r\right)\\
&= 0.5\left(r_+-r^2\right).
\end{aligned}$$

Therefore,

$$r=\sqrt{r_+-\frac{2s_+}{\left(1+\sqrt{r_+}\right)^2}}\ ,\ s=\sqrt{r_+}-\sqrt{r_+-\frac{2s_+}{\left(1+\sqrt{r_+}\right)^2}}\ ,\ \beta=\frac{\beta_+}{1+\sqrt[4]{r_+-\dfrac{2s_+}{\left(1+\sqrt{r_+}\right)^2}}}\ .$$

Hence, for any solution $x(r\,,s)$ of (2.41), $x(0,0)=\beta_\infty=f_\infty(0,0)$ implies $\underset{n}{\&}\, x(r_n\,,s_n)=f_\infty(r_n\,,s_n)$ and, in particular, $x(r_0\,,s_0)=f_\infty(r_0\,,s_0)$ for any $(r_0\,,s_0)$ in the past history of $(0\,,0)$.

2°

$$\begin{aligned} I(\beta_+,r_+,s_+) &= \frac{(1-r_+)^2-2s_+}{\beta_+^2\left(1+\sqrt{r_+}\right)^2} = \frac{\left(1-(r+s)^2\right)^2-2s(1+r+s)^2(0.5s+r)}{\beta_+^2(1+r+s)^2} \\ &= \frac{(1-r-s)^2-2s(0.5s+r)}{\beta_+^2} \\ &= \frac{(1-r)^2-2s(1-r)+s^2-s^2-2rs}{\beta_+^2} \\ &= \frac{(1-r)^2-2s}{\beta^2\left(1+\sqrt{r}\right)^2} = I(\beta,\gamma,s)\ . \end{aligned}$$

3° By 2°,

$$I(\beta_0\,,r_0\,,s_0)=I(\beta_\infty\,,0\,,0) \iff \frac{\beta_\infty}{1+\sqrt{r_0}}\sqrt{(1-r_0)^2-2s_0}=\beta_\infty\,.$$

This implies $r_0\geq 0\leq s_0\leq 0.5(1-r_0)^2$, which is the claim. □

The last two equations of the generator (2.40) do not depend on β and so the truncated generator

$$r_+ := (r+s)^2\ ,\ \ s_+ := s(1+r+s)^2(0.5s+r)\,, \tag{2.42}$$

can be studied independently. In contrast with the generator (2.40), whose fixed points constitute an interval $\{(\beta\,,0\,,0)\mid 0\leq\beta\leq\beta_\infty\}$ of the β-axis, the generator (2.42) has (in the first quadrant) two: $(0\,,0)$, $(a\,,\sqrt{a}-a)$, where $a=0.29...$ (the root of the equation $\sqrt{r}\left(1+\sqrt{r}\right)^3=2$). The attraction basin of the fixed point $(0\,,0)$ is rather complicated, but if $r\geq a\ \&\ s\geq\sqrt{a}-a$, then $r_+=(r+s)^2\geq\left(a+\sqrt{a}-a\right)^2=a$ and so (keep in mind (2.42))

$$\begin{aligned} s_+ \geq \left(\sqrt{a}-a\right)\left(1+\sqrt{a}\right)^2\left(0.5\left(\sqrt{a}-a\right)+a\right) &= 0.5\left(\sqrt{a}-a\right)\left(1+\sqrt{a}\right)^3\sqrt{a} \\ &= \sqrt{a}-a \end{aligned}$$

(because of definition of a). Therefore, the condition $\underset{n}{\&}\, r_n<a$ is necessary for convergence of the sequence $(r_n\,,s_n)$, generated by the generator (2.42) from a starter $(r_0\,,s_0)$, to $(0\,,0)$. Is it sufficient? The answer is provided by

Proposition 2.13. 1° $\underset{n}{\&}\, r_n<a \iff s_0\leq f_\infty(r_0)$, *where* $f_0(r):=\sqrt{a}-r$ *and* $f_{n+1}(r)$ *is the (unique) solution for* s *of the equation*

$$f_n\left((r+s)^2\right)=s(1+r+s)^2(0.5s+r)\,.$$

2° *The function* f_∞ *is the only solution of the system*

$$x\Big(\big(r+x(r)\big)^2\Big)=x(r)\big(1+r+x(r)\big)^2\big(0.5x(r)+r\big)\ \ \&\ \ x(0)=0\,. \tag{2.43}$$

Proof. 1° Let $\mathop{\&}\limits_{k=0}^{n} r_n < a$. Then

$$r_{n+1} < a \Longleftrightarrow (r_n + s_n)^2 < a \Longleftrightarrow s_n < \sqrt{a} - r_n =: f_o(r_n) .$$

Suppose that, for some $k \geq 0$,

$$r_{n+1} < a \Longleftrightarrow s_{n-k} < f_k(r_{n-k}) ,$$

where f_k is decreasing in $(0, \infty)$ and $f_k(1) < 0 < f_k(a)$. Let b_k be the zero of f_k: $f_k(b_k) = 0$. Clearly, $a < b_k < 1$. Using (2.42), rewrite the last inequality as

$$F_k(r_{n-k-1}, s_{n-k-1}) > 0 , \tag{2.44}$$

where

$$F_k(r, s) := f_k\big((r+s)^2\big) - s(1+r+s)^2(0.5s + r) . \tag{2.45}$$

As f_k is decreasing by the induction hypothesis, the function $s \mapsto F_k(r_{n-k-1}, s)$ is decreasing in $(0, 1)$ from $F_k(r_{n-k-1}, 0) = f_k\big(r_{n-k-1}^2\big) > f_k(b_k) = 0$ $\big($since $r_{n-k-1} < a \Longrightarrow r_{n-k-1}^2 < a^2 < a < b_k\big)$ to

$$\begin{aligned} F_k(r_{n-k-1}, 1) &= f_k\big((r_{n-k-1} + 1)^2\big) - (2 + r_{n-k-1})^2(0.5 + r_{n-k-1}) \\ &< f_k(1) - (2 + r_{n-k-1})^2(0.5 + r_{n-k-1}) \\ &< -(2 + r_{n-k-1})^2(0.5 + r_{n-k-1}) < 0 . \end{aligned}$$

So, the equation $F_k(r_{n-k-1}, s) = 0$ is uniquely solvable for s in $(0, 1)$. Denote the solution $f_{k+1}(r_{n-k-1})$:

$$F_k\big(r, f_{k+1}(r)\big) = 0 , \ \forall r > 0 . \tag{2.46}$$

In particular, $F_k\big(r_{n-k-1}, f_{k+1}(r_{n-k-1})\big) = 0$. Inasmuch as F_k is decreasing with respect to the second argument, comparison with (2.44) shows that $s_{n-k-1} < f_{k+1}(r_{n-k-1})$. The function f_{k+1} is decreasing in $(0, \infty)$:

$$\begin{aligned} 0 < r < r' &\Longrightarrow F_k\big(r, f_{k+1}(r)\big) = 0 = F_k\big(r', f_{k+1}(r')\big) < F_k\big(r, f_{k+1}(r')\big) \\ &\Longrightarrow f_{k+1}(r) > f_{k+1}(r') , \end{aligned}$$

since F_k is decreasing in each of its two arguments. Besides,

$$F_k\big(1, f_{k+1}(1)\big) = 0 > f_k(1) = F_k(1, 0) \Longrightarrow f_{k+1}(1) < 0$$

and $f_{k+1}(a) > 0$, since the converse $f_{k+1}(a) \leq 0$ implies (due to (2.46) and monotonicity of F_k with respect to the second argument)

$$0 = F_k\big(a, f_{k+1}(a)\big) \geq F_k(a, 0) = f_k\left(a^2\right) > f_k(a) > 0$$

(by the induction hypothesis), which is a contradiction. Thus, $r_{n+1} < a \Longleftrightarrow$

$s_{n-k} < f_k(r_{n-k})$ implies $r_{n+1} < a \iff s_{n-k-1} < f_{k+1}(r_{n-k-1})$. By induction, $r_{n+1} < a \iff s_0 < f_n(r_0)$ and $\underset{n}{\&}\, r_n < a \iff s_0 \le \inf_n f_n(r_0)$. The sequence f_n is pointwise decreasing in $(0\,,a)$:

$$\underset{n}{\&}\, f_{n+1}(r) < f_n(r)\ ,\ \forall r \in (0\,,a)\,. \tag{2.47}$$

This is verified inductively. First, we have to show that $0 < r < a \Longrightarrow f_1(r) < f_0(r)$ or, as F_0 is decreasing with respect to the second argument, that $0 = F_0(r\,,f_1(r)) > F_0(r\,,f_0(r))$. By definition (see (2.45)),

$$F_0(r\,,f_0(r)) = f_0\Big(\big(r + f_0(r)\big)^2\Big) - f_0(r)\big(1 + r + f_0(r)\big)^2\big(0.5f_0(r) + r\big)\,,$$

where $f_0(r) = \sqrt{a} - r$. So,

$$\begin{aligned} F_0(r\,,f_0(r)) &= f_0(a) - \left(\sqrt{a} - r\right)\left(1 + \sqrt{a}\right)^2\left(0.5\left(\sqrt{a} - r\right) + r\right) \\ &= f_0(a) - 0.5\left(\sqrt{a} - r\right)\left(1 + \sqrt{a}\right)^2\left(\sqrt{a} + r\right) \\ &= \sqrt{a} - a - 0.5\left(1 + \sqrt{a}\right)^2\left(a - r^2\right) \end{aligned}$$

and

$$\begin{aligned} 0 < r < a \Longrightarrow F_0(r\,,f_0(r)) &< \sqrt{a} - a - 0.5\left(1 + \sqrt{a}\right)^2\left(a - a^2\right) \\ &= \sqrt{a} - a - 0.5a\left(1 + \sqrt{a}\right)^2\left(1 + \sqrt{a}\right)\left(1 - \sqrt{a}\right) \\ &= \sqrt{a}\left(1 - \sqrt{a} - 0.5\sqrt{a}\left(1 + \sqrt{a}\right)^3\left(1 - \sqrt{a}\right)\right) \\ &= \sqrt{a}\left(1 - \sqrt{a}\right)\left(1 - 0.5\sqrt{a}\left(1 + \sqrt{a}\right)^3\right) = 0\,, \end{aligned}$$

because $\sqrt{a}\left(1 + \sqrt{a}\right)^3 = 2$ by definition of a. Thus, $0 < r < a \Longrightarrow f_1(r) < f_0(r)$. Suppose now that $0 < r < a \Longrightarrow f_n(r) < f_{n-1}(r)$ for some $n \ge 1$. Then

$$\begin{aligned} &F_{n-1}(r, f_n(r)) = 0 = F_n(r, f_{n+1}(r)) \\ &= f_n\Big(\big(r + f_{n+1}(r)\big)^2\Big) - f_{n+1}(r)\left(1 + r + f_{n+1}(r)\right)\left(0.5f_{n+1}(r) + r\right) \\ &< f_{n-1}\Big(\big(r + f_{n+1}(r)\big)^2\Big) - f_{n+1}(r)\left(1 + r + f_{n+1}(r)\right)\left(0.5f_{n+1}(r) + r\right) \\ &= F_{n-1}(r\,,f_{n+1}(r))\,. \end{aligned}$$

Inasmuch as F_{n-1} is decreasing with respect to the second argument,

$$F_{n-1}(r, f_n(r)) < F_{n-1}(r\,,f_{n+1}(r)) \Longrightarrow f_n(r) > f_{n+1}(r)\,.$$

By induction, (2.48) is proved. Then $\inf_n f_n = \lim_n f_n = f_\infty$ and $\underset{n}{\&}\, r_n < a \iff s_0 \le f_\infty(r_0)$.

2° Taking limits in (2.46) results in $F_\infty(r\,,f_\infty(r)) = 0$, i.e., f_∞ solves the functional equation (2.43). Besides,

$$0 = F_\infty(0\,,f_\infty(0)) \Longrightarrow f_\infty\left(f_\infty(0)^2\right) = 0.5f_\infty(0)^2\left(1+f_\infty(0)\right)^2 \Longrightarrow f_\infty(0) = 0\,,$$

for $t := f_\infty(0)^2 > 0 \Longrightarrow f_\infty(t) = 0.5t\left(1+\sqrt{t}\right)^2$ and this function is not a solution of that functional equation. Indeed, $x(t) = 0.5t\left(1+\sqrt{t}\right)^2$

$$\Longrightarrow t + x(t) = t\left(1 + 0.5\left(1+\sqrt{t}\right)^2\right)$$

$$\Longrightarrow \left(t+x(t)\right)^2 = t^2\left(1+0.5\left(1+\sqrt{t}\right)^2\right)^2$$

$$\Longrightarrow x\left(\left(t+x(t)\right)^2\right) = 0.5t^2\left(1+0.5\left(1+\sqrt{t}\right)^2\right)\left(1+t+0.5t\left(1+\sqrt{t}\right)^2\right)^2,$$

while, for the same $x(t)$, $x(t)\left(1+t+x(t)\right)^2(0.5x(t)+t)$

$$= 0.5t\left(1+\sqrt{t}\right)^2\left(1+t+0.5t\left(1+\sqrt{t}\right)^2\right)^2\left(0.25t\left(1+\sqrt{t}\right)^2+t\right)$$

$$= 0.125t^2\left(1+\sqrt{t}\right)^2\left(1+t+0.5t\left(1+\sqrt{t}\right)^2\right)^2\left(\left(1+\sqrt{t}\right)^2+4\right).$$

So, if $x(t) = 0.5t\left(1+\sqrt{t}\right)^2$, then

$$\left(\left(t+x(t)\right)^2\right) = x(t)\left(1+t+x(t)\right)^2(0.5x(t)+t)$$

$$\Longleftrightarrow \left(1+0.5\left(1+\sqrt{t}\right)^2\right)^2 = 0.25\left(1+\sqrt{t}\right)^2\left(\left(1+\sqrt{t}\right)^2+4\right)$$

$$\Longleftrightarrow 1 + \left(1+\sqrt{t}\right)^2 + 0.25\left(1+\sqrt{t}\right)^4 = 0.25\left(1+\sqrt{t}\right)^4 + \left(1+\sqrt{t}\right)^2,$$

which is impossible. It follows that f_∞ is a solution of the system (2.43).

To see the uniqueness of this solution, let x be a solution and consider the generator

$$p_+ := \left(p + x(p)\right)^2. \tag{2.48}$$

For it, $r = p$ & $s = x(p)$

$$\Longrightarrow r_+ := (r+s)^2 = \left(p+x(p)\right)^2 = p_+$$

$$\Longrightarrow s_+ := s(1+r+s)^2(0.5s+r) = x(p)\left(1+p+x(p)\right)^2(0.5x(p)+p)$$

$$= x\left(\left(p+x(p)\right)^2\right) = x(p_+)$$

(because x is a solution of (2.43)). It follows (by induction) that

$$r_0 = p_0 \;\; \& \;\; s_0 = x(p_0) \Longrightarrow \underset{n}{\&}\Big(r_n = p_n \;\; \& \;\; s_n = x(p_n)\Big)$$
$$\Longrightarrow \underset{n}{\&}\Big(r_n = p_n \;\; \& \;\; s_n = x(r_n)\Big),$$

i.e., $p_+ := \big(p+x(p)\big)^2 \;\&\; s = x(p)$ is equivalent to (2.42). Hence, $x = f_\infty$. □

The knowledge of an invariant allows us to get an explicit expression for the convergence domain of the generator (2.39).

Lemma 2.14. *Let the sequence* $(\beta_n, \gamma_n, \delta_n)$ *be generated by the generator (2.39) from the starter* $(\beta_0, \gamma_0, \delta_0)$ *with* $\beta_0 > 0$ *. It converges if and only if*

$$\gamma_0 \geq 0 \leq \delta_0 < \frac{(1-\gamma_0)^2(1+\gamma_0)}{2c\beta_0} .$$

In this case,

$$\beta_n = \frac{1}{\sqrt{I_0}}\sqrt{(1-\gamma_n)^2 - \frac{2s_n}{(1+\gamma_n)}},$$

$$t_n = c^{-1}\left(\frac{1-\gamma_0}{\beta_0} - \frac{1-\gamma_n}{\beta_n}\right) , \; t_\infty = c^{-1}\left(\frac{1-\gamma_0}{\beta_0} - \sqrt{I_0}\right),$$

where

$$I_0 := \frac{1}{\beta_0^2}\left((1-\gamma_0)^2 - \frac{2s_0}{(1+\gamma_0)^2}\right) , \; s_0 := c\beta_0\delta_0(1+\gamma_0).$$

Proof. If the sequence $(\beta_n, \gamma_n, \delta_n)$ converges (that is $\beta_\infty < \infty$), then $\gamma_n = \beta_{n+1}/\beta_n - 1 \to 0$ and $s_n = \gamma_{n+1} - \gamma_n^2 \to 0$. So, $\gamma_\infty = s_\infty = 0$ and $\underset{n}{\&}\, I(\beta_n, r_n, s_n) = I(\beta_\infty, 0, 0) = \beta_\infty^{-2} > 0$. In particular,

$$\frac{1}{\beta_0^2}\left((1-\gamma_0)^2 - \frac{2s_0}{(1+\gamma_0)^2}\right) = \frac{1}{\beta_\infty^2} > 0,$$

i.e., $s_0 = c\beta_0\delta_0(1+\gamma_0) < 0.5\left(1-\gamma_0^2\right)^2$ or, equivalently,

$$\delta_0 < \frac{(1-\gamma_0)^2(1+\gamma_0)}{2c\beta_0}.$$

Conversely, if $s_0 < 0.5\big(1-\gamma_0^2\big)^2$, then $I(\beta_\infty, r_\infty, s_\infty) = I_0 > 0$, that is,

$$\frac{1}{\beta_\infty^2}\left((1-\gamma_\infty)^2 - \frac{2s_\infty}{(1+\gamma_\infty)^2}\right) > 0.$$

So, $\beta_\infty < \infty$, $\gamma_\infty = s_\infty = 0$, and $\underset{n}{\&} I(\beta_n, r_n, s_n) = I(\beta_\infty, 0, 0) = \beta_\infty^{-2}$. Solving the equation $I(\beta_n, r_n, s_n) = \beta_\infty^{-2}$ for β_n, we obtain

$$\beta_n = \beta_\infty \sqrt{\left(1 - \sqrt{r_n}\right)^2 - \frac{2s_n}{(1 + \sqrt{r_n}}} = \frac{1}{\sqrt{I_0}} \sqrt{(1 - \gamma_n)^2 - \frac{2s_n}{(1 + \gamma_n)}}.$$

Besides, (2.39) implies

$$t_n = \sum_0^{n-1} \delta_k = \sum_0^{n-1} \frac{\gamma_{k+1} - \gamma_k^2}{c\beta_{k+1}} = c^{-1} \sum_0^{n-1} \left(\frac{1 - \gamma_k^2}{\beta_k(1 + \gamma_k)} - \frac{1 - \gamma_{k+1}}{\beta_{k+1}} \right)$$

$$= c^{-1} \sum_0^{n-1} \left(\frac{1 - \gamma_k}{\beta_k} - \frac{1 - \gamma_{k+1}}{\beta_{k+1}} \right) = \frac{1 - \gamma_0}{c\beta_0} - \frac{1 - \gamma_n}{c\beta_n}$$

and

$$t_\infty = \frac{1 - \gamma_0}{c\beta_0} - \frac{1}{c\beta_\infty} = \frac{1 - \gamma_0}{c\beta_0} - \frac{1}{c}\sqrt{I_0}.$$

□

We are well prepared now to state the convergence theorem for Ulm's method. In fact, to get the theorem, it suffices to replace the condition $\beta < \infty \ \& \ t_\infty < \infty$ in Lemma 2.8 by its equivalent found in Lemma 2.9 and Proposition 2.11.

Theorem 2.15. *Let the operator* $\mathbf{f}$ *be* ω*-regularly smooth on* D. *If the starters* $\mathbf{A}_0, x_0, \alpha_0, \beta_0, \gamma_0$ *are such that*

$$\|\mathbf{A}_0\| \le \beta_0 \ \& \ \left\|\mathbf{I} - \mathbf{A}_0\mathbf{f}'(x_0)\right\| \le \gamma_0 < 1 \ \& \ \left\|\mathbf{A}_1\mathbf{f}(x_0\right\| \le \delta_0$$

and

$$\beta_0 \le f_\infty(\alpha_0, \gamma_0, \delta_0) \ ,$$

where f_∞ *is the function of Proposition 2.11, then*

1° *the sequence* t_n *converges to the limit*

$$t_\infty \begin{cases} = \alpha_0 - \omega^{-1}\left(\omega(\alpha_0) - \dfrac{1-\gamma_0}{\beta_0} + \dfrac{1}{\beta_\infty}\right) , & \text{if } \underset{n}{\&} \alpha_n > 0 , \\ \le \omega^{-1}\left(\dfrac{1-\gamma_0}{\beta_0} - \dfrac{1}{\beta_\infty} + 2\omega(\alpha_0/2) - \omega(\alpha_0)\right) , & \text{if } \alpha_k > 0 = \alpha_{k+1} \text{ for some } k; \end{cases}$$

2° *the sequence* $(\mathbf{A}_n, x_n)$ *generated by Ulm's method* (2.19) *from the starter* $(\mathbf{A}_0 , x_0)$ *remains in the ball*

$$B\big((\mathbf{A}_0, x_0), (\beta_\infty - \beta_0, t_\infty)\big) := \left\{(\mathbf{A} , x) \ \middle| \ \|\mathbf{A} - \mathbf{A}_0\| \le \beta_\infty - \beta_0 \ \& \ \|x - x_0\| \le t_\infty \right\},$$

and converges to a limit $(\mathbf{A}_\infty, x_\infty)$*;*

3° *this limit solves the system*

$$\mathbf{X}\mathbf{f}'(x) = \mathbf{I} \quad \& \quad \mathbf{A}_0\mathbf{f}(x) = 0 \tag{2.49}$$

for $(\mathbf{X}, x) \in \mathcal{L}(\mathbb{Y}, \mathbb{X}) \times \mathbb{X}$ *;*

4° x_∞ *is the only solution of the equation* $\mathbf{A}_0\mathbf{f}(x) = 0$ *in the ball* $B(x_0, R)$*, where* R *is the unique solution for* t *of the equation*

$$\frac{\Psi(\bar{\alpha}_0, t) - \Psi(\bar{\alpha}_0, t_\infty)}{t - t_\infty} = \frac{1 - \gamma_0}{\beta_0}.$$

5° *for all* $n = 0, 1, \ldots$

$$\Delta_n := \|x_\infty - x_n\| \le t_\infty - t_n\,,$$
$$\|\mathbf{A}_\infty - \mathbf{A}_n\| \le \beta_\infty - \beta_n\,,$$
$$\frac{\Delta_{n+1}}{\Delta_n} \le \beta_{n+1}\frac{w(\Delta_n)}{\Delta_n} + \gamma_n^2.$$

6° *these error bounds are exact in the sense that they are attained for a scalar* ω*-regularly smooth function.*

In the special case when $\mathbf{f}$ is Lipschitz smooth (ω is linear: $\omega(t) = ct$) and so Lemma 2.14 applies, the theorem takes on the more rigorous form of

Corollary 2.16. *Let the operator* $\mathbf{f}$ *be Lipschitz smooth on* D*:*

$$\|\mathbf{f}'(x) - \mathbf{f}'(x_0)\| \le c\|x - x_0\|\ ,\ \forall\, x, x_0 \in D\,.$$

If the starters $(\beta_0, \gamma_0, \delta_0)$ *and* $(\mathbf{A}_0, x_0)$ *satisfy*

$$\|\mathbf{A}_0\| \le \beta_0 \quad \& \quad \|\mathbf{I} - \mathbf{A}_0\mathbf{f}'(x_0)\| \le \gamma_0 < 1 \quad \& \quad \|\mathbf{A}_1\mathbf{f}(x_0)\| \le \delta_0 < \frac{(1-\gamma_0)^2(1+\gamma_0)}{2c\beta_0}\,,$$

then

1° *the sequence* $(\mathbf{A}_n, x_n)$ *generated by Ulm's method* (2.19) *from* $(\mathbf{A}_0, x_0)$ *remains in the ball* $B\big((\mathbf{A}_0, x_0), (\beta_\infty - \beta_0, t_\infty)\big)$*, where*

$$\beta_\infty = I_0 := \frac{\beta_0(1+\gamma_0)}{\sqrt{(1-\gamma_0^2)^2 - 2c\beta_0\delta_0(1+\gamma_0)}}\ ,\ t_\infty = \frac{1-\gamma_0}{c\beta_0} - \frac{1}{cI_0}\,,$$

and converges to a limit $(\mathbf{A}_\infty, x_\infty)$*;*

2° *this limit solves the system* (2.49)*;*

3° x_∞ *is the only solution of the equation* $\mathbf{A}_0\mathbf{f}(x) = 0$ *in the ball* $B(x_0, R)$*, where*

$$R := \frac{1-\gamma_0}{c\beta_0} + \frac{1}{cI_0} > t_\infty\,;$$

4° *for all* $n = 0, 1, \dots$,

$$\left\|\mathbf{I} - \mathbf{A}_n \mathbf{f}'(x_n)\right\| \le \gamma_n \ ,$$
$$\Delta_n \le \frac{1-\gamma_n}{c\beta_n} - \frac{1}{cI_0} \ ,$$
$$\frac{\Delta_{n+1}}{\Delta_n} \le 0.5c\beta_{n+1}\Delta_n + \gamma_n^2;$$

5° *these error bounds are exact.*

If the operator $\mathbf{f}'(x_0)$ is invertible, then the condition $\|\mathbf{I} - \mathbf{A}_0\mathbf{f}'(x_0)\| < 1$ forces $\mathbf{A}_0$ to be invertible too (Proposition 2.1, 1°). In this case, $\mathbf{A}_0\mathbf{f}(x_\infty) = 0 \Longrightarrow \mathbf{f}(x_\infty) = 0$, i.e., x_∞ solves the equation $\mathbf{f}(x) = 0$. Moreover, $\underset{n}{\&} \mathcal{N}(\mathbf{A}_n) = \mathcal{N}(\mathbf{A}_0) = \{0\}$, that is, all $\mathbf{A}_n$ are left-invertible and so invertible (Proposition 2.1, 2°). Then $\mathbf{f}'(x_n)$ and $\mathbf{f}'(x_\infty)$ are invertible too and the manifold of left inverses of $\mathbf{f}'(x_\infty)$ consists of $\mathbf{f}'(x_\infty)^{-1}$ only. So, $\mathbf{A}_\infty = \mathbf{f}'(x_\infty)^{-1}$.

2.5 Rate of convergence

According to Theorem 2.15, $\Delta_{n+1}/\Delta_n \le \beta_{n+1}w(\Delta_n)/\Delta_n + \gamma_n^2$, where $\beta_{n+1} < \beta_\infty$ and (due to convexity of w) $w(\Delta_n)/\Delta_n \le w'(\Delta_n) = \omega(\Delta_n) \to 0$. The rate of convergence of a sequence a_n to its limit a_∞ characterized by the relation $|a_{n+1} - a_\infty| = o(|a_n - a_\infty|)$ is called in the literature *superlinear*. So, we can say that the sequence x_n generated by Ulm's method converges to the solution x_∞ of the equation $\mathbf{A}_0\mathbf{f}(x) = 0$ superlinearly. However, the bound stated by the theorem allows us to estimate the convergence rate along the whole process, not only asymptotically.

Now we are going to demonstrate that, under the conditions of Theorem 2.15, the convergence rate of the sequence $(\mathbf{A}_n, x_n)$ is quadratic:

$$\|\mathbf{A}_{n+1} - \mathbf{A}_\infty\| + \|x_{n+1} - x_\infty\| = O\big(\|\mathbf{A}_n - \mathbf{A}_\infty\| + \|x_n - x_\infty\|\big)^2 .$$

Proposition 2.17. *If* $\beta_\infty < \infty$ & $\omega'(\alpha_\infty) < \infty$ *(in particular, when* ω *is linear), then*

$$\beta_\infty\omega'(\alpha_\infty)\Delta_{n+1} + \bar\gamma_{n+1}^2 \le \Big(\big(2 + \beta_\infty\omega(t_\infty)\big)^2 + 1\Big)\big(\beta_\infty\omega'(\alpha_\infty)\Delta_n + \bar\gamma_n^2\big)^2 .$$

Proof. Let $r_n := \bar\gamma_n^2$ and $s_n := \beta_\infty\omega'(\alpha_\infty)\Delta_n$. By Theorem 2.15, $s_{n+1} \le s_n(r_n + s_n)$. Besides, as follows from Lemma 2.6,

$$\bar\gamma_{n+1} \le r_n + \beta_\infty\left(\omega\left(\alpha_{n+1} + \bar\delta_n\right) - \omega(\alpha_{n+1})\right) \ ,$$

where $\omega\left(\alpha_{n+1}+\bar{\delta}_n\right)-\omega(\alpha_{n+1})\le\omega'(\alpha_{n+1})\bar{\delta}_n\le\omega'(\alpha_\infty)\bar{\delta}_n$ due to concavity of ω. Therefore, $\bar{\gamma}_{n+1}\le r_n+\beta_\infty\omega'(\alpha_\infty)\bar{\delta}_n$. Inasmuch as $\bar{\delta}_n=\|x_{n+1}-x_\infty+x_\infty-x_n\|\le\Delta_n+\Delta_{n+1}$, we have by (2.30)

$$\frac{\bar{\delta}_n}{\Delta_n}\le 1+\frac{\Delta_{n+1}}{\Delta_n}\le 1+\bar{\beta}_{n+1}\frac{w(\Delta_n)}{\Delta_n}+\bar{\gamma}_n^2<1+\beta_\infty\omega(\Delta_n)+r_n\,,$$

so that

$$\bar{\gamma}_{n+1}\le r_n+\beta_\infty\omega'(\alpha_\infty)\Delta_n\left(1+\beta_\infty\omega(\Delta_n)+r_n\right)=r_n+s_n\left(1+\beta_\infty\omega(\Delta_n)+r_n\right)$$

and $r_{n+1}+s_{n+1}\le\left(r_n+s_n(1+\beta_\infty\omega(\Delta_n)+r_n)\right)^2+s_n(r_n+s_n)$. It follows that

$$\frac{r_{n+1}+s_{n+1}}{(r_n+s_n)^2}\le\left(1+\frac{\beta_\infty\omega(\Delta_n)s_n+r_ns_n}{r_n+s_n}\right)^2+\frac{s_n}{r_n+s_n}\,,$$

where $\Delta_n<t_\infty<\infty$ and $r_n<1$ by Theorem 2.15. So,

$$\frac{r_{n+1}+s_{n+1}}{(r_n+s_n)^2}\le\left(1+(\beta_\infty\omega(t_\infty)+1)\frac{s_n}{r_n+s_n}\right)^2+\frac{s_n}{r_n+s_n}\le(2+\beta_\infty\omega(t_\infty))^2+1.$$

□

2.6 A posteriori error bounds

The error bounds established by Theorem 2.15 are based on information available at the starter $(\mathbf{A}_0,x_0)$. Such bounds are called a priori ones. The theorem allows us to obtain also a posteriori bounds which take into account the information acquired when n-th iteration has already been computed. Namely, having $(\mathbf{A}_n,x_n)$ in hand, one can feed it into the algorithm that has supplied q_0 at $(\mathbf{A}_0,x_0)$ and get a better than q_n approximation $q_0^{(n)}$ to $\bar{q}_n$. By Lemma 2.7, $\underset{n}{\&}\,\bar{q}_{n+k}\prec q_k^{(n)}\prec q_{n+k}$. As a result, all error bounds of Theorem 2.15 tighten up.

There is a possibility of further improvement of a posteriori error bounds. To realize it, one should exploit the fact that the smaller is the set on which the regular smoothness modulus has been evaluated, the more accurately it characterizes the operator's behavior on the set. Symbolically,

$$D'\subset D\Longrightarrow\omega_{D'}(t)\le\omega_D(t)\ ,\ \forall t\ge 0.$$

This implication can be deduced formally from Definition 2.2 (deduce).

2.7 An application: Chandrasekhar's integral equation

Convergence theorems of the type of Theorem 2.15 characterize a method's performance for a worst representative of a (more or less) broad class of operators. So, naturally their conclusions are usually too conservative to be effectively used in practice when one deals with a specific member of the class typically far from being the worst. Therefore, in an application, there is no real need for an initial approximation to satisfy the convergence condition stated by the theorem in order to produce a convergent sequence of successive approximations (though, normally, it helps to accelerate convergence). This point is illustrated by our application of Ulm's method to Chandrasekhar's integral equation (2.17), described below.

Differentiation of Chandrasekhar's operator gives the following result:

$$(\mathbf{f}'(x)u)(s) = \left(1 - 0.5s\int_0^1 \frac{x(t)}{s+t}dt\right)u(s) - 0.5sx(s)\int_0^1 \frac{u(t)}{s+t}dt\ .$$

It agrees with the general form of a linear integral operator

$$(\mathbf{A}u)(s) = v(s)u(s) + \int_0^1 w(s,t)u(t)dt\ . \tag{2.50}$$

This form is invariant with respect to the operations of summation and composition used in (2.19) for construction of $\mathbf{A}_+$ from $\mathbf{A}$ and $\mathbf{f}'(x)$:

$$\big((\mathbf{A}_1 + \mathbf{A}_2)u\big)(s) = \big(v_1(s) + v_2(s)\big)u(s) + \int_0^1 \big(w_1(s,t) + w_2(s,t)\big)u(t)dt\ , \tag{2.51}$$

$$\big((\mathbf{A}_1\mathbf{A}_2)u\big)(s) = v_1(s)v_2(s)u(s) + \int_0^1\left(v_1(s)w_2(s,t) + v_2(t)w_1(s,t) + \int_0^1 w_1(s,\sigma)w_2(\sigma,t)d\sigma\right)u(t)dt\ . \tag{2.52}$$

So, if we take $\mathbf{A}_0$ in the form (2.50), then all $\mathbf{A}_n$ will have this form:

$$(\mathbf{A}_n u)(s) = v_n(s)u(s) + \int_0^1 w_n(s,t)u(t)dt\ .$$

Therefore, evaluation of $\mathbf{A}_{n+1}$ means evaluation of its functional parameters $v_{n+1}\,, w_{n+1}$, given the corresponding parameters

$$p_n(s) := 1 - 0.5s\int_0^1 \frac{x_n(t)}{s+t}dt\ ,\ \ q_n(s,t) := -0.5\frac{sx_n(s)}{s+t}$$

of $\mathbf{f}'(x_n)$ and $v_n\,, w_n$ of $\mathbf{A}_n$. Using the rules (2.51) and (2.52), we obtain that

$$v_{n+1}(s) = \tilde{v}_n(s)v_n(s) \ ,$$

$$w_{n+1}(s,t) = \tilde{v}_n(s)w_n(s,t) + v_n(t)\tilde{w}_n(s,t) + \int_0^1 \tilde{w}_n(s,\sigma)w_n(\sigma,t)d\sigma \ ,$$

where

$$\tilde{v}_n(s) := 2 - v_n(s)p_n(s) \ ,$$

$$\tilde{w}_n(s,t) := -v_n(s)q_n(s,t) - p_n(t)w_n(s,t) - \int_0^1 w_n(s,\sigma)q_n(\sigma,t)d\sigma \ .$$

Having v_{n+1}, w_{n+1} in hand, we compute $x_{n+1}(s)$ as

$$x_n(s) - \mathbf{A}_{n+1}\mathbf{f}(x_n)(s) = x_n(s) - v_{n+1}(s)\mathbf{f}(x_n)(s) - \int_0^1 w_{n+1}(s,t)\mathbf{f}(x_n)(t)dt \ .$$

The functions x_n, p_n, q_n, v_n, w_n are represented in computer memory by their respective splines on a Tchebyshev mesh on the segment $[0,1]$ and the corresponding grid on the square $[0,1] \times [0,1]$.

For a given pair $(\mathbf{A}_0, x_0)$, the norm of the operator $\mathbf{I} - \mathbf{A}_0\mathbf{f}'(x_0)$ acting on the space $\mathbb{C}[0,1]$ of functions continuous on $[0,1]$

$$\|\mathbf{I} - \mathbf{A}_0\mathbf{f}'(x_0)\| = \max_{0 \le s \le 1} \left(|1 - v_0(s)p_0(s)| + \int_0^1 |\tilde{w}_0(s,t)|dt \right) .$$

In particular, for $\mathbf{A}_0 := \mathbf{I}$ and $x_0(s) := 2$, $\forall\, s$, $\bar{\gamma}_0 := \|\mathbf{I} - \mathbf{A}_0\mathbf{f}'(x_0)\| = 1.38...$, i.e., the convergence condition of Theorem 2.15 is violated. Nevertheless, the sequence of Ulm's iterations converges to the solution of the system

$$\mathbf{X}\mathbf{f}'(x) = \mathbf{I} \ \ \& \ \ \mathbf{f}(x) = 0.$$

The progress is presented in Table 2.1. The approximate solution of the equation $\mathbf{f}(x) = 0$ obtained after 10 iterations is tabulated in the last two columns. Figure 2.1 shows the plots of the initial (dotted line) and the final (solid line) approximations.

The approach we have used for application of Ulm's method to Chandrasekhar's integral equation (iterative computation of functional parameters of operators $\mathbf{A}_n$) works also for Hammerstein integral operators

$$\mathbf{H}(x)(s) := \int_0^1 K(s,t)\varphi(t\,,x(t))\,dt\,,$$

where K and φ are given continuous functions and φ is differentiable with respect to the second argument. Differentiability of φ implies differentiability of $\mathbf{H}$:

$$\left(\mathbf{H}'(x)u\right)(s) = \int_0^1 K(s,t)\frac{\partial\varphi}{\partial x}(t,x(t))\,dt\,.$$

TABLE 2.1: Chandrasekhar's equation

n	$\bar{\beta}_n$	$\bar{\gamma}_n$	$\bar{\delta}_n$	$\|\mathbf{f}(x_n)\|$	t	$x_{10}(t)$
0	1.0000	1.3862E-0	1.0339E-0	1.0000E-0	0.0	1.000000
1	2.3862	1.8950E-0	2.5459E-1	1.1148E-1	0.1	1.243222
2	6.2707	2.3317E-0	3.0787E-2	1.3104E-2	0.2	1.440975
3	15.730	2.4363E-0	1.1826E-2	6.6802E-4	0.3	1.626778
4	35.154	2.2941E-0	2.2316E-2	8.1955E-4	0.4	1.806078
5	70.985	1.7570E-0	2.8533E-2	6.1497E-4	0.5	1.981069
6	122.91	5.0050E-1	1.1286E-2	1.9449E-4	0.6	2.152868
7	146.80	2.4883E-1	3.2069E-4	8.7694E-6	0.7	2.322129
8	131.61	3.3662E-2	3.6798E-6	9.0824E-8	0.8	2.489264
9	133.44	5.5102E-4	1.9202E-8	5.8083E-10	0.9	2.654552
10	133.47	6.3061E-7	9.3529E-11	1.2400E-11	1.0	2.818189

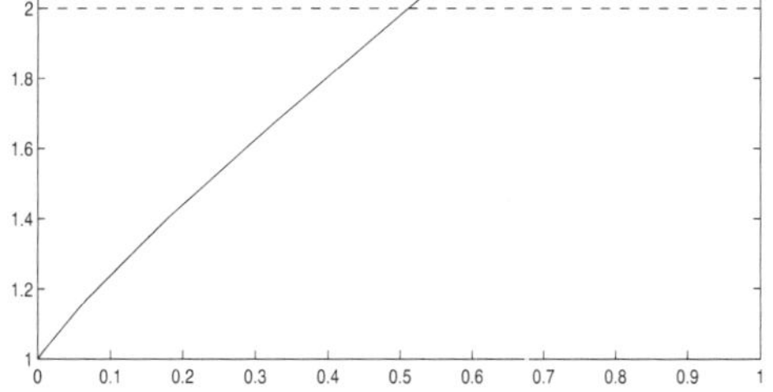

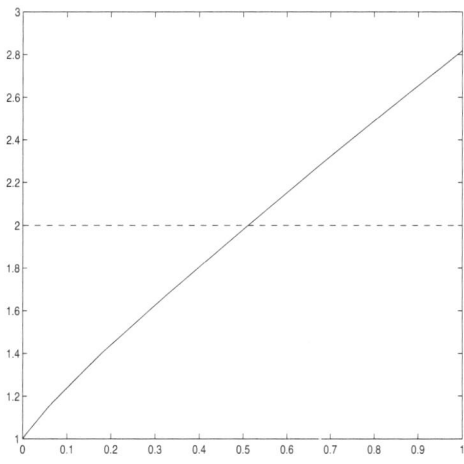

FIGURE 2.1: Chandrasekhar's equation, plots

The parameters of $\mathbf{H}'(x)$ are $p(s) = 0$ and $q(s,t) = K(s,t)\dfrac{\partial \varphi}{\partial x}(t, x(t))$. Correspondingly, the parameters of $\mathbf{A}_{n+1}$ are $v_{n+1}(s) = 2v_n(s)$ and

$$w_{n+1}(s,t) = 2w_n(s,t) + v_n(t)\tilde{w}_n(s,t) + \int_0^1 \tilde{w}_n(s,\sigma)w_n(\sigma,t)d\sigma\,,$$

where

$$\tilde{w}_n(s,t) := -v_n(s)q_n(s,t) - \int_0^1 w_n(s,\sigma)q_n(\sigma,t)d\sigma\,.$$

2.8 Research projects

The most problematic component of the convergence condition stated by Theorem 2.15 is the requirement that $\|\mathbf{I} - \mathbf{A}_0\mathbf{f}'(x_0)\| < 1$. I know no satisfactory solution to the problem of finding a starting pair $(\mathbf{A}_0, x_0)$ that would meet this requirement. This problem is closely related to the one of the left-invertibility of a given linear operator $\mathbf{A} \in \mathcal{L}(\mathbb{X}, \mathbb{Y})$. Clearly,

$$\inf_{\mathbf{B}} \|\mathbf{I} - \mathbf{BA}\| = \begin{cases} 0\,, & \text{if } \mathbf{A} \text{ is left-invertible}\,,\\ 1\,, & \text{otherwise}\,,\end{cases}$$

since, if $\mathbf{A}$ is left-invertible, then its left-inverse can be taken for $\mathbf{B}$, and if it is not, then no $\mathbf{B} \in \mathcal{L}(\mathbb{Y}, \mathbb{X})$ can satisfy $\|\mathbf{I} - \mathbf{BA}\| < 1$ by Proposition 2.1, while $\inf_{\mathbf{B}} \|\mathbf{I} - \mathbf{BA}\| \le \|\mathbf{I}\| = 1$. Thus, the existence of a proinverse of $\mathbf{A}$, that is, a $\mathbf{B}$ with $\|\mathbf{I} - \mathbf{BA}\| < 1$, is equivalent to the left-invertibility of $\mathbf{A}$. The question of the existence of a proinverse is not trivial even for matrices with nondominant diagonals. Inversion of a general linear operator in infinite dimensions is a formidable task. So, these problems present a real challenge that would constitute the subject of a serious research project.

Another interesting situation arises when the equation $\mathbf{f}(x) = 0$ has no solutions in D, $\inf_{x\in D} \|\mathbf{f}(x)\| > 0$. In this case the sequence x_n generated by Ulm's method from a starter $(\mathbf{A}_0\,, x_0)$ cannot converge. How will it behave then? Will it diverge, enter into a cycle, become chaotic, or, perhaps, some average of $x_n, x_{n+1}, \ldots, x_{n+m}$ will cluster around a minimizer of $\|\mathbf{f}(x)\|$ over D? These questions are similar to those asked in the theory of finite-dimensional discrete dynamical systems [50], [56], whose problems are motivated by examples from mathematical biology, economics, finance, genetics, and other fields. Analysis of iterative methods for solving operator equations provides a new source of discrete dynamical systems with its own specifics.

The application of Section 2.8 shows that that Ulm's method (2.1) is easily applied to operators whose derivative admits parametric form invariant with respect to summation and composition. However, many interesting operators do not have this property. One example is the Riccati matrix operator (2.18). It is related to the Riccati matrix equation arising frequently in estimation and control. The derivative

$$\mathbf{f}'(x)u = u(ax + c) + (xa + b)u$$

is invariant under summation:

$$\begin{aligned}\mathbf{A}_i u := up_i + q_i u\,, i = 1, 2\,, \Longrightarrow (\mathbf{A}_1 + \mathbf{A}_2)u &= \mathbf{A}_1 u + \mathbf{A}_2 u \\ &= up_1 + q_1 u + up_2 + q_2 u \\ &= u(p_1 + p_2) + (q_1 + q_2)u\,,\end{aligned}$$

but not under composition:

$$\begin{aligned}\mathbf{A}_1\mathbf{A}_2 u &= \mathbf{A}_1(up_2 + q_2 u) = (up_2 + q_2 u)p_1 + q_1(up_2 + q_2 u) \\ &= up_2 p_1 + q_2 u p_1 + q_1 u p_2 + q_1 q_2 u \neq ua + bu\end{aligned}$$

for any a and b (prove it). In contrast, the operator $\mathbf{A}u := puq$ is invariant under composition: $\mathbf{A}_1\mathbf{A}_2 u = \mathbf{A}_1(p_2 u q_2) = p_1 p_2 u q_2 q_1$, but not under summation:

$$(\mathbf{A}_1 + \mathbf{A}_2)u = \mathbf{A}_1 u + \mathbf{A}_2 u = p_1 u q_1 + p_2 u q_2 \neq aub\,.$$

So, it is important from the practical point of view to identify classes of differentiable operators whose derivative is invariant (as are diagonal operators mentioned in Section 1.8) under both summation and composition.

Chapter 3

Ulm's method without derivatives

3.1 Motivation

With all its attractive properties, Ulm's method (2.1) has a serious shortcoming: the derivative $\mathbf{f}'(x)$ has to be evaluated at each iteration. This makes it inapplicable to equations with nondifferentiable operators and in situations when evaluation of the derivative is too costly. Nondifferentiable operators arise each time the operator $\mathbf{f}$ under consideration is defined not on the whole space $\mathbb{X}$, but on some proper subset $D \subset \mathbb{X}$ of it and so each new iteration x_+ generated by an iterative method must belong to this subset. Clearly, no general iterative method can satisfy this requirement. A natural way to deal with the problem is to force x_+ onto D (for example, by metric projection) before evaluating $\mathbf{f}(x_+)$. In other words, the operator $\mathbf{f}$ must be *globalized* to make it suitable for application of a general iterative method. Even differentiable operators most likely become nondifferentiable after their globalization.

The obvious idea to free Ulm's method of this shortcoming is to replace the derivative in (2.1) with some approximation. One such approximation is the so-called *divided difference operator* (briefly, dd). It is named by analogy with one-dimensional approximation of the derivative, the divided difference

$$f'(x) \approx \frac{f(x_1) - f(x)}{x_1 - x} ,$$

where x_1 is close to x.

3.2 The divided difference operator

Let $\mathbf{f}$ be a continuous operator from a convex subset D of the Banach space $\mathbb{X}$ into another Banach space $\mathbb{Y}$.

Definition 3.1. *A linear bounded operator* $\mathbf{A}$ *from* $\mathbb{X}$ *to* $\mathbb{Y}$ *is called a divided difference operator of* $\mathbf{f}$*, if, for any given pair of points* (x, x_1) *of* D*, it satisfies the (secant) equation*

$$\mathbf{A}(x_1 - x) = \mathbf{f}(x_1) - \mathbf{f}(x) \ . \tag{3.1}$$

To emphasize its dependence on x, x_1 and $\mathbf{f}$, we denote it by the symbol $[x, x_1 \,|\, \mathbf{f}]$.

For given $x \in \mathbb{X}$ and $y \in \mathbb{Y}$, linear operators satisfying the equation $\mathbf{A}x = y$ constitute an affine manifold in the space $\mathcal{L}(\mathbb{X}, \mathbb{Y})$ of all bounded linear operators between $\mathbb{X}$ and $\mathbb{Y}$:

$$\mathbf{A}_0 x = y \ \ \& \ \ \mathbf{A}x = y \Longrightarrow (\mathbf{A} - \mathbf{A}_0)x = 0 \Longrightarrow \mathbf{A} = \mathbf{A}_0 + \mathcal{L}_x \ ,$$

where $\mathcal{L}_x \subset \mathcal{L}(\mathbb{X}, \mathbb{Y})$ is the subspace of operators vanishing at x. So, the symbol $[x, x_1 \,|\, \mathbf{f}]$ should be understood as the notation for this manifold or, more precisely, as its particular representative selected according to a certain rule specified in advance. If $[x, x_1 \,|\, \mathbf{f}]$ is selected to be continuous at x with respect to x_1, then

$$\begin{aligned} \mathbf{f}(x+h) &= \mathbf{f}(x) + [x, x+h \,|\, \mathbf{f}]h \\ &= \mathbf{f}(x) + [x, x \,|\, \mathbf{f}]h + \big([x, x+h \,|\, \mathbf{f}] - [x, x \,|\, \mathbf{f}]\big)h \\ &= \mathbf{f}(x) + [x, x \,|\, \mathbf{f}]h + o(h) \ , \end{aligned}$$

which means that $[x, x \,|\, \mathbf{f}] = \mathbf{f}'(x)$. Otherwise,

$$\lim_{t \to +0} t[x, x+th \,|\, \mathbf{f}]h = \lim_{t \to +0} \big(\mathbf{f}(x+th) - \mathbf{f}(x)\big)$$

(if it exists) may vary depending on h, $\|h\| = 1$. In this case, this limit is the directional derivative $\mathbf{f}'(x, h)$ of $\mathbf{f}$ at x in the direction h. The following proposition lists for the record some properties of the set-valued map $\mathbf{f} \mapsto [x, x_1 \,|\, \mathbf{f}]$.

Proposition 3.2. 1° $[x, x_1 \,|\, \mathbf{f}] = \mathbf{f}$ *if and only if* $\mathbf{f}$ *is linear.*

2° $[x, x_1 \,|\, \alpha\mathbf{f}_1 + \beta\mathbf{f}_2\,] = \alpha[x, x_1 \,|\, \mathbf{f}_1\,] + \beta[x, x_1 \,|\, \mathbf{f}_2\,]$.

3° *If* $\mathbf{f}$ *is a composition of the operators* $\mathbf{f}_1$, $\mathbf{f}_2$ *:* $\mathbf{f} = \mathbf{f}_1 \circ \mathbf{f}_2$*, then*

$$[x, x_1 \,|\, \mathbf{f}] = [\mathbf{f}_2(x), \mathbf{f}_2(x_1) \,|\, \mathbf{f}_1\,] \cdot [x, x_1 \,|\, \mathbf{f}_2\,] \ .$$

4° *If* $\mathbf{A}$ *and* $\mathbf{B}$ *are two dd's of* $\mathbf{f}$*, then any their convex combination* $(1-\alpha)\mathbf{A} + \alpha\mathbf{B}$, $0 \le \alpha \le 1$*, is a dd too.*

Proof. 1° If $\mathbf{f}$ is linear, then $\mathbf{f}(x) - \mathbf{f}(x_1) = \mathbf{f}(x - x_1)$, so that $\mathbf{f} = [x, x_1 \,|\, \mathbf{f}]$. Conversely, the equality $\mathbf{f} = [x, x_1 \,|\, \mathbf{f}]$ implies linearity of $\mathbf{f}$, since $[x, x_1 \,|\, \mathbf{f}]$ is linear by definition.

2° By definition,

$$\begin{aligned}[x, x_1 \,|\, \alpha_1\mathbf{f}_1 + \alpha_2\mathbf{f}_2\,](x - x_1) &= (\alpha_1\mathbf{f}_1 + \alpha_2\mathbf{f}_2)(x) - (\alpha_1\mathbf{f}_1 + \alpha_2\mathbf{f}_2)(x_1)\\ &= \alpha_1\mathbf{f}_1(x) + \alpha_2\mathbf{f}_2(x) - \alpha_1\mathbf{f}_1(x_1) - \alpha_2\mathbf{f}_2)(x_1)\\ &= \alpha_1\big(\mathbf{f}_1(x) - \mathbf{f}_1(x_1)\big) + \alpha_2\big(\mathbf{f}_2(x_1) - \mathbf{f}_2(x_1)\big)\\ &= \alpha_1[x, x_1 \,|\, \mathbf{f}_1\,](x - x_1) + \alpha_2[x, x_1 \,|\, \mathbf{f}_2\,](x - x_1)\\ &= \Big(\alpha_1[x, x_1 \,|\, \mathbf{f}_1\,] + \alpha_2[x, x_1 \,|\, \mathbf{f}_2\,]\Big)(x - x_1)\,.\end{aligned}$$

Thus, for all $z \in \mathbb{X}$, $[x, x_1 \,|\, \alpha_1\mathbf{f}_1 + \alpha_2\mathbf{f}_2\,]z = \Big(\alpha_1[x, x_1 \,|\, \mathbf{f}_1\,] + \alpha_2[x, x_1 \,|\, \mathbf{f}_2\,]\Big)z$, so that

$$[x, x_1 \,|\, \alpha_1\mathbf{f}_1 + \alpha_2\mathbf{f}_2\,] = \alpha_1[x, x_1 \,|\, \mathbf{f}_1\,] + \alpha_2[x, x_1 \,|\, \mathbf{f}_2\,]\,.$$

3° By definition,

$$\begin{aligned}[x, x_1 \,|\, \mathbf{f}_1\circ\mathbf{f}_2\,](x - x_1) &= \big(\mathbf{f}_1\circ\mathbf{f}_2\big)(x) - \big(\mathbf{f}_1\circ\mathbf{f}_2\big)(x_1)\\ &= \mathbf{f}_1\big(\mathbf{f}_2(x)\big) - \mathbf{f}_1\big(\mathbf{f}_2(x_1)\big)\\ &= \big[\mathbf{f}_2(x), \mathbf{f}_2(x_1) \,\big|\, \mathbf{f}_1\,\big]\big(\mathbf{f}_2(x) - \mathbf{f}_2(x_1)\big)\\ &= \big[\mathbf{f}_2(x), \mathbf{f}_2(x_1) \,\big|\, \mathbf{f}_1\,\big]\,[x, x_1 \,|\, \mathbf{f}_2\,](x - x_1)\,.\end{aligned}$$

Thus, for all $z \in \mathbb{X}$, $[x, x_1 \,|\, \mathbf{f}_1\circ\mathbf{f}_2\,]z = \big[\mathbf{f}_2(x), \mathbf{f}_2(x_1) \,\big|\, \mathbf{f}_1\,\big]\,[x, x_1 \,|\, \mathbf{f}_2\,]z$, so that

$$[x, x_1 \,|\, \mathbf{f}_1\circ\mathbf{f}_2\,] = \big[\mathbf{f}_2(x), \mathbf{f}_2(x_1) \,\big|\, \mathbf{f}_1\,\big]\,[x, x_1 \,|\, \mathbf{f}_2\,]\,.$$

4° As $\mathbf{A}$ and $\mathbf{B}$ are dd's of $\mathbf{f}$, they are linear, so that

$$\big((1-\alpha)\mathbf{A} + \alpha\mathbf{B}\big)(x - x_1) = (1-\alpha)\mathbf{A}(x - x_1) + \alpha\mathbf{B}(x - x_1)\,.$$

By the same reasoning, $\mathbf{A}(x - x_1) = \mathbf{f}(x) - \mathbf{f}(x_1)$ and $\mathbf{B}(x - x_1) = \mathbf{f}(x) - \mathbf{f}(x_1)$. So,

$$\begin{aligned}\big((1-\alpha)\mathbf{A} + \alpha\mathbf{B}\big)(x - x_1) &= (1-\alpha)\big(\mathbf{f}(x) - \mathbf{f}(x_1)\big) + \alpha\big(\mathbf{f}(x) - \mathbf{f}(x_1)\big)\\ &= (1 - \alpha + \alpha)\big(\mathbf{f}(x) - \mathbf{f}(x_1)\big) = \mathbf{f}(x) - \mathbf{f}(x_1)\,,\end{aligned}$$

i.e., $(1-\alpha)\mathbf{A} + \alpha\mathbf{B}$ is a dd of $\mathbf{f}$. □

For the quadratic operator (2.12), the secant equation (3.1) becomes

$$\begin{aligned}\mathbf{Q}(x_1) - \mathbf{Q}(x) &= 0.5\big(\mathbf{B}(x_1, x_1) - \mathbf{B}(x_1, x) + \mathbf{B}(x_1, x) - \mathbf{B}(x, x)\big) + \mathbf{A}(x_1 - x)\\ &= 0.5\big(\mathbf{B}(x_1, x_1 - x) + \mathbf{B}(x_1 - x, x)\big) + \mathbf{A}(x_1 - x)\,.\end{aligned}$$

Since we can assume (with no loss of generality) $\mathbf{B}$ to be symmetric,

$$\mathbf{B}(x_1, x_1 - x) + \mathbf{B}(x_1 - x, x) = \mathbf{B}(x_1, x_1 - x) + \mathbf{B}(x, x_1 - x) = \mathbf{B}(x_1 + x, x_1 - x)\,,$$

so that $\mathbf{Q}(x_1) - \mathbf{Q}(x) = \Big(0.5\mathbf{B}(x + x_1, \cdot) + \mathbf{A}\Big)(x_1 - x)$. Hence, the operator

$0.5\mathbf{B}(x + x_1, \cdot) + \mathbf{A}$ is a $[x, x_1 \,|\, \mathbf{Q}]$. In particular, for the operator (2.18), $\mathbf{B}(x\,,y) = 0.5xay$, $\mathbf{A}x = bx+xc$ and so $[x, x_1 \,|\, \mathbf{f}]h = 0.25(x+x_1)ah+bh+hc$.

As another example, consider the operator

$$\mathbf{f}(x)(r,s) := x\Big(p(r,s)\,,q(r,s)\Big) - f(r,s)x(r,s)\,,$$

where p, q, f are given functions of two variables (cf. (2.41)). It is linear. Indeed, by the operator's definition

$$\begin{aligned}\mathbf{f}(\alpha x + \beta y)(r,s) &= (\alpha x + \beta y)\big(p(r,s)\,,q(r,s)\big) - f(r,s)(\alpha x + \beta y)(r,s)\\ &= \alpha x\big(p(r,s)\,,q(r,s)\Big) + \beta y\big(p(r,s)\,,q(r,s)\Big) - \\ &\qquad\qquad \alpha f(r,s)x(r,s) - \beta f(r,s)\beta y(r,s)\\ &= \alpha\Big(x\big(p(r,s)\,,q(r,s)\big) - f(r,s)x(r,s)\Big) + \\ &\qquad\qquad \beta\Big(x\big(p(r,s)\,,q(r,s)\big) - f(r,s)x(r,s)\Big)\\ &= \alpha\mathbf{f}(x)(r,s) + \beta\mathbf{f}(y)(r,s) = \big(\alpha\mathbf{f}(x) + \beta\mathbf{f}(y)\big)(r,s)\end{aligned}$$

and so $\mathbf{f}(\alpha x+\beta y) = \alpha\mathbf{f}(x)+\beta\mathbf{f}(y)$. Then, by Proposition 3.2, 1°, the operator $h(r,s) \mapsto \mathbf{f}(h)(r,s)$ is the $[x_1, x \,|\, \mathbf{f}]$.

It is not uncommon that a nonlinear operator $\mathbf{f}$ corresponding to the equation under consideration is a composition of a linear operator and another nonlinear one. For example, the Hammerstein integral operator

$$\mathbf{H}(x)(t) := \int_0^1 K(s,t)\varphi(s, x(s))\,ds\ , \quad \forall x \in \mathbb{C}[0,1], \tag{3.2}$$

where K and φ are given continuous functions, is the composition of the linear integral operator

$$(\mathbf{A}x)(t) := \int_0^1 K(s,t)x(s)\,ds\ , \quad \forall x \in \mathbb{C}[0,1]$$

and of the nonlinear operator $\mathfrak{f}(x)(t) := \varphi(t, x(t))$. By Proposition 3.2,

$$[x_1, x_2 \,|\, \mathbf{H}] = \mathbf{A}[x_1, x_2 \,|\, \mathfrak{f}]\ ,$$

so that selection of a dd for $\mathbf{H}$ reduces to the same question for $\mathfrak{f}$. The selection of $[x_1, x_2 \,|\, \mathfrak{f}]$ should take into account specific properties of the function φ. If, for example, φ is a polynomial in the second argument,

$$\varphi(t,x) = \sum_{i=0}^{m} a_i(t)x^i\ ,$$

then

$$\varphi(t,x_1) - \varphi(t,x_2) = \sum_{i=0}^{m} a_i(t)\big(x_1^i - x_2^i\big) = \sum_{i=0}^{m} a_i(t)\sum_{j=0}^{i-1} x_1^{i-1-j}x_2^j(x_1 - x_2)\,,$$

so that $\sum_{i=0}^{m} a_i(t) \sum_{j=0}^{i-1} x_1^{i-1-j} x_2^j$ is a possible choice for $[x_1, x_2 \,|\, \mathfrak{f}]$.

When $\mathbf{f}$ is acting on a space of real-valued functions, such as $\mathbb{C}[0,1]$ or $\mathbb{L}_2[0,1]$, its dd can be defined as follows:

$$\big([x, x_1 \,|\, \mathbf{f}]h\big)(t) = \begin{cases} \dfrac{\mathbf{f}(x)(t) - \mathbf{f}(x_1)(t)}{x(t) - x_1(t)} h(t)\,, & \text{if } x_1(t) \neq x(t) \\ 0\,, & \text{otherwise}. \end{cases} \tag{3.3}$$

(Do you agree?) Note that this operator is *diagonal*:

$$\big([x, x_1 \,|\, \mathbf{f}]h\big)(t) = p(t)h(t)\ ,\ p(t) := \begin{cases} \dfrac{\mathbf{f}(x)(t) - \mathbf{f}(x_1)(t)}{x(t) - x_1(t)}\,, & \text{if } x_1(t) \neq x(t) \\ 0\,, & \text{otherwise}. \end{cases}$$

The variant of Ulm's method we discuss in this chapter is proposed in [20]. It is obtained by replacing the derivative $\mathbf{f}'$ in (2.1) by a dd $[x_+, x|\,\mathbf{f}\,]$ of $\mathbf{f}$, where x denotes (as in (2.1)) the current approximation and x_+ is the next one:

$$x_+ := x - \mathbf{A}\mathbf{f}(x)\ ,\ \mathbf{A}_+ := 2\mathbf{A} - \mathbf{A}[x_+, x|\,\mathbf{f}\,]\mathbf{A}\ . \tag{3.4}$$

The convergence analysis of any iterative method that involves evaluation of the selected dd has to be based on one or another assumption about the continuity properties of that dd. For example, Potra in [48] and [47] assumes the dd to be a *consistent approximation* to the derivative:

$$\big\|[x, y \,|\, \mathbf{f}] - \mathbf{f}'(u)\big\| \leq c\big(\|x-u\| + \|y-u\|\big)\ ,\quad \forall\, x, y, u \in D\,.$$

In [2] and [31], the inequality

$$\big\|[x, y \,|\, \mathbf{f}] - [u, v \,|\, \mathbf{f}]\big\| \leq c\big(\|x-u\| + \|y-v\|\big)\ ,\quad \forall\, x, y, u, v \in D, \tag{3.5}$$

(*Lipschitz continuity* of dd) is required. In [27] Hernández and Rubio replace Lipschitz continuity by the more general *Hölder continuity*, which means that

$$\big\|[x, y \,|\, \mathbf{f}] - [u, v \,|\, \mathbf{f}]\big\| \leq c\big(\|x-u\|^p + \|y-v\|^p\big)\ ,\quad \forall\, x, y, u, v \in D$$

for some $p \in (0,1]$. In [28],[29],[30] these authors relax this requirement still further, assuming that a continuous nondecreasing function $\omega : [0,\infty) \times [0,\infty) \to [0,\infty)$ is known such that

$$\big\|[x, y \,|\, \mathbf{f}] - [u, v \,|\, \mathbf{f}]\big\| \leq \omega\big(\|x-u\|, \|y-v\|\big)\ ,\quad \forall\, x, y, u, v \in D. \tag{3.6}$$

Assumptions of the type of ω-*continuity*

$$\|[x, y \,|\, \mathbf{f}] - [u, v \,|\, \mathbf{f}]\| \leq \omega\big(\|x-u\| + \|y-v\|\big)\ ,\quad \forall\, x, y, u, v \in D, \tag{3.7}$$

or (3.6) are too coarse a tool for convergence analysis. First we note that the least ω satisfying (3.7),

$$\underline{\omega}(t) := \sup_{x,y,u,v} \Big\{ \|[x, y \,|\, \mathbf{f}] - [u, v \,|\, \mathbf{f}]\| \;\Big|\; (x, y, u, v) \in D^4\ \&\ \|x-u\| + \|y-v\| \leq t \Big\}\,,$$

in addition to being continuous and nondecreasing, is zero at zero and subadditive: $\underline{\omega}(s+t) \le \underline{\omega}(s) + \underline{\omega}(t)$, $\forall\, s > 0,\ t > 0$. To prove subadditivity, take any four points x, y, u, v in D with $\|x-u\| + \|y-v\| \le s+t$ and define

$$(w, w') := \left(x + \frac{s}{s+t}(u-x)\,,\, y + \frac{s}{s+t}(v-y) \right) .$$

Then

$$\|x-w\| + \|y-w'\| = \frac{s}{s+t}\big(\|x-u\| + \|y-v\|\big) \le s$$

and

$$\|w-u\| + \|w'-v\| = \frac{t}{s+t}\big(\|x-u\| + \|y-v\|\big) \le t\ ,$$

so that $\big\|[x,y\,|\,\mathbf{f}\,] - [w,w'\,|\,\mathbf{f}\,]\big\| \le \underline{\omega}(s)$, $\big\|[w,w'\,|\,\mathbf{f}\,] - [u,v\,|\,\mathbf{f}\,]\big\| \le \underline{\omega}(t)$, and

$$\begin{aligned}\big\|[x,y\,|\,\mathbf{f}\,] - [u,v\,|\,\mathbf{f}\,]\big\| &\le \big\|[x,y\,|\,\mathbf{f}\,] - [w,w'\,|\,\mathbf{f}\,]\big\| + \big\|[w,w'\,|\,\mathbf{f}\,] - [u,v\,|\,\mathbf{f}\,]\big\| \\ &\le \underline{\omega}(s) + \underline{\omega}(t).\end{aligned}$$

Inasmuch as this is true for any quadruple of points x, y, u, v of D satisfying $\|x-u\| + \|y-v\| \le s+t$, this is true also for the corresponding supremum $\underline{\omega}(s+t)$.

The functions ω possessing all four properties
(i) $\omega(0) = 0$,
(ii) continuity on $[0, \infty)$,
(iii) monotonicity,
(iv) subadditivity,
are called in [57] continuity moduli, because each such function is a continuity modulus of itself [42]. So, there is no sense in allowing for ω's which are not continuity moduli, for such an ω can be replaced by a pointwise lesser continuity modulus and the replacement will result in immediate improvement of all parameters describing convergence properties of the iterative method in question.

Analyzing the method (3.4), we assume that the dd involved is ω*-regularly continuous* in the sense of the following definition, where

$$\underline{h}([x_1,x_2\,|\,\mathbf{f}\,]) := \inf_{x_1,x_2} \Big\{ \|[x_1,x_2\,|\,\mathbf{f}\,]\| \ \Big|\ (x_1,x_2) \in D^2 \Big\}$$

and Ω is as in Definition 2.2.

Definition 3.3. *[18] The dd $[x_1,x_2\,|\,\mathbf{f}\,]$ is said to be ω-regularly continuous on D, if an $\omega \in \Omega$ (call it a regularity modulus) and a constant $\underline{h} \in [0, \underline{h}([x_1,x_2\,|\,\mathbf{f}\,])]$ are known such that $\forall\, x_1, x_2, u_1, u_2 \in D$*

$$\omega^{-1}\Big(\min\big\{\|[x_1,x_2\,|\,\mathbf{f}\,]\|, \|[u_1,u_2\,|\,\mathbf{f}\,]\|\big\} - \underline{h} + \big\|[x_1,x_2\,|\,\mathbf{f}\,] - [u_1,u_2\,|\,\mathbf{f}\,]\big\|\Big)$$

$$-\omega^{-1}\Big(\min\{\big\|[x_1,x_2\,|\,\mathbf{f}\,]\big\|,\big\|[u_1,u_2\,|\,\mathbf{f}\,]\big\|\}-\underline{h}\Big) \qquad (3.8)$$
$$\le \|x_1-u_1\|+\|x_2-u_2\|.$$

We say also that it is regularly continuous on D if it has there a regularity modulus.

Using this definition, it is easy to prove

Proposition 3.4. *An* $\omega\in\Omega$ *is a regularity modulus of the dd* $\mathbf{B}\big(0.5(x_1+x_2),\cdot\big)+\mathbf{A}$ *of the quadratic operator* (2.12) *on* $\mathbb{X}$ *if and only if* $\omega(t)\ge 0.5\|\mathbf{B}\|t\ ,\ \forall t\ge 0$.

The proof is very similar to that of Proposition 2.5. We leave it as an exercise for the reader.

A general approach to constructing a regularity modulus of a selected dd of a given operator $\mathbf{f}$ is pointed out by the following proposition. To state it, we define the sets

$$D_h:=\Big\{(x_1,x_2)\in D^2\ \Big|\ \big\|[x_1,x_2\,|\,\mathbf{f}\,]\big\|\ge h\Big\}\ ,\quad h>0\ .$$

$$\mathcal{P}(h,t\,|\,\mathbf{f}):=\left\{(x_1,x_2,u_1,u_2)\in D^4\ \middle|\ \begin{array}{c}(x_1,x_2)\in D_h\ \ \&\ \ (u_1,u_2)\in D_h\\ \|x_1-u_1\|+\|x_2-u_2\|\le t\end{array}\right\}$$

$$\mathcal{Q}(h,\tau\,|\,\mathbf{f}):=\left\{(x_1,x_2,u_1,u_2)\in D^4\ \middle|\ \begin{array}{c}(x_1,x_2)\in D_h\ \ \&\ \ (u_1,u_2)\in D_h\\ \big\|[x_1,x_2\,|\,\mathbf{f}\,]-[u_1,u_2\,|\,\mathbf{f}]\big\|\ge\tau\end{array}\right\}$$

and the functions

$$\xi(h,t\,|\,\mathbf{f}):=\sup_{x_1,x_2,u_1,u_2}\left\{\ \big\|[x_1,x_2\,|\mathbf{f}\,]-[u_1,u_2\,|\,\mathbf{f}\,]\big\|\ \middle|\ (x_1,x_2,u_1,u_2)\in\mathcal{P}(h,t\,|\,\mathbf{f})\right\}.$$

$$\eta(h,\tau\,|\,\mathbf{f}):=\inf_{x_1,x_2,u_1,u_2}\left\{\ \|x_1-u_1\|+\|x_2-u_2\|\ \middle|\ (x_1,x_2,u_1,u_2)\in\mathcal{Q}(h,\tau\,|\,\mathbf{f})\right\}.$$

It can be shown [13] that the functions $t\mapsto\xi(h,t\,|\,\mathbf{f})$ and $\tau\mapsto\eta(h,\tau\,|\,\mathbf{f})$ are mutually inverse.

Proposition 3.5. *Let* $\omega\in\Omega$. *The following statements are equivalent.*

1° ω *is a regularity modulus of the dd* $[x_1,x_2\,|\,\mathbf{f}\,]$;

$$2^\circ\ \omega\left(\omega^{-1}(h)+t\right)-h\ge\xi_h(t\,|\,\mathbf{f})\ ,\ \forall h\ge 0\ ,\ t\ge 0\,; \qquad (3.9)$$

$$3^\circ\ \omega^{-1}(h+\tau)-\omega^{-1}(h)\le\eta_h(\tau\,|\,\mathbf{f})\ ,\ \ \forall h\ge 0\ ,\ \tau\ge 0\ . \qquad (3.10)$$

Proof. It is enough to prove equivalence of 1° and 2°. The equivalence $1^\circ\iff 3^\circ$ is proved analogously. Let an $\omega\in\Omega$ satisfy (3.8), so that $\big\|[x_1,x_2\,|\,\mathbf{f}\,]-[u_1,u_2\,|\,\mathbf{f}\,]\big\|$

$$\le\omega\big(\omega^{-1}(\min\{\big\|[x_1,x_2\,|\,\mathbf{f}\,]\big\|,\big\|[u_1,u_2\,|\,\mathbf{f}\,]\big\|\}-\underline{h})+\|x_1-u_1\|+\|x_2-u_2\|\big)$$
$$-\min\{\big\|[x_1,x_2\,|\,\mathbf{f}\,]\big\|,\big\|[u_1,u_2\,|\,\mathbf{f}\,]\big\|\}+\underline{h}\ . \qquad (3.11)$$

For $(x_1, x_2) \in D_h$ and $(u_1, u_2) \in D_h$, $\min\{\|[x_1, x_2 \,|\, \mathbf{f}\,]\|, \|[u_1, u_2 \,|\, \mathbf{f}\,]\|\} - \underline{h} \geq h$,

$$\omega^{-1}(\min\{\|[x_1, x_2 \,|\, \mathbf{f}\,]\|, \|[u_1, u_2 \,|\, \mathbf{f}\,]\|\} - \underline{h}) \geq \omega^{-1}(h) \ ,$$

and (in view of the concavity of ω) (3.11) $\leq \omega(\omega^{-1}(h) + \|x_1 - u_1\| + \|x_2 - u_2\|) - h$. If, in addition, $\|x_1 - u_1\| + \|x_2 - u_2\| \leq t$, then $(x_1, x_2, u_1, u_2) \in \mathcal{P}(h, t \,|\, \mathbf{f})$ and

$$\|[x_1, x_2 \,|\, \mathbf{f}\,] - [u_1, u_2 \,|\, \mathbf{f}\,]\| \leq \omega\left(\omega^{-1}(h) + t\right) - h \ .$$

Conversely, if this inequality holds $\forall\,(x_1, x_2, u_1, u_2) \in \mathcal{P}(h, t \,|\, \mathbf{f})$, whatever $h \geq 0$ and $t \geq 0$, then it is true, in particular, for $h := \min\{\|[x_1, x_2 \,|\, \mathbf{f}\,]\|, \|[u_1, u_2 \,|\, \mathbf{f}\,]\|\} - \underline{h}$ and $t := \|x_1 - u_1\| + \|x_2 - u_2\|$. Thus, we have (3.11), which is equivalent to (3.8). □

Corollary 3.6. *Let ω be a regularity modulus of the dd $[x_1, x_2 \,|\, \mathbf{f}]$ and let $\mathbf{A}$ be a bounded linear operator from $\mathbb{Y}$ into another Banach space Z . Then, $\forall\,\lambda \geq \|\mathbf{A}\|$, the function $\omega_1 := \lambda\omega$ is a regularity modulus of the dd $[x_1, x_2 \,|\, \mathbf{Af}]$.*

Proof. As

$$\eta(h/\lambda\,, \tau/\lambda \,|\, \mathbf{f}) \geq \omega^{-1}\big((h + \tau)/\lambda\big) - \omega^{-1}(h/\lambda) = \omega_1^{-1}(h + \tau) - \omega_1^{-1}(h),$$

by the proposition, it suffices to prove that $\eta(h\,, \tau \,|\, \mathbf{Af}) \geq \eta(h/\lambda\,, \tau/\lambda \,|\, \mathbf{f})$. Using Proposition 3.2, we see that

$$\|[x_1, x_2 \,|\, \mathbf{Af}]\| \leq \|\mathbf{A}\|\|[x_1, x_2 \,|\, \mathbf{f}]\| \leq \lambda\|[x_1, x_2 \,|\, \mathbf{f}]\|$$

and so

$$\|[x_1, x_2 \,|\, \mathbf{Af}]\| \geq h \Longrightarrow \|[x_1, x_2 \,|\, \mathbf{f}]\| \geq \frac{h}{\lambda}.$$

Hence, $D(h \,|\, \mathbf{Af}) \subset D(h/\lambda \,|\, \mathbf{f})$. Similarly,

$$\|[x_1, x_2 \,|\, \mathbf{Af}] - [u_1, u_2 \,|\, \mathbf{Af}]\| \geq \tau \Longrightarrow \|[x_1, x_2 \,|\, \mathbf{f}] - [u_1, u_2 \,|\, \mathbf{f}]\| \geq \frac{\tau}{\lambda}.$$

So, $\mathcal{Q}(h\,, \tau \,|\, \mathbf{Af}) \subset \mathcal{Q}(h/\lambda\,, \tau/\lambda \,|\, \mathbf{f})$ and $\eta(h\,, \tau \,|\, \mathbf{Af}) \geq \eta(h/\lambda\,, \tau/\lambda \,|\, \mathbf{f})$. □

This corollary reduces the task of finding a regularity modulus of the dd

$$([x_1, x_2 \,|\, \mathbf{H}\,]h)(t) = \int_0^1 K(s, t)\frac{\varphi(s, x_1(s)) - \varphi(s, x_2(s))}{x_1(s) - x_2(s)}h(s)\,ds$$

of the Hammerstein integral operator (3.2) to the same problem for the dd $[x_1, x_2 \,|\, \mathfrak{f}\,]$. The search for its regularity modulus inevitably involves the use of specific properties of the function φ. If, for example, the function $x \mapsto \varphi'_x(t, x)$ belongs to the class Ω, $\forall\,t \in [0, 1]$, then the best regularity modulus of the dd $[x_1, x_2 \,|\, \mathfrak{f}\,]$ is given by the following proposition.

Proposition 3.7. *Suppose that* $\forall\, s \in [0,1]$ *the function* $x \mapsto \varphi'_x(s,x)$ *is defined, nondecreasing, and concave in* $[0,\infty)$, *and* $\varphi'_x(s,0) = 0$. *Then the function*

$$\underline{\omega}(t) := \int_0^1 \max_{0\le s\le 1} \varphi'_x(s, t\max\{\tau, 1-\tau\})\, d\tau$$

is a regularity modulus of the dd $[x, y\,|\,\mathfrak{f}\,]$.

Proof. As

$$([x_1, x_2, \mathfrak{f}\,]h)(s) = \frac{\varphi(s, x_1(s)) - \varphi(s, x_2(s))}{x_1(s) - x_2(s)} h(s) =: \big[x_1(s), x_2(s)\,\big|\,\varphi\,\big] h(s)$$

for short, we have $\big\|[x_1, x_2\,|\,\mathfrak{f}\,]\big\| = \max\limits_{0\le s\le 1} \big[x_1(s), x_2(s)\,\big|\,\varphi\,\big]$ and

$$\big\|[x_1, x_2\,|\,\mathfrak{f}\,] - [u_1, u_2\,|\,\mathfrak{f}\,]\big\| = \max_{0\le s\le 1} \Big|\big[x_1(s), x_2(s)\,\big|\,\varphi\,\big] - \big[u_1(s), u_2(s)\,\big|\,\varphi\,\big]\Big|\,,$$

so that $\xi(h\,, t\,|\,\mathfrak{f}\,) = \sup\limits_{x_1, x_2, u_1, u_2} \max\limits_{0\le s\le 1} \Big|\big[x_1(s), x_2(s)\,\big|\,\varphi\,\big] - \big[u_1(s), u_2(s)\,\big|\,\varphi\,\big]\Big|$ subject to

$$\begin{aligned} &\max_{0\le s\le 1} \big[x_1(s), x_2(s)\,\big|\,\varphi\,\big] \ge h \\ &\max_{0\le s\le 1} \big[u_1(s), u_2(s)\,\big|\,\varphi\,\big] \ge h \\ &\|x_1 - u_1\| + \|x_2 - u_2\| \le t\,. \end{aligned}$$

Concavity of φ'_x with respect to the second argument implies that

$$\big|\varphi'_x(s, x_1) - \varphi'_x(s, x_2)\big| \le \varphi'_x\big(s, |x_1 - x_2|\big)\,.$$

So, $\Big|\big[x_1(s), x_2(s)\,\big|\,\varphi\big] - \big[u_1(s), u_2(s)\,\big|\,\varphi\big]\Big|$

$$= \left|\int_0^1 \left[\varphi'_x(s, x_1(s)+\tau(x_2(s)-x_1(s))) - \varphi'_x(s, u_1(s)+\tau(u_2(s)-u_1(s)))\right] d\tau\right|$$

$$\leq \int_0^1 \varphi'_x\big(s, \big|x_1(s)+\tau(x_2(s)-x_1(s)) - u_1(s) - \tau\big(u_2(s)-u_1(s)\big)\big|\big)\, d\tau$$

$$\leq \int_0^1 \varphi'_x\big(s, (1-\tau)\big|x_1(s)-u_1(s)\big| + \tau\big|x_2(s)-u_2(s)\big|\big)\, d\tau$$

$$\leq \int_0^1 \varphi'_x\big(s, (1-\tau)\|x_1-u_1\| + \tau\|x_2-u_2\|\big)\, d\tau$$

$$\leq \int_0^1 \varphi'_x\big(s, \max\{\tau, 1-\tau\}\big(\|x_1-u_1\| + \|x_2-u_2\|\big)\big)\, d\tau$$

$$\leq \int_0^1 \varphi'_x\big(s, t\max\{\tau, 1-\tau\}\big)\, d\tau\ ,$$

due to the constraint $\|x_1 - u_1\| + \|x_2 - u_2\| \leq t$. Hence,

$$\begin{aligned} \xi(h\,, t\,|\,\mathfrak{f}) &\leq \max_{0\leq s\leq 1} \int_0^1 \varphi'_x(s, t\max\{\tau, 1-\tau\})\, d\tau \\ &= \int_0^1 \max_{0\leq s\leq 1} \varphi'_x(s, t\max\{\tau, 1-\tau\})\, d\tau =: \underline{\omega}(t)\ . \end{aligned} \tag{3.12}$$

On the other hand, for any pair (x_1, x_2) with $\max\limits_{0\leq\sigma\leq 1} [x_1(\sigma), x_2(\sigma)\,|\,\varphi] \geq h$, any $\lambda \in [0, 1]$, and $u_1(s) := x_1(s) + \lambda t$, $u_2(s) := x_2(s) + (1-\lambda)t$, we see that $\big[u_1(\sigma), u_2(\sigma)\,\big|\,\varphi\big]$

$$\begin{aligned} &= \int_0^1 \varphi'_x\big(\sigma, x_1(\sigma) + \lambda t + \tau\big(x_2(\sigma) + (1-\lambda)t - x_1(\sigma) - \lambda t\big)\big) d\tau \\ &= \int_0^1 \varphi'_x\big(\sigma, (1-\tau)\big(x_1(\sigma) + \lambda t\big) + \tau\big(x_2(\sigma) + (1-\lambda)t\big)\big) d\tau \\ &\geq \int_0^1 \varphi'_x\big(\sigma, (1-\tau)x_1(\sigma) + \tau x_2(\sigma)\big) d\tau = \big[x_1(\sigma), x_2(\sigma)\,\big|\,\varphi\big]\ , \end{aligned}$$

for all $\sigma \in [0, 1]$, $\max\limits_{0\leq\sigma\leq 1} \big[u_1(\sigma), u_2(\sigma)\,\big|\,\varphi\big] \geq \max\limits_{0\leq\sigma\leq 1} \big[x_1(\sigma), x_2(\sigma)\,\big|\,\varphi\big] \geq h$ and

$\|x_1 - u_1\| + \|x_2 - u_2\| = t$, so that $\xi(h\,, t\,|\,\mathfrak{f}\,)$

$$\geq \sup_{x_1,x_2}\left\{\max_{0\leq s\leq 1}\Big(\big[x_1(s)+\lambda t, x_2(s)+(1-\lambda)t\,\big|\,\varphi\big] - \big[x_1(s), x_2(s)\,\big|\,\varphi\big]\Big)\,\Bigg|\,\max_{0\leq\sigma\leq 1}\big[x_1(\sigma), x_2(\sigma)\,\big|\,\varphi\big]\geq h\right\}$$

$$= \max_{0\leq s\leq 1}\left[\sup_{x,y}\Big\{\big[x_1(s)+\lambda t, x_2(s)+(1-\lambda)t\,\big|\,\varphi\big] - \big[x_1(s), x_2(s)\,\big|\,\varphi\big]\Big|\max_{0\leq\sigma\leq 1}\big[x_1(\sigma), x_2(\sigma)\,\big|\,\varphi\big]\geq h\Big\}\right]. \tag{3.13}$$

Here $\big[x_1(s)+\lambda t\,, x_2(s)+(1-\lambda)t\,\big|\,\varphi\big] - \big[x_1(s)\,, x_2(s)\,\big|\,\varphi\big]$

$$= \int_0^1\Big[\varphi'_x\big(s, x_1(s)+\lambda t+\tau\big(x_2(s)+(1-\lambda)t-x_1(s)-\lambda t\big)\big) - \varphi'_x\big(s, x_1(s)+\tau\big(x_2(s)-x_1(s)\big)\big)\Big]\,d\tau\,,$$

so that, due to the concavity of φ'_x with respect to the second argument, the supremum in (3.13) is attained when $x_1(s)+\tau\big(x_2(s)-x_1(s)\big)$ attains its infimum subject to the constraint in (3.13). To evaluate this infimum, note that, for all feasible pairs (x_1, x_2), the objective $x_1(s)+\tau\big(x_2(s)-x_1(s)\big)\geq 0$ and so the infimum ≥ 0. At the same time, for any $\overline{x}_1$ with $\max\limits_{0\leq\sigma\leq 1}\varphi'_x\big(\sigma, \overline{x}_1(\sigma)\big)\geq h$ and $\tilde{x}_1(\sigma) := \tilde{x}_2(\sigma) := \overline{x}_1(\sigma)$, $\forall\,\sigma\in[0,1]$, we have $\max\limits_{0\leq\sigma\leq 1}\big[\tilde{x_1}(\sigma), \tilde{x_2}(\sigma)\,\big|\,\varphi\big]$

$$= \max_{0\leq\sigma\leq 1}\int_0^1 \varphi'_x\big(\sigma, \tilde{x}_1(\sigma)+\tau\big(\tilde{x}_2(\sigma)-\tilde{x}_1(\sigma)\big)\big)\,d\tau = \max_{0\leq\sigma\leq 1}\varphi'_x\big(\sigma, \overline{x}_1\sigma\big)\geq h$$

and so the infimum $\leq \tilde{x}_1(s)$, $\forall\,s\in[0,1]$, i.e., $\leq \min\limits_{0\leq s\leq 1}\tilde{x}_1(s) = 0$. Thus, the supremum in (3.13) $\geq \displaystyle\int_0^1 \varphi'_x(s, t(\lambda+\tau(1-2\lambda)))\,d\tau$, $\forall\,\lambda\in[0,1]$, that is,

$$\geq \max_{0\leq\lambda\leq 1}\int_0^1 \varphi'_x\big(s, t(\lambda+\tau(1-2\lambda))\big)\,d\tau = \int_0^1 \varphi'_x\big(s, t\max_{0\leq\lambda\leq 1}\big(\lambda+\tau(1-2\lambda)\big)\big)\,d\tau$$
$$= \int_0^1 \varphi'_x\big(s, t\max\{\tau, 1-\tau\}\big)\,d\tau\,,$$

and $\xi(h\,, t\,|\,\mathfrak{f}\,)\geq \max\limits_{0\leq s\leq 1}\displaystyle\int_0^1 \varphi'_x\big(s, t\max\{\tau, 1-\tau\}\big)\,d\tau = \underline{\omega}(t)$. Comparing this with (3.12), we conclude that $\xi(h\,, t\,|\,\mathfrak{f}\,) = \underline{\omega}(t)$. It follows by Proposition 3.5 that any regularity modulus ω of $[x_1, x_2\,|\,\mathfrak{f}\,]$ must satisfy the inequality $\inf\limits_{h\geq 0}\big[\omega\left(\omega^{-1}(h)+t\right)-h\big]\geq\underline{\omega}(t)$. Taking into account the concavity of ω, we see that this infimum $=\omega(t)$. As $\underline{\omega}\in\Omega$, it is clear now that $\underline{\omega}$ is a regularity modulus of $[x_1, x_2\,|\,\mathfrak{f}\,]$ (pointwise the least one). □

We conclude this section with two immediate consequences of ω-regular continuity of a dd $[x_1, x_2 \,|\, \mathbf{f}]$, which will be referred to repeatedly.

Lemma 3.8. *If dd* $[x_1, x_2 \,|\, \mathbf{f}]$ *is ω-regularly continuous on D, then for all* $x_1, x_2, u_1, u_2 \in D$ $\quad \big\| [x_1, x_2 \,|\, \mathbf{f}] - [u_1, u_2 \,|\, \mathbf{f}] \big\|$

$$\le \omega\Big(\min\{\omega^{-1}(\|[x_1, x_2|\mathbf{f}]\| - \underline{h}), \omega^{-1}(\|[u_1, u_2 \,|\, \mathbf{f}]\| - \underline{h})\} + \|x_1 - u_1\| + \|x_2 - u_2\|\Big)$$
$$- \omega\Big(\min\{\omega^{-1}(\|[x_1, x_2 \,|\, \mathbf{f}]\| - \underline{h}), \omega^{-1}(\|[u_1, u_2 \,|\, \mathbf{f}]\| - \underline{h})\}\Big).$$

Lemma 3.9. *If dd* $[x_1, x_2 \,|\, \mathbf{f}]$ *is ω-regularly continuous on D, then for all* $x_1, x_2, u_1, u_2 \in D$

$$\Big|\omega^{-1}(\|[x_1, x_2 \,|\, \mathbf{f}]\| - \underline{h}) - \omega^{-1}(\|[u_1, u_2 \,|\, \mathbf{f}]\| - \underline{h})\Big| \le \|x_1 - u_1\| + \|x_2 - u_2\| \,.$$

It follows that

$$\omega^{-1}(\|[x_1, x_2 \,|\, \mathbf{f}]\| - \underline{h}) \ge \Big(\omega^{-1}(\|[u_1, u_2 \,|\, \mathbf{f}]\| - \underline{h}) - \|x_1 - u_1\| - \|x_2 - u_2\|\Big)^+ . \tag{3.14}$$

3.3 Majorant generator and convergence lemma

Beginning the convergence analysis of the method (3.4), we note that, if the operator $\mathbf{A}_0$ is invertible, then the operators $\mathbf{f}$ and $\mathbf{A}$ in (3.4) can be replaced by their respective normalizations $\mathbf{A}_0\mathbf{f}$ and $\mathbf{A}\mathbf{A}_0^{-1}$ without affecting either the set of solutions of the equation $\mathbf{f}(x) = 0$ or the method. Indeed, $\mathbf{A}\mathbf{f}(x) = \mathbf{A}\mathbf{A}_0^{-1}\mathbf{A}_0\mathbf{f}(x)$, $[x_+, x \,|\, \mathbf{A}_0\mathbf{f}] = \mathbf{A}_0[x_+, x \,|\, \mathbf{f}]$ (Proposition 3.2, 3°, 1°), and

$$\mathbf{A}_+ = 2\mathbf{A} - \mathbf{A}[x_+, x \,|\, \mathbf{f}]\mathbf{A} \iff \mathbf{A}_+\mathbf{A}_0^{-1} = 2\mathbf{A}\mathbf{A}_0^{-1} - \mathbf{A}\mathbf{A}_0^{-1}[x_+, x \,|\, \mathbf{A}_0\mathbf{f}]\mathbf{A}\mathbf{A}_0^{-1}.$$

To avoid introducing additional notation, we assume (with a negligible loss of generality) that $\mathbb{Y} = \mathbb{X}$ (i.e., $\mathbf{f}$ acts on $\mathbb{X}$) and $\mathbf{f}$ and $\mathbf{A}$ are already normalized:

$$\mathbf{A}_0 = \mathbf{I}\,. \tag{3.15}$$

This $\mathbf{A}_0$ determines $\mathbf{A}_1 = 2\mathbf{I} - \big[x_0 - \mathbf{f}(x_0), x_0 \,\big|\, \mathbf{f}\big]$.

The current iteration $(x\,, \mathbf{A})$ of the method (3.4) induces the quadruple $\bar{q} = \big(\bar{t}, \bar{\beta}, \bar{\delta}, \bar{\gamma}\big)$ of reals

$$\bar{t} := \|x - x_0\| \;,\; \bar{\beta} := \|\mathbf{A}\| \;,\; \bar{\delta} := \|x_+ - x\| \;,\; \bar{\gamma} := \big\|\mathbf{I} - \mathbf{A}[x_+, x \,|\, \mathbf{f}]\big\| \,.$$

The following lemma relates the next quadruple $\bar{q}_+ = \big(\bar{t}_+, \bar{\beta}_+, \bar{\delta}_+, \bar{\gamma}_+\big)$ with $\bar{q}$.

Lemma 3.10.

$$1^\circ\ \bar{t}_+ := \|x_+ - x_0\| \le \bar{t} + \bar{\delta};$$
$$2^\circ\ \bar{\beta}_+ := \|\mathbf{A}_+\| \le \bar{\beta}(1+\bar{\gamma});$$
$$3^\circ\ \bar{\delta}_+ := \|x_{++} - x_+\| \le \bar{\delta}\bar{\gamma}(1+\bar{\gamma}).$$

If the selected dd $[x_1, x_2 \,|\, \mathbf{f}]$ *of* $\mathbf{f}$ *is* ω*-regularly continuous on* D*, then*
$4^\circ\ \bar{\gamma}_+ := \big\|\mathbf{I} - \mathbf{A}_+[x_{++}, x_+ \,|\, \mathbf{f}]\,\big\| \le$

$$\le \bar{\gamma}^2 + \bar{\beta}_+\Big[\omega\Big(\min\{(\bar{a} - \bar{t}_{++} - \bar{t}_+)^+, (\bar{a} - \bar{t}_+ - \bar{t})^+\} + \bar{\delta}_+ + \bar{\delta}\Big) -$$
$$\omega\Big(\min\{(\bar{a} - \bar{t}_{++} - \bar{t}_+)^+, (\bar{a} - \bar{t}_+ - \bar{t})^+\}\Big)\Big].$$

where $\bar{a} := \omega^{-1}\big(\big\|[x_1, x_0 \,|\, \mathbf{f}]\big\| - \underline{h}\big) - \|x_1 - x_0\|$.

5° *All these upper bounds are exact: they hold as equalities for the scalar quadratic polynomial*

$$p(x) := x^2 + 2c_1 x - c_2\ ,\ \ c_1 > 0\ ,\ \ c_2 > 0\,.$$

Proof. 1° is a trivial consequence of the triangle inequality.
2° $\bar{\beta}_+ = \big\|\big(2\mathbf{I} - \mathbf{A}[x_+, x \,|\, \mathbf{f}]\big)\mathbf{A}\big\| \le \|\mathbf{A}\|\big(1 + \big\|\mathbf{I} - \mathbf{A}[x_+, x \,|\, \mathbf{f}]\big\|\big) = \bar{\beta}(1+\bar{\gamma})$.
3° By the secant equation, $\mathbf{f}(x_+) - \mathbf{f}(x) = [x_+, x \,|\, \mathbf{f}](x_+ - x)$ and so

$$\begin{aligned}
\bar{\delta}_+ &= \big\|\mathbf{A}_+\mathbf{f}(x_+)\big\| = \big\|\big(2\mathbf{I} - \mathbf{A}[x_+, x \,|\, \mathbf{f}]\big)\mathbf{A}\big(\mathbf{f}(x_+) - \mathbf{f}(x) + \mathbf{f}(x)\big)\big\| \\
&= \big\|\big(2\mathbf{I} - \mathbf{A}[x_+, x \,|\, \mathbf{f}]\big)\mathbf{A}\big([x_+, x \,|\, \mathbf{f}](x_+ - x) + \mathbf{f}(x)\big)\big\| \\
&\le \big\|2\mathbf{I} - \mathbf{A}[x_+, x \,|\, \mathbf{f}]\big\| \cdot \big\|\mathbf{A}[x_+, x \,|\, \mathbf{f}](x_+ - x) + \mathbf{A}\mathbf{f}(x)\big\| \\
&\le \big(1 + \big\|\mathbf{I} - \mathbf{A}[x_+, x \,|\, \mathbf{f}]\big\|\big)\big\|\mathbf{A}[x_+, x \,|\, \mathbf{f}](x_+ - x) - (x_+ - x)\big\| \\
&\le \bar{\delta}\bar{\gamma}(1+\bar{\gamma})\,.
\end{aligned}$$

4°

$$\begin{aligned}
\bar{\gamma}_+ &= \big\|\mathbf{I} - \mathbf{A}_+[x_{++}, x_+ \,|\, \mathbf{f}]\big\| \\
&= \big\|\mathbf{I} - \mathbf{A}_+[x_+, x \,|\, \mathbf{f}] - \mathbf{A}_+\big([x_{++}, x_+ \,|\, \mathbf{f}] - [x_+, x \,|\, \mathbf{f}]\big)\big\| \\
&\le \big\|\mathbf{I} - \mathbf{A}_+[x_+, x \,|\, \mathbf{f}]\big\| + \bar{\beta}_+\big\|[x_{++}, x_+ \,|\, \mathbf{f}] - [x_+, x \,|\, \mathbf{f}]\big\|\,.
\end{aligned}$$

Here

$$\mathbf{I} - \mathbf{A}_+[x_+, x \,|\, \mathbf{f}] = \mathbf{I} - \big(2\mathbf{I} - \mathbf{A}[x_+, x \,|\, \mathbf{f}]\big)\mathbf{A}[x_+, x \,|\, \mathbf{f}] = \big(\mathbf{I} - \mathbf{A}[x_+, x \,|\, \mathbf{f}]\big)^2,$$

so that

$$\big\|\mathbf{I} - \mathbf{A}_+[x_+, x \,|\, \mathbf{f}]\big\| \le \big\|\mathbf{I} - \mathbf{A}[x_+, x \,|\, \mathbf{f}]\big\|^2 = \bar{\gamma}^2.$$

Besides, by Lemma 3.8, $\big\|[x_{++}, x_+ \,|\, \mathbf{f}] - [x_+, x \,|\, \mathbf{f}]\big\|$

$$\le \omega\Big(\min\{\omega^{-1}\big(\big\|[x_{++}, x_+ \,|\, \mathbf{f}]\big\| - \underline{h}\big)\,, \omega^{-1}\big(\big\|[x_+, x \,|\, \mathbf{f}]\big\| - \underline{h}\big)\} + \bar{\delta}_+ + \bar{\delta}\Big) -$$
$$\omega\Big(\min\{\omega^{-1}\big(\big\|[x_{++}, x_+ \,|\, \mathbf{f}]\big\| - \underline{h}\big)\,, \omega^{-1}\big(\big\|[x_+, x \,|\, \mathbf{f}]\big\| - \underline{h}\big)\}\Big).$$

By (3.14), $\omega^{-1}(\|[x_{++}, x_+ \,|\, \mathbf{f}]\| - \underline{h})$

$$\begin{aligned}&\geq \Big(\omega^{-1}(\|[x_1, x_0 \,|\, \mathbf{f}]\| - \underline{h}) - \|x_{++} - x_1\| - \|x_+ - x_0\|\Big)^+ \\ &\geq \Big(\omega^{-1}(\|[x_1, x_0 \,|\, \mathbf{f}]\| - \underline{h}) - \bar{t}_{++} - \bar{t}_+ - \|x_1 - x_0\|\Big)^+ \\ &= (\bar{a} - \bar{t}_{++} - \bar{t}_+)^+,\end{aligned}$$

and, analogously, $\omega^{-1}(\|[x_+, x \,|\, \mathbf{f}]\| - \underline{h}) \geq (\bar{a} - \bar{t}_+ - \bar{t})^+$. So,

$$\begin{aligned}\|[x_{++}, x_+ \,|\, \mathbf{f}] - [x_+, x \,|\, \mathbf{f}]\| \leq \omega\Big(\min\{(\bar{a} - \bar{t}_{++} - \bar{t}_+)^+, (\bar{a} - \bar{t}_+ - \bar{t})^+\} + \bar{\delta}_+ + \bar{\delta}\Big) - \\ \omega\Big(\min\{(\bar{a} - \bar{t}_{++} - \bar{t}_+)^+, (\bar{a} - \bar{t}_+ - \bar{t})^+\}\Big)\end{aligned} \tag{3.16}$$

and

$$\begin{aligned}\bar{\gamma}_+ \leq \bar{\gamma}^2 + \bar{\beta}_+ \Big[\omega\Big(\min\{(\bar{a} - \bar{t}_{++} - \bar{t}_+)^+, (\bar{a} - \bar{t}_+ - \bar{t})^+\} + \bar{\delta}_+ + \bar{\delta}\Big) - \\ \omega\Big(\min\{(\bar{a} - \bar{t}_{++} - \bar{t}_+)^+, (\bar{a} - \bar{t}_+ - \bar{t})^+\}\Big)\Big].\end{aligned}$$

5° First we note that the mapping (the iteration of the method (3.4))

$$\mathbf{U} : \mathbb{R}^2 \to \mathbb{R}^2 \ , \ \mathbf{U}(x, A) := \big(x_+ \, , (2 - A[x_+, x \,|\, p])A\big) \ , \ x_+ := x - Ap(x)$$

maps the set

$$M := \big\{(x, A) \;\big|\; x^* < x < x_0 \ \ \& \ \ A > 0 \ \ \& \ \ A[x_+, x \,|\, p] < 1\big\},$$

where $x^* := \sqrt{c_1^2 + c_2} - c_1$ (the zero of p), into itself: $\mathbf{U}(M) \subset M$. Indeed, $(x, A) \in M \Longrightarrow A_+ = A + A(1 - A[x_+, x \,|\, p]) > A > 0$ and

$$1 - A_+[x_+, x \,|\, p] = 1 - (2 - A(1 - A[x_+, x \,|\, p])A[x_+, x \,|\, p] = (1 - A[x_+, x \,|\, p])^2 > 0,$$

i.e., $A_+[x_+, x \,|\, p] < 1$. Moreover,

$$\begin{aligned}(x, A) \in M &\Longrightarrow p(x) > 0 \ \ \& \ \ A > 0 \Longrightarrow Ap(x) > 0 \\ &\Longrightarrow p(x_+) = p(x) + [x_+, x \,|\, p](x_+ - x) = p(x) - Ap(x)[x_+, x \,|\, p],\end{aligned}$$

where $A[x_+, x \,|\, p] < 1$. Therefore, $p(x_+) > p(x) - p(x) = 0 = p(x^*)$ and so $x_+ > x^*$ $\big(p(x)$ is negative in $(0\,, x^*)\big)$. Thus,

$$(x\,, A) \in M \Longrightarrow \mathbf{U}(x\,, A) \Longrightarrow x^* < x_+ < x_0 \ \ \& \ \ A_+ > 0 \ \ \& \ \ A_+[x_+, x \,|\, p] < 1.$$

It follows that $\bar{t}_+ = |x_+ - x_0| = x_0 - x + Ap(x) = \bar{t} + \bar{\delta}$,

$$\bar{\beta}_+ = |A_+| = \big|(2 - A_+[x_+, x \,|\, p])A\big| = |A|(1 + 1 - A_+[x_+, x \,|\, p] = \bar{\beta}(1 + \bar{\gamma}),$$

$$\begin{aligned}\bar\delta_+ &= |A_+ p(x_+)| = \big|(2 - A[x_+, x\,|\,p])A\big(p(x_+) - p(x) + p(x)\big)\big|\\ &= (1+\bar\gamma)A\big|[x_+, x\,|\,p](x_+ - x) + p(x)\big|\\ &= (1+\bar\gamma)\big|A[x_+, x\,|\,p](x_+ - x) + Ap(x)\big|\\ &= (1+\bar\gamma)\big|A[x_+, x\,|\,p](x_+ - x) - (x_+ - x)\big| = \bar\delta\bar\gamma(1+\bar\gamma)\,,\end{aligned}$$

and

$$\bar\gamma_+ = \big|1 - A_+[x_{++}, x_+\,|\,p]\big| = \big|1 - A_+[x_+, x\,|\,p] - A_+\big([x_{++}, x_+\,|\,p] - [x_+, x\,|\,p]\big)\big|\,.$$

As we know already, $1 - A_+[x_+, x\,|\,p] = \big(1 - A[x_+, x\,|\,p]\big)^2 = \bar\gamma^2$. Besides, $[x', x\,|\,p] = \big(p(x') - p(x)\big)/(x' - x) = x + x' + 2c_1$, so that $[x_{++}, x_+\,|\,p] - [x_+, x\,|\,p] = x_{++} - x_+ + x_+ - x = -\bar\delta_+ - \bar\delta$. Hence, $\bar\gamma_+ = \bar\gamma^2 + \bar\beta_+\big(\bar\delta_+ + \bar\delta\big)$. We see that the inequality 4° can become an equality if and only if

$$\Big(\omega\big((a - 2\bar t_+ - \bar\delta_+)^+ + \bar\delta_+ + \bar\delta\Big) - \omega\big((a - 2\bar t_+ - \bar\delta_+)^+\big)\Big) = \bar\delta_+ + \bar\delta \tag{3.17}$$

for all $\bar t_+$ and $\bar\delta_+$. In particular, this must hold when $2\bar t_+ + \bar\delta_+ > a$. Thus, (3.17) $\Longrightarrow \omega(t) = t$, which is a regularity modulus of the dd $[x_+, x\,|\,p]$ (verify it). □

The lemma suggests the following majorant generator $q_+ = \mathbf{g}(q)$:

$$t_+ := t+\delta\,,\ \beta_+ := \beta(1+\gamma)\,,\ \delta_+ := \delta\gamma(1+\gamma)\,,\ \gamma_+ := \gamma^2+\beta_+ e(a-2t-\delta\,,\delta+\delta_+)\,, \tag{3.18}$$

where (cf. (2.8))

$$e(u\,,t) := \omega\big((u-t)^+ + t\big) - \omega\big((u-t)^+\big) = \begin{cases}\omega(u) - \omega(u-t)\,, & \text{if } 0 \le t \le u\,,\\ \omega(t)\,, & \text{if } t \ge u\,.\end{cases}$$

and a is the best lower bound for $\bar a$ available: $\bar a \ge a$. This generator has the same monotonicity property $q \prec q' \Longrightarrow \mathbf{g}(q) \prec \mathbf{g}(q')$ as the generator (2.21). As stated by Lemma 3.10, $\bar q_+ \prec \mathbf{g}(\bar q)$. So, as in the preceding chapter,

$$\bar q_0 \prec q_0 \Longrightarrow \mathop{\&}_{n} \bar q_n \prec q_n\,. \tag{3.19}$$

Using this fact and taking into account that (3.15) implies $\beta_0 = 1$, we obtain the convergence lemma for the method (3.4) (an analog of Lemma 2.8).

Lemma 3.11. *If* q_0 *causes the sequences* β_n *and* t_n *to converge:*

$$\beta_\infty < \infty \ \ \&\ \ t_\infty < \infty,,$$

then

1° *the successive iterations* $(x_n\,,\mathbf{A}_n)$ *of the method* (3.4) *remain in the ball*

$$B\big((x_0\,,\mathbf{I})\,,(t_\infty\,,\beta_\infty - 1)\big) := \Big\{(x\,,\mathbf{A})\ \Big|\ \|x - x_0\| < t_\infty\ \ \&\ \ \|\mathbf{A} - \mathbf{I}\| < \beta_\infty - 1\Big\}$$

and converge to a limit $(x_\infty, \mathbf{A}_\infty)$;

2° *this limit solves for* $(x, \mathbf{X}) \in \mathbb{X} \times \mathcal{L}(\mathbb{Y}, \mathbb{X})$ *the system*

$$\mathbf{f}(x) = 0 \quad \& \quad \mathbf{X}[x, x \,|\, \mathbf{f}] = \mathbf{I}\,; \tag{3.20}$$

3° x_∞ *is the only solution of the equation* $\mathbf{f}(x) = 0$ *in the ball* $B(x_0, R)$, *where*

$$R := \omega^{-1}(1 - \gamma_0) - \delta_0 - t_\infty\,;$$

4° *for all* $n = 0, 1, \ldots$,

$$\begin{aligned}
&\Delta_n := \|x_\infty - x_n\| \le t_\infty - t_n\,,\\
&\|\mathbf{A}_\infty - \mathbf{A}_n\| \le \beta_\infty - \beta_n\,,\\
&\frac{\Delta_{n+1}}{\Delta_n} \le \gamma_n + \beta_n\Big(\omega\big((a - t_n - t_\infty)^+ + \delta_n + \Delta_n\big) - \omega\big((a - t_n - t_\infty)^+\big)\Big);
\end{aligned} \tag{3.21}$$

5° *these bounds are exact in the sense that they are attained for the quadratic polynomial*

$$q(x) := x^2 + 2c_1 x - c_2\,,\ c_1 > 0\,,\ c_1^2 > c_2 > 0.$$

Proof. 1° By (3.19), we have $\underset{n}{\&}\, \bar{\delta}_n \le \delta_n$. So,

$$\begin{aligned}
\|x_{m+n} - x_n\| &\le \sum_{k=n}^{m+n-1} \|x_{k+1} - x_k\| = \sum_{k=n}^{m+n-1} \bar{\delta}_k \le \sum_{k=n}^{n+m-1} \delta_k\\
&= \sum_{k=n}^{n+m-1} (t_{k+1} - t_k) = t_{m+n} - t_n < t_\infty - t_n\,.
\end{aligned} \tag{3.22}$$

Similarly, $\underset{n}{\&}\,(\bar{\beta}_n \le \beta_n \ \ \& \ \ \bar{\gamma}_n \le \gamma_n)$ implies

$$\begin{aligned}
\|\mathbf{A}_{m+n} - \mathbf{A}_n\| &\le \sum_{k=n}^{m+n-1} \|\mathbf{A}_{k+1} - \mathbf{A}_k\| \le \sum_{k=n}^{m+n-1} \|\mathbf{A}_k\| \cdot \|\mathbf{I} - \mathbf{A}_k[x_{k+1}, x_k \,|\, \mathbf{f}]\|\\
&\le \sum_{k=n}^{m+n-1} \beta_k \gamma_k = \sum_{k=n}^{m+n-1} (\beta_{k+1} - \beta_k) = \beta_{m+n} - \beta_n < \beta_\infty - \beta_n.
\end{aligned} \tag{3.23}$$

Since $\beta_\infty < \infty$ and $t_\infty < \infty$ by assumption, it follows that x_n and $\mathbf{A}_n$ are Cauchy sequences in respective Banach spaces and so converge to limits $x_\infty \in \mathbb{X}$ and $\mathbf{A}_\infty \in \mathcal{L}(\mathbb{Y}, \mathbb{X})$. Setting $n = 0$ in (3.22) and (3.23) shows that $\underset{m}{\&}\,(x_m, \mathbf{A}_m) \in B\big((x_0, \mathbf{I}), (t_\infty, \beta_\infty - 1)\big)$, while forcing m to ∞ yields the first two inequalities in (3.21).

2° The assumption $\beta_\infty < \infty$ implies by (3.18) that $\gamma_n = \beta_{n+1}/\beta_n - 1 \to 0$, and so $\underset{n}{\&}\gamma_n < 1$, for $\gamma_n \geq 1 \Longrightarrow \gamma_{n+1} \geq 1 \Longrightarrow \underset{k\geq n}{\&} \gamma_k \geq 1$. It follows that all $\mathbf{A}_n$ have the same null space, namely zero:

$$\underset{n}{\&}\mathcal{N}(\mathbf{A}_n) = \mathcal{N}(\mathbf{A}_0) = \{0\} \tag{3.24}$$

due to (3.15). Indeed, as seen from (3.4), $\mathbf{A}_{n+1}y = 0$

$$\begin{aligned}
&\Longleftrightarrow -\mathbf{A}_n y = (\mathbf{I} - \mathbf{A}_n[x_{n+1}, x_n \,|\, \mathbf{f}])\mathbf{A}_n y \\
&\Longrightarrow \|\mathbf{A}_n y\| \leq \|\mathbf{I} - \mathbf{A}_n[x_{n+1}, x_n \,|\, \mathbf{f}]\| \cdot \|\mathbf{A}_n y\| = \bar{\gamma}_n \|\mathbf{A}_n y\| \leq \gamma_n \|\mathbf{A}_n y\| \\
&\Longrightarrow \mathbf{A}_n y = 0 \ ,
\end{aligned}$$

for $\gamma_n < 1$. So, $\mathcal{N}(\mathbf{A}_{n+1}) \subset \mathcal{N}(\mathbf{A}_n)$. On the other hand, by (3.4),

$$y \in \mathcal{N}(\mathbf{A}_n) \Longleftrightarrow \mathbf{A}_n y = 0 \Longrightarrow \mathbf{A}_{n+1}y = 0 \Longleftrightarrow y \in \mathcal{N}(\mathbf{A}_{n+1}) \ ,$$

that is, $\mathcal{N}(\mathbf{A}_n) \subset \mathcal{N}(\mathbf{A}_{n+1})$. Thus, $\underset{n}{\&}\mathcal{N}(\mathbf{A}_{n+1}) = \mathcal{N}(\mathbf{A}_n)$ and so (3.24). Inasmuch as $\bar{\gamma}_n := \big\|\mathbf{I} - \mathbf{A}_n[x_{n+1}, x_n \,|\, \mathbf{f}]\big\| \leq \gamma_n < 1$, the operator $\mathbf{A}_n[x_{n+1}, x_n \,|\, \mathbf{f}]$ is boundedly invertible,

$$\left\|\big(\mathbf{A}_n[x_{n+1}, x_n \,|\, \mathbf{f}]\big)^{-1}\right\| \leq (1 - \bar{\gamma}_n)^{-1} \leq (1 - \gamma_n)^{-1} \,,$$

and $[x_{n+1}, x_n \,|\, \mathbf{f}]\big(\mathbf{A}_n[x_{n+1}, x_n \,|\, \mathbf{f}]\big)^{-1}$ is a right-inverse of $\mathbf{A}_n$ (Proposition 2.1). Therefore,

$$\begin{aligned}
y \in \Big(\mathbf{I} - [x_{n+1}, x_n \,|\, \mathbf{f}]\,(\mathbf{A}_n[x_{n+1}, x_n \,|\, \mathbf{f}])^{-1}\,\mathbf{A}_n\Big)\,\mathbb{Y} \Longrightarrow& \\
y = \Big(\mathbf{I} - [x_{n+1}, x_n \,|\, \mathbf{f}]\,(\mathbf{A}_n[x_{n+1}, x_n \,|\, \mathbf{f}])^{-1}\,\mathbf{A}_n\Big)\,z \,,\ & z \in \mathbb{Y} \\
\Longrightarrow \mathbf{A}_n y = 0 \Longrightarrow y \in \mathcal{N}(\mathbf{A}_n) \,,&
\end{aligned}$$

i.e., $\Big(\mathbf{I} - [x_{n+1}, x_n \,|\, \mathbf{f}]\,(\mathbf{A}_n[x_{n+1}, x_n \,|\, \mathbf{f}])^{-1}\,\mathbf{A}_n\Big)\,\mathbb{Y} \subset \mathcal{N}(\mathbf{A}_n)$. Conversely, (3.24) implies $\mathcal{N}(\mathbf{A}_n) \subset \Big(\mathbf{I} - [x_{n+1}, x_n \,|\, \mathbf{f}]\,(\mathbf{A}_n[x_{n+1}, x_n \,|\, \mathbf{f}])^{-1}\,\mathbf{A}_n\Big)\,\mathbb{Y}$. Thus,

$$y \in \mathcal{N}(\mathbf{A}_n) \Longleftrightarrow y \in \Big(\mathbf{I} - [x_{n+1}, x_n \,|\, \mathbf{f}]\big(\mathbf{A}_n[x_{n+1}, x_n \,, |\, \mathbf{f}]\big)^{-1}\mathbf{A}_n\Big)\,\mathbb{Y}.$$

So,

$$\begin{aligned}
\|\mathbf{f}(x_n)\| &= dist\big(\mathbf{f}(x_n), \mathcal{N}(\mathbf{A}_0)\big) = dist\big(\mathbf{f}(x_n), \mathcal{N}(\mathbf{A}_n)\big) \\
&= \inf_{y\in\mathbb{Y}} \left\|\mathbf{f}(x_n) - \Big(\mathbf{I} - [x_{n+1}, x_n \,, |\, \mathbf{f}]\big(\mathbf{A}_n[x_{n+1}, x_n \,, |\, \mathbf{f}]\big)^{-1}\mathbf{A}_n\Big)\,y\right\| \\
&\leq \left\|\mathbf{f}(x_n) - \Big(\mathbf{I} - [x_{n+1}, x_n \,|\, \mathbf{f}]\big(\mathbf{A}_n[x_{n+1}, x_n \,|\, \mathbf{f}]\big)^{-1}\mathbf{A}_n\Big)\,\mathbf{f}(x_n)\right\| \\
&\leq \|[x_{n+1}, x_n \,|\, \mathbf{f}]\| \cdot \left\|\big(\mathbf{A}_n[x_{n+1}, x_n, \mathbf{f}]\big)^{-1}\right\| \cdot \|\mathbf{A}_n \mathbf{f}(x_n)\| \\
&\leq \delta_n \frac{\|[x_{n+1}, x_n \,|\, \mathbf{f}]\|}{1 - \gamma_n} \,.
\end{aligned}$$

The norm in the numerator converges to $\big\|[x_\infty, x_\infty \,|\, \mathbf{f}]\big\|$, whereas $\delta_n \to 0$ by 1°. Hence, $\big\|\mathbf{f}(x_\infty)\big\| = \lim\big\|\mathbf{f}(x_n)\big\| = 0$. Besides, as $\big\|\mathbf{I} - \mathbf{A}_n[x_{n+1}, x_n \,|\, \mathbf{f}]\big\| \le \gamma_n < 1$ and $\gamma_n \to 0$, $\mathbf{A}_n[x_{n+1}, x_n \,|\, \mathbf{f}] \to \mathbf{I}$. At the same time,

$$\mathbf{A}_n[x_{n+1}, x_n \,|\, \mathbf{f}] \to \mathbf{A}_\infty[x_\infty, x_\infty \,|\, \mathbf{f}]\,.$$

Consequently, $\mathbf{A}_\infty[x_\infty, x_\infty \,|\, \mathbf{f}] = \mathbf{I}$.

3° Let $x^* \in D$ be another solution of the equation $\mathbf{f}(x) = 0$ and let $\bar{R}$ be its distance from x_0: $\bar{R} := \|x_0 - x^*\|$. Then

$$0 = \mathbf{f}(x^*) - \mathbf{f}(x_\infty) = [x^*, x_\infty \,|\, \mathbf{f}](x^* - x_\infty),$$

so that the operator $[x^*, x_\infty \,|\, \mathbf{f}]$ is not invertible. It follows that $\big\|\mathbf{I} - [x^*, x_\infty \,|\, \mathbf{f}]\big\| \ge 1$ (otherwise, $\sum_{k=0}^{\infty}\big(\mathbf{I} - [x^*, x_\infty \,|\, \mathbf{f}]\big)^k = [x^*, x_\infty \,|\, \mathbf{f}]^{-1}$). On the other hand,

$$\big\|\mathbf{I} - [x^*, x_\infty \,|\, \mathbf{f}]\big\| \le \big\|\mathbf{I} - [x_1, x_0 \,|\, \mathbf{f}]\big\| + \big\|[x_1, x_0 \,|\, \mathbf{f}] - [x^*, x_\infty \,|\, \mathbf{f}]\big\|\,,$$

where the first norm is $\bar{\gamma}_0$ in view of (3.15), while the second

$$\begin{aligned}
&\le \omega\Big(\min\{\omega^{-1}(\|[x_1, x_0 \,|\, \mathbf{f}]\| - \underline{h})\,, \omega^{-1}(\|[x^*, x_\infty \,|\, \mathbf{f}]\| - \underline{h})\} + \\
&\qquad\qquad \|x_1 - x^*\| + \|x_0 - x_\infty\|\Big) - \\
&\qquad \omega\Big(\min\{\omega^{-1}(\|[x_1, x_0 \,|\, \mathbf{f}]\| - \underline{h})\,, \omega^{-1}(\|[x^*, x_\infty \,|\, \mathbf{f}]\| - \underline{h})\}\Big) \\
&\le \omega\Big(\min\{\omega^{-1}(\|[x_1, x_0 \,|\, \mathbf{f}]\| - \underline{h})\,, \omega^{-1}(\|[x^*, x_\infty \,|\, \mathbf{f}]\| - \underline{h})\} + \\
&\qquad\qquad \|x_1 - x_0\| + \|x_0 - x^*\| + \bar{t}_\infty\Big) \\
&\qquad \omega\Big(\min\{\omega^{-1}(\|[x_1, x_0 \,|\, \mathbf{f}]\| - \underline{h})\,, \omega^{-1}(\|[x^*, x_\infty \,|\, \mathbf{f}]\| - \underline{h})\}\Big)\,.
\end{aligned} \tag{3.25}$$

According to (3.14), $\omega^{-1}\big(\big\|[x^*, x_\infty \,|\, \mathbf{f}]\big\| - \underline{h}\big)$

$$\begin{aligned}
&\ge \big(\omega^{-1}\big(\big\|[x_1, x_0 \,|\, \mathbf{f}]\big\| - \underline{h}\big) - \|x^* - x_1\| - \|x_\infty - x_0\|\big)^+ \\
&\ge \big(\omega^{-1}\big(\big\|[x_1, x_0 \,|\, \mathbf{f}]\big\| - \underline{h}\big) - \|x_1 - x_0\| - \|x_0 - x^*\| - \bar{t}_\infty\big)^+ \\
&= \big(\bar{a} - \bar{R} - \bar{t}_\infty\big)^+.
\end{aligned}$$

Hence, the minimum in (3.25) $\ge \big(\bar{a} - \bar{R} - \bar{t}_\infty\big)^+$, so that (because of the concavity of ω)

$$\big\|[x_1, x_0 \,|\, \mathbf{f}] - [x^*, x_\infty \,|\, \mathbf{f}]\big\| \le \omega\Big(\big(\bar{a} - \bar{t}_\infty - \bar{R}\big)^+ + \bar{\delta}_0 + \bar{t}_\infty + \bar{R}\Big) - \omega\Big(\big(\bar{a} - \bar{t}_\infty - \bar{R}\big)^+\Big).$$

Thus,

$$1 \le \bar{\gamma}_0 + \omega\Big(\big(\bar{a} - \bar{t}_\infty - \bar{R}\big)^+ + \bar{\delta}_0 + \bar{t}_\infty + \bar{R}\Big) - \omega\Big(\big(\bar{a} - \bar{t}_\infty - \bar{R}\big)^+\Big) \le \bar{\gamma}_0 + \omega\big(\bar{\delta}_0 + \bar{t}_\infty + \bar{R}\big)\,.$$

Solving this inequality for $\bar{R}$ yields

$$\bar{R} \geq \omega^{-1}(1-\bar{\gamma}_0) - \bar{\delta}_0 - \bar{t}_\infty \geq \omega^{-1}(1-\gamma_0) - \delta_0 - t_\infty .$$

4° To get the third bound in (3.21), observe that $x_{n+1} - x_\infty$

$$= x_n - x_\infty - \mathbf{A}_n\big(\mathbf{f}(x_n) - \mathbf{f}(x_\infty)\big) = x_n - x_\infty - \mathbf{A}_n[x_n, x_\infty \,|\, \mathbf{f}](x_n - x_\infty)$$
$$= \Big(\mathbf{I} - \mathbf{A}_n[x_{n+1}, x_n \,|\, \mathbf{f}] + \mathbf{A}_n\big([x_{n+1}, x_n \,|\, \mathbf{f}] - [x_n, x_\infty \,|\, \mathbf{f}]\big)\Big)(x_n - x_\infty) ,$$

whence

$$\begin{aligned}\Delta_{n+1} &\leq \Delta_n\Big(\bar{\gamma}_n + \bar{\beta}_n\big\|[x_{n+1}, x_n \,|\, \mathbf{f}] - [x_n, x_\infty \,|\, \mathbf{f}]\big\|\Big) \\ &\leq \Delta_n\Big(\gamma_n + \beta_n\big\|[x_{n+1}, x_n \,|\, \mathbf{f}] - [x_n, x_\infty \,|\, \mathbf{f}]\big\|\Big) .\end{aligned}$$

By Lemma 3.8, $\big\|[x_{n+1}, x_n \,|\, \mathbf{f}] - [x_n, x_\infty \,|\, \mathbf{f}]\big\|$

$$\begin{aligned}= \omega\Big(&\min\big\{\omega^{-1}\big(\|[x_{n+1}, x_n \,|\, \mathbf{f}]\| - \underline{h}\big) , (\omega^{-1}\big(\|[x_n, x_\infty \,|\, \mathbf{f}]\| - \underline{h}\big)\big\} + \\ &\bar{\delta}_n + \Delta_n\Big) - \\ &\omega\Big(\min\big\{\omega^{-1}\big(\|[x_{n+1}, x_n \,|\, \mathbf{f}]\| - \underline{h}\big) , \omega^{-1}\big(\|[x_n, x_\infty \,|\, \mathbf{f}]\| - \underline{h}\big)\big\}\Big),\end{aligned}$$

where, by (3.14), $\omega^{-1}\big(\|[x_{n+1}, x_n \,|\, \mathbf{f}]\| - \underline{h}\big)$

$$\begin{aligned}&\geq \Big(\omega^{-1}\big(\|[x_0, x_{-1} \,|\, \mathbf{f}]\| - \underline{h}\big) - \|x_{n+1} - x_0\| - \|x_n - x_{-1}\|\Big)^+ \\ &\geq \big(\omega^{-1}(1-\underline{h}) - \bar{t}_{n+1} - \|x_n - x_0\| - \|x_0 - x_{-1}\|\big)^+ \\ &= \big(\bar{a} - \bar{t}_{n+1} - \bar{t}_n\big)^+\end{aligned}$$

and $\omega^{-1}\big(\|[x_n, x_\infty \,|\, \mathbf{f}]\| - \underline{h}\big)$

$$\begin{aligned}&\geq \Big(\omega^{-1}\big(\|[x_0, x_{-1} \,|\, \mathbf{f}]\| - \underline{h}\big) - \|x_n - x_0\| - \|x_\infty - x_{-1}\|\Big)^+ \\ &\geq \big(\omega^{-1}(1-\underline{h}) - \bar{t}_n - \bar{t}_\infty - \|x_0 - x_{-1}\|\big)^+ \geq \big(\bar{a} - \bar{t}_n - \bar{t}_\infty\big)^+ .\end{aligned}$$

So, $\big\|[x_{n+1}, x_n \,|\, \mathbf{f}] - [x_n, x_\infty \,|\, \mathbf{f}]\big\|$

$$\begin{aligned}&\leq \omega\Big(\min\big\{(a - \bar{t}_n - \bar{t}_{n+1})^+, (a - \bar{t}_n - \bar{t}_\infty)^+\big\} + \bar{\delta}_n + \Delta_n\Big) - \\ &\qquad \omega\Big(\min\big\{(a - \bar{t}_n - \bar{t}_{n+1})^+, (a - \bar{t}_n - \bar{t}_\infty)^+\big\}\Big) \\ &= \omega\Big((a - \bar{t}_n - \bar{t}_\infty)^+ + \bar{\delta}_n + \Delta_n\Big) - \omega\Big((a - \bar{t}_n - \bar{t}_\infty)^+\Big) \\ &\leq \omega\Big((a - t_n - t_\infty)^+ + \delta_n + \Delta_n\Big) - \omega\Big((a - t_n - t_\infty)^+\Big)\end{aligned}$$

and

$$\frac{\Delta_{n+1}}{\Delta_n} \leq \gamma_n + \beta_n\Big(\omega(a - t_n - t_\infty)^+ + \delta_n + \Delta_n\Big) - \omega(a - t_n - t_\infty)^+\Big).$$

□

Now we have to answer the same question we asked about the generator (2.21): precisely which starter q_0 causes the sequence q_n to converge? As then, we see immediately that the condition

$$\underset{n}{\&}\, \gamma_n < 1 \tag{3.26}$$

is necessary for convergence: $\gamma \geq 1 \Longrightarrow \gamma_+ \geq 1 \Longrightarrow \delta_+ \geq 2\delta \Longrightarrow \delta_n \to \infty$. As then, it is also sufficient. While proving it, we use the abbreviation

$$e_n := e\big(a - 2t_n - \delta_n\,, \delta_n + \delta_{n+1}\big)$$
$$= \begin{cases} \omega(a - t_{n+1} - t_n) - \omega(a - t_{n+2} - t_{n+1})\,, & \text{if } t_{n+2} + t_{n+1} \leq a\,, \\ \omega(\delta_{n+1} + \delta_n)\,, \text{ if } t_{n+2} + t_{n+1} \geq a\,. \end{cases}$$

Lemma 3.12. *Suppose that* (3.26) *is true. Then*

$$t_{n+1} + t_n \leq a \Longrightarrow t_{n+1} + t_n = a - \omega^{-1}\left(\omega(a - \delta_0) - \frac{1-\gamma_0}{\beta_0} + \frac{1-\gamma_n}{\beta_n}\right),$$

$$t_{n+1} + t_n \leq a < t_{n+2} + t_{n+1} \Longrightarrow \underset{m>n}{\&}\, t_{m+2} + t_{m+1} < t_{n+1} + t_n + \omega^{-1}\left(\frac{1-\gamma_n}{\beta_n}\right).$$

Proof. By (3.18), $\underset{k}{\&}(1-\gamma_{k+1})/\beta_{k+1} = (1-\gamma_k)/\beta_k - e_k$ and so

$$\sum_0^{n-1} e_k = \sum_0^{n-1}\left(\frac{1-\gamma_k}{\beta_k} - \frac{1-\gamma_{k+1}}{\beta_{k+1}}\right) = \frac{1-\gamma_0}{\beta_0} - \frac{1-\gamma_n}{\beta_n}\,. \tag{3.27}$$

If $t_{n+1} + t_n \leq a$, then $\forall\, k = 0, 1, \ldots, n-1$ $e_k = \omega(a - t_{k+1} - t_k) - \omega(a - t_{k+2} - t_{k+1})$ and (with $t_0 = 0$ in mind)

$$\sum_0^{n-1} e_k = \sum_0^{n-1}\big(\omega(a - t_{k+1} - t_k) - \omega(a - t_{k+2} - t_{k+1})\big) = \omega(a - t_1) - \omega(a - t_{n+1} - t_n)\,.$$

Together with (3.27), this gives the equality

$$\frac{1-\gamma_0}{\beta_0} - \frac{1-\gamma_n}{\beta_n} = \omega(a - \delta_0) - \omega(a - t_{n+1} - t_n)\,,$$

whence

$$t_{n+1} + t_n = a - \omega^{-1}\left(\omega(a - \delta_0) - \frac{1-\gamma_0}{\beta_0} + \frac{1-\gamma_n}{\beta_n}\right).$$

If $t_{n+1}+t_n \le a < t_{n+2}+t_{n+1}$, then $\underset{k\ge n}{\&}\ e_k = \omega(\delta_{k+1}+\delta_k)$ and

$$\sum_{k=n}^{m} e_k = \sum_{k=n}^{m} \omega(\delta_{k+1}+\delta_k) \ge \omega\left(\sum_{k=n}^{m}(\delta_{k+1}+\delta_k)\right),$$

because of the subadditivity of ω. On the other hand, similarly to (3.27),

$$\sum_{k=n}^{m} e_k = \frac{1-\gamma_n}{\beta_n} - \frac{1-\gamma_{m+1}}{\beta_{m+1}}.$$

Hence, $\omega\left(\sum_{k=n}^{m}(\delta_{k+1}+\delta_k)\right) \le (1-\gamma_n)/\beta_n - (1-\gamma_{m+1})/\beta_{m+1}$ and

$$\sum_{k=n}^{m}(\delta_{k+1}+\delta_k) \le \omega^{-1}\left(\frac{1-\gamma_n}{\beta_n} - \frac{1-\gamma_{m+1}}{\beta_{m+1}}\right) < \omega^{-1}\left(\frac{1-\gamma_n}{\beta_n}\right).$$

By (3.18), $\delta_{k+1}+\delta_k = t_{k+2}-t_k$ and

$$\sum_{k=n}^{m}(t_{k+2}-t_k) = t_{m+2}+t_{m+1}-t_{n+1}-t_n.$$

It follows that

$$\underset{m>n}{\&}\ t_{m+2}+t_{m+1} < t_{n+1}+t_n+\omega^{-1}\left(\frac{1-\gamma_n}{\beta_n}\right).$$

□

Corollary 3.13. *$t_\infty < \infty \Longleftrightarrow \underset{n}{\&}\ \gamma_n < 1$. In this case,*

$$\underset{n}{\&}\ t_{n+1}+t_n \le a \Longrightarrow t_\infty = \frac{1}{2}\left(a-\omega^{-1}(a-\delta_0) - \frac{1-\gamma_0}{\beta_0} + \frac{1}{\beta_\infty}\right),$$

$$\exists n \text{ with } t_{n+1}+t_n \le a < t_{n+2}+t_{n+1} \Longrightarrow t_\infty \le \frac{1}{2}\left(t_{n+1}+t_n+\omega^{-1}\left(\frac{1-\gamma_n}{\beta_n}\right)\right).$$

The set of all starters $q_0 = (0,1,\gamma_0,\delta_0)$ resulting in $\underset{n}{\&}\ \gamma_n < 1$ is determined by the next proposition, where

$$e(t,\gamma,\delta) := e\Big(a-2t-\delta\,,\,\delta(1+\gamma+\gamma^2)\Big)$$
$$= \begin{cases} \omega\Big(a-2t-\delta\Big) - \omega\Big(a-2t-\delta(2+\gamma+\gamma^2)\Big), & \text{if } 2t+\delta\left(2+\gamma+\gamma^2\right) \le a, \\ \omega\Big(\delta(1+\gamma+\gamma^2)\Big), & \text{if } 2t+\delta\left(2+\gamma+\gamma^2\right) \ge a. \end{cases}$$

Proposition 3.14. 1° $\underset{n}{\&}\gamma_n < 1 \Longleftrightarrow f_\infty(0,\gamma_0,\delta_0) \geq 1$, *where* f_∞ *is defined recursively:* $f_0(t,\gamma,\delta) := (1-\gamma)/e(t,\gamma,\delta)$ *and* $f_{n+1}(t,\gamma,\delta)$ *is the solution for* β *of the equation*

$$f_n\Big(t+\delta,\ \gamma^2+\beta(1+\gamma)e(t,\gamma,\delta),\ \delta\gamma(1+\gamma)\Big) = \beta(1+\gamma)\,.$$

2° *The function* f_∞ *is a solution of the system (a functional equation with an end condition)*

$$\begin{gathered} x\Big(t+\delta\,,\gamma^2+x(t,\gamma,\delta)(1+\gamma)e(t,\gamma,\delta),\ \delta\gamma(1+\gamma)\Big) = x(t,\gamma,\delta)(1+\gamma) \\ x(t,0,0) = \beta_\infty. \end{gathered} \tag{3.28}$$

Proof. By (3.18),

$$\begin{aligned} \gamma_{n+1} < 1 &\Longleftrightarrow \gamma_n^2 + \beta_{n+1}e(t_n,\gamma_n,\delta_n) < 1 \\ &\Longleftrightarrow \beta_n < \frac{1-\gamma_n}{e(t_n,\gamma_n,\delta_n)} =:\ f_0(t_n,\gamma_n,\delta_n)\,. \end{aligned}$$

As seen from the definition of $e(t,\gamma,\delta)$, it is increasing in γ and δ and not increasing in t. So, f_0 is positive, decreasing in γ and δ and not decreasing in t. Suppose that, for some $k \geq 0$, $\beta_{n-k} < f_k(t_{n-k},\gamma_{n-k},\delta_{n-k})$, where f_k is positive, decreasing in γ and δ and not increasing in t. Using (3.18), rewrite this inequality as

$$F_k\big(t_{n-k-1}\,,\beta_{n-k-1}\,,\gamma_{n-k-1}\,,\delta_{n-k-1}\big) > 0\,,$$

where

$$F_k(t,\beta,\gamma,\delta) := f_k\Big(t+\delta\,,\ \gamma^2+\beta(1+\gamma)e(t,\gamma,\delta)\ ,\ \delta\gamma(1+\gamma)\Big) - \beta(1+\gamma)\,. \tag{3.29}$$

Since f_k is decreasing in the second argument,

$$f_k(t+\delta\,,\ \gamma^2+\beta(1+\gamma)e(t\,,\gamma\,,\delta)\ ,\ \delta\gamma(1+\gamma)) \leq f_k(t+\delta\,,\ \gamma^2\ ,\ \delta\gamma(1+\gamma))$$

and so

$$\begin{aligned} &\lim_{\beta\to\infty}\Big(f_k(t+\delta\,,\ \gamma^2+\beta(1+\gamma)e(t\,,\gamma\,,\delta)\ ,\ \delta\gamma(1+\gamma)) - \beta(1+\gamma)\Big) \\ &\leq \lim_{\beta\to\infty}\Big(f_k(t+\delta\,,\ \gamma^2\,,\ \delta\gamma(1+\gamma)) - \beta(1+\gamma)\Big) = -\infty\,. \end{aligned}$$

Hence, the function $\beta \mapsto F_k(t,\beta,\gamma,\delta)$ is decreasing in the interval $(0,\infty)$ from $f_k\Big(t+\delta\,,\ \gamma^2\ ,\ \delta\gamma(1+\gamma)\Big) > 0$ to

$$\lim_{\beta\to\infty}\Big(f_k(t+\delta\,,\ \gamma^2+\beta(1+\gamma)e(t\,,\gamma\,,\delta)\ ,\ \delta\gamma(1+\gamma)) - \beta(1+\gamma)\Big) = -\infty\,.$$

Therefore, the equation $F_k(t,\beta,\gamma,\delta) = 0$ for β has a unique positive solution, which we denote $f_{k+1}(t,\gamma,\delta)$:

$$F_k\big(t, f_{k+1}(t,\gamma,\delta),\gamma,\delta\big) = 0 \ . \tag{3.30}$$

Moreover, like f_k, f_{k+1} is decreasing in γ and δ and not increasing in t. Indeed, by the induction hypothesis, f_k is not decreasing in the first and the third arguments and decreasing in the second. Hence, F_k is decreasing in β and γ, so that

$$\begin{aligned}\gamma < \gamma' \Longrightarrow F_k\big(t, f_{k+1}(t,\gamma,\delta),\gamma,\delta\big) = 0 = F_k\big(t\,, f_{k+1}(t\,,\gamma'\,,\delta)\,,\gamma'\,,\delta\big)\\ < F_k\big(t, f_{k+1}(t\,,\gamma'\,,\delta),\gamma,\delta\big)\\ \Longrightarrow f_{k+1}(t,\gamma',\delta) < f_{k+1}(t,\gamma,\delta) \ .\end{aligned}$$

Similarly,

$$\begin{aligned}\delta < \delta' \Longrightarrow F_k(t, f_{k+1}(t,\gamma,\delta),\gamma,\delta) = 0 = F_k(t, f_{k+1}(t,\gamma,\delta'),\gamma,\delta')\\ < F_k(t, f_{k+1}(t,\gamma,\delta'),\gamma,\delta)\\ \Longrightarrow f_{k+1}(t,\gamma,\delta') < f_{k+1}(t,\gamma,\delta)\end{aligned}$$

and

$$\begin{aligned}t < t' \Longrightarrow F_k(t, f_{k+1}(t,\gamma,\delta),\gamma,\delta) = 0 = F_k(t', f_{k+1}(t',\gamma,\delta),\gamma,\delta)\\ \le F_k(t, f_{k+1}(t',\gamma,\delta),\gamma,\delta)\\ \Longrightarrow f_{k+1}(t',\gamma,\delta) \le f_{k+1}(t,\gamma,\delta) \, .\end{aligned}$$

Thus,

$$\beta_{n-k} < f_k(t_{n-k},\gamma_{n-k},\delta_{n-k}) \Longrightarrow \beta_{n-k-1} < f_{k+1}(t_{n-k-1},\gamma_{n-k-1},\delta_{n-k-1}) \ .$$

By induction, $\gamma_{n+1} < 1 \Longleftrightarrow \beta_0 < f_n(t_0,\gamma_0,\delta_0)$. It follows that

$$\underset{n}{\&}\ \gamma_n < 1 \Longleftrightarrow \beta_0 \le \inf_n f_n(t_0,\gamma_0,\delta_0) \ .$$

The sequence f_n is pointwise decreasing:

$$\underset{n}{\&}\ f_{n+1}(t,\beta,\gamma) < f_n(t,\beta,\gamma) \ . \tag{3.31}$$

This is proved inductively. First, we have to verify that

$$f_1(t,\beta,\gamma) < f_0(t,\beta,\gamma) \ .$$

Since $F_0(t,\beta,\gamma,\delta) = (1-\gamma)/e(t,\gamma,\delta) - \beta(1+\gamma)$ is decreasing with respect to its second argument, it is enough to show that $F_0\big(t, f_0(t,\gamma,\delta),\gamma,\delta\big) < F_0\big(t, f_1(t,\gamma,\delta),\gamma,\delta\big) = 0$ (see (3.30). By (3.29), $F_0(t, f_0(t,\gamma,\delta),\gamma,\delta)$

$$= f_0\Big(t{+}\delta\,,\ \gamma^2{+}\ f_0(t,\gamma,\delta)(1+\gamma)e(t,\gamma,\delta)\ ,\ \delta\gamma(1+\gamma)\Big) - f_0(t,\gamma,\delta)(1+\gamma)\ ,$$

where $\gamma^2 + f_0(t,\gamma,\delta)(1+\gamma)e(t,\gamma,\delta) = 1$. So, $F_0\big(t, f_0(t,\gamma,\delta),\gamma,\delta\big)$

$$= f_0\,(t+\delta\,,1\,,\delta\gamma(1+\gamma)) - \frac{1-\gamma}{e(t,\gamma,\delta)}(1+\gamma) = -\frac{1-\gamma^2}{e(t,\gamma,\delta)} < 0\,.$$

Suppose now that $f_n(t,\gamma,\delta) < f_{n-1}(t,\gamma,\delta)$ for some $n \geq 1$. Then

$$F_{n-1}\big(t, f_n(t,\gamma,\delta),\gamma,\delta\big) = 0 = F_n\big(t, f_{n+1}(t,\gamma,\delta),\gamma,\delta\big)$$

$$= f_n\big(t+\delta\,,\,\gamma^2 + f_{n+1}(t,\gamma,\delta)(1+\gamma)e(t,\gamma,\delta)\,,\delta\gamma(1+\gamma)\big) - f_{n+1}(t,\gamma,\delta)(1+\gamma)$$

$$< f_{n-1}\big(t+\delta\,,\,\gamma^2 + f_{n+1}(t,\gamma,\delta)(1+\gamma)e(t,\gamma,\delta)\,,\delta\gamma(1+\gamma)\big) - f_{n+1}(t,\gamma,\delta)(1+\gamma)$$

$$= F_{n-1}\big(t, f_{n+1}(t,\gamma,\delta),\gamma,\delta\big) \Longrightarrow f_{n+1}(t,\gamma,\delta) < f_n(t,\gamma,\delta)\ ,$$

because F_{n-1} is decreasing with respect to the second argument. By induction, (3.31) is proved. Now $\inf_n f_n = f_\infty$. Taking limits in (3.30) yields $F_\infty\big(t, f_\infty(t,\gamma,\delta),\gamma,\delta\big) = 0$, i.e.,

$$f_\infty\Big(t+\delta\,,\gamma^2 + f_\infty(t,\gamma,\delta)(1+\gamma)e(t,\gamma,\delta)\,,\delta\gamma(1+\gamma)\Big) = f_\infty(t,\gamma,\delta)(1+\gamma).$$

Besides, (3.30) implies $f_\infty(t,0,0) = \beta_\infty$. Thus, the function $f_\infty(t,\gamma,\delta)$ is a solution of the system (3.28). □

For linear ω ($\omega(t) = ct$) the generator (3.18) reduces to

$$\beta_+ := \beta(1+\gamma)\ ,\ \gamma_+ := \gamma^2 + c\beta_+\delta\left(1+\gamma+\gamma^2\right)\ ,\ \delta_+ := \delta\gamma(1+\gamma)\ . \qquad (3.32)$$

An invariant of this simplified generator is the function

$$I(\beta\,,\gamma\,,\delta) := \delta^2 - 2\delta\frac{1+\gamma}{c\beta} + \left(\frac{1-\gamma}{c\beta}\right)^2.$$

Indeed, $I(\beta_+,\gamma_+,\delta_+)$

$$= \delta_+^2 - 2\,\delta_+\frac{1+\gamma_+}{c\beta_+} + \left(\frac{1-\gamma_+}{c\beta_+}\right)^2$$

$$= \delta_+^2 - 2\,\delta_+\frac{1+\gamma^2+c\beta_+(\delta+\delta_+)}{c\beta_+} + \left(\frac{1-\gamma^2-c\beta(\delta+\delta_+)}{c\beta_+}\right)^2$$

$$= \delta_+^2 - 2\,\delta_+\frac{1+\gamma^2}{c\beta_+} - 2\delta\delta_+ - 2\,\delta_+^2 + \left(\frac{1-\gamma^2}{c\beta_+}\right)^2 - 2\frac{1-\gamma^2}{c\beta_+}(\delta+\delta_+) + (\delta+\delta_+)^2.$$

As $(1-\gamma^2)/(c\beta_+) = (1-\gamma)(1+\gamma)/\big(c\beta(1+\gamma)\big) = (1-\gamma)/(c\beta)$, we obtain that $I(\beta_+\,,\gamma_+\,,\delta_+)$

$$= \delta_+^2 - 2\,\delta_+\frac{1+\gamma^2}{c\beta_+} - 2\,\delta\delta_+ - 2\delta_+^2 + \left(\frac{1-\gamma}{c\beta}\right)^2 - 2\frac{1-\gamma}{c\beta}(\delta+\delta_+) + \delta^2 + 2\delta\delta_+ + \delta_+^2$$

$$= -2\,\delta_+\frac{1+\gamma^2}{c\beta_+} + \left(\frac{1-\gamma}{c\beta}\right)^2 - 2\frac{1-\gamma}{c\beta}(\delta+\delta_+) + \delta^{\,2}$$

$$= \delta^{\,2} - 2\,\frac{\delta}{c\beta}\Big(\gamma(1+\gamma^2) + (1-\gamma)(1+\gamma+\gamma^2)\Big) + \left(\frac{1-\gamma}{c\beta}\right)^2$$

$$= \delta^{\,2} - 2\,\delta\frac{1+\gamma}{c\beta} + \left(\frac{1-\gamma}{c\beta}\right)^2 = I(\beta,\gamma,\delta)\,.$$

The system (3.28) for the generator (3.32) takes the form

$$x\big(\gamma^2 + cx(\gamma,\delta)\delta(1+\gamma)(1+\gamma+\gamma^2)\,,\,\delta\gamma(1+\gamma)\big) = x(\gamma,\delta)(1+\gamma) \;\;\&\;\; x(0\,,0) = \beta_\infty\,. \tag{3.33}$$

Its solution

$$\begin{aligned}(\gamma\,,\delta) \mapsto\; & \frac{\sqrt{4\gamma\delta^{\,2} + (c\beta_\infty)^{-2}(1-\gamma)^2} - \delta(1+\gamma)}{c\big((c\beta_\infty)^{-2} - \delta^2\big)} \\ &= c^{-1}\frac{(1-\gamma)^2}{\sqrt{4\gamma\delta^{\,2} + (c\beta_\infty)^{-2}(1-\gamma)^2} + \delta(1+\gamma)}\end{aligned} \tag{3.34}$$

is obtained by solving the equation $I(\beta\,,\gamma\,,\delta) = I(\beta_\infty\,,0\,,0)$ for β. According to Proposition 3.14, the convergence domain of the generator (3.32) is described by the inequality

$$(\gamma\,,\delta) \mapsto c^{-1}\frac{(1-\gamma)^2}{\sqrt{4\gamma\delta^{\,2} + (c\beta_\infty)^{-2}(1-\gamma)^2} + \delta(1+\gamma)} \geq 1\,,$$

which is equivalent to $(1-\gamma)^2 - \delta(1+\gamma) \geq \sqrt{4\gamma\delta^{\,2} + (c\beta_\infty)^{-2}(1-\gamma)^2}$

$$\begin{aligned}&\Longleftrightarrow (1-\gamma)^4 - 2\delta(1-\gamma)^2(1+\gamma) + \delta^2(1+\gamma)^2 \geq 4\gamma\delta^{\,2} + (c\beta_\infty)^{-2}(1-\gamma)^2 \\ &\Longleftrightarrow (1-\gamma)^4 - 2\delta(1-\gamma)^2(1+\gamma) + \delta^{\,2}(1-\gamma)^2 \geq (c\beta_\infty)^{-2}(1-\gamma)^2 \\ &\Longleftrightarrow \delta^{\,2} - 2\delta(1+\gamma) + (1-\gamma)^2 - (c\beta_\infty)^{-2} \geq 0 \\ &\Longleftrightarrow \delta \leq 1+\gamma - \sqrt{4\gamma + (c\beta_\infty)^{-2}} \leq 1+\gamma - 2\sqrt{\gamma} = (1-\sqrt{\gamma})^2\,.\end{aligned}$$

The generator (3.32) can be simplified even further by changing δ for $s := c\beta_+\delta(1+\gamma)$. After the change, it becomes

$$\beta_+ := \beta(1+\gamma)\,,\;\; \gamma_+ := \gamma^2 + s\frac{1+\gamma+\gamma^2}{1+\gamma}\,,\;\; s_+ := s\gamma(1+\gamma_+)^2\,. \tag{3.35}$$

In contrast with (3.32), β does not appear in the second and third equations. Fixed points of this generator fill up the nonnegative half $\{(\beta,0,0) \mid \beta \geq 0\}$ of the β-axis. So, convergence of the sequence (β_n, γ_n, s_n), generated by (3.32) from the starter $(1, \gamma_0, s_0)$, is equivalent to convergence to $(0,0)$ of the sequence (γ_n, s_n), generated by the truncated generator

$$\gamma_+ := \gamma^2 + s\frac{1+\gamma+\gamma^2}{1+\gamma}\ , \quad s_+ := s\gamma(1+\gamma_+)^2\ , \tag{3.36}$$

defined by the last two equations in (3.35). This generator has two fixed points

$$(0,0)\ ,\quad (0.465...,0.216...)\ ,$$

of which only the origin is attractive. Its basin of attraction (as well as the convergence domain of the generator (3.35)) is determined by the following.

Proposition 3.15. 1° *The sequence (γ_n, s_n), generated by the generator (3.36) from the starter (γ_0, s_0), converges to $(0,0)$ if and only if*

$$s_0 \leq f_\infty(\gamma_0)\ ,$$

where $f_0(\gamma) := \big((1-\gamma^2)(1+\gamma)\big)/(1+\gamma+\gamma^2)$ and $f_{n+1}(\gamma)$ is the (unique) solution for s of the equation

$$f_n\left(\gamma^2 + s\frac{1+\gamma+\gamma^2}{1+\gamma}\right) = s\gamma\left(1+\gamma^2+s\frac{1+\gamma+\gamma^2}{1+\gamma}\right)^2. \tag{3.37}$$

2° *The function f_∞ is the only solution of the system*

$$x\left(\gamma^2 + x(\gamma)\frac{1+\gamma+\gamma^2}{1+\gamma}\right) = x(\gamma)\,\gamma\left(1+\gamma^2+x(\gamma)\frac{1+\gamma+\gamma^2}{1+\gamma}\right)^2 \quad \& \quad x(1)=0\,. \tag{3.38}$$

3° $f_\infty(\gamma) = (1+\gamma)^2(1-\sqrt{\gamma})^2$.
4° *The function $I(\gamma,s) := f_\infty(\gamma) - s$ is an invariant of the generator* (3.36).
5° *If $s_0 \leq f_\infty(\gamma_0)$, then, for all $n = 0,1,\ldots,$*

$$\beta_n = \prod_0^{n-1}(1+\gamma_k)\,,$$

$$t_n = \frac{1}{2}\left(\frac{1-\gamma_0^2+s_0}{c(1+\gamma_0)} - \frac{1-\gamma_n^2+s_n}{c\beta_n(1+\gamma_n)}\right).$$

Proof. 1° Let $\mathop{\&}\limits_{k=0}^{n} \gamma_k < 1$ and (for short) $g(\gamma) := (1+\gamma+\gamma^2)/(1+\gamma)$. By (3.36),

$$\gamma_{n+1} < 1 \Longleftrightarrow \gamma_n^2 + s_n g(\gamma) < 1 \Longleftrightarrow s_n < \frac{(1-\gamma_n^2)}{g(\gamma_n} =: f_0(\gamma_n).$$

f_0 is positive and decreasing in $(0\,,1)$ (its derivative

$$f_0'(\gamma) = -\frac{4\gamma + 5\gamma^2 + 2\gamma^3 + \gamma^4}{(1+\gamma+\gamma^2)^2}$$

is negative $\forall\,\gamma > 0$), $f_0(0) = 1$, and $f_0(1) = 0$. Suppose that, for some $k \ge 0$,

$$\gamma_{n+1} < 1 \iff s_{n-k} < f_k(\gamma_{n-k})\,, \tag{3.39}$$

where f_k is positive and decreasing in $(0\,,1)$, $f_k(0) = 1$, and $f_k(1) = 0$. Using (3.36), rewrite this inequality as

$$F_k(\gamma_{n-k-1}\,,\, s_{n-k-1}) > 0\,, \tag{3.40}$$

where

$$F_k(\gamma\,, s) := f_k(\gamma^2 + sg(\gamma)) - s\gamma\left(1 + \gamma^2 + sg(\gamma)\right)^2\,. \tag{3.41}$$

As f_k is decreasing and $f_k(1) = 0$ by the induction hypothesis, F_k is decreasing with respect to $s \in \left[0\,, (1-\gamma^2)\,/g(\gamma)\right]$ from $F_k(\gamma\,,0) = f_k(\gamma^2) > 0$ to

$$F_k\left(\gamma\,, \frac{(1-\gamma^2)}{g(\gamma)}\right)$$

$$= f_k\left(\gamma^2 + \frac{(1-\gamma^2)}{g(\gamma)}\,g(\gamma)\right) - \frac{(1-\gamma^2)}{g(\gamma)}\,\gamma\left(1+\gamma^2+\frac{(1-\gamma^2)}{g(\gamma)}\,g(\gamma)\right)^2$$

$$= f_k(1) - 4\gamma\,\frac{(1-\gamma^2)}{g(\gamma)} = -4\gamma\,\frac{(1-\gamma^2)}{g(\gamma)} < 0.$$

Therefore, the equation $F_k(\gamma\,,s) = 0$ is uniquely solvable for $s \in \left(0\,, (1-\gamma^2)/g(\gamma)\right)$. We denote its solution $f_{k+1}(\gamma)$:

$$F_k\left(\gamma\,, f_{k+1}(\gamma)\right) = 0\ ,\ \forall\,\gamma \in [0\,,1]\,. \tag{3.42}$$

In particular,

$$F_k\left(\gamma_{n-k-1}\,, f_{k+1}(\gamma_{n-k-1})\right) = 0\,. \tag{3.43}$$

Comparison with (3.40) shows that

$$s_{n-k-1} < f_{k+1}(\gamma_{n-k-1})\,.$$

The function f_{k+1} is decreasing in $[0\,,1]$. Indeed, since F_k is decreasing in each of its two arguments,

$$\begin{aligned} 0 \le \gamma < \gamma' \le 1 \Longrightarrow F_k\left(\gamma\,, f_{k+1}(\gamma)\right) = 0 = F_k\left(\gamma'\,, f_{k+1}(\gamma')\right) \\ < F_k\left(\gamma\,, f_{k+1}(\gamma')\right) \\ \Longrightarrow f_{k+1}(\gamma') < f_{k+1}(\gamma)\,. \end{aligned}$$

Besides, $f_{k+1}(1) = 0$, for by definition $0 \leq f_{k+1}(\gamma) \leq (1-\gamma^2)/g(\gamma)$. Moreover, by (3.42) $F_k(0, f_{k+1}(0)) = 0$, while according to (3.41) $F_k(0,1) = f_k(g(0)) = f_k(1) = 0$. Hence, $F_k(0, f_{k+1}(0)) = 0 = F_k(0,1)$ and so $f_{k+1}(0) = 1$, for F_k has only one solution in $(0,1)$. Thus, (3.39) implies

$$\gamma_{n+1} < 1 \Longleftrightarrow s_{n-k-1} < f_{k+1}(\gamma_{n-k-1}) .$$

By induction, $\gamma_{n+1} < 1 \Longleftrightarrow s_0 < f_n(\gamma_0)$ and $\underset{n}{\&} \gamma_n < 1 \Longleftrightarrow s_0 \leq \inf_n f_n(\gamma_0)$. The sequence f_n is pointwise decreasing:

$$\underset{n}{\&} f_{n+1}(\gamma) < f_n(\gamma) , \ \forall \gamma \in [0,1) . \tag{3.44}$$

This is verified inductively. First, we have to show that $f_1(\gamma) < f_0(\gamma)$ or, equivalently, $F_0(\gamma, f_1(\gamma)) = 0 > F_0(\gamma, f_0(\gamma))$. By the definitions of F_0 and f_0,

$$F_0(\gamma, f_0(\gamma)) = f_0\big(\gamma^2 + f_0(\gamma)g(\gamma) - f_0(\gamma)\gamma(1+\gamma^2 + f_0(\gamma)g(\gamma)\big)^2$$

$$= f_0\left(\gamma^2 + \frac{(1-\gamma^2)}{g(\gamma)}\right) - f_0(\gamma)\gamma\left(1+\gamma^2 + \frac{(1-\gamma^2)}{g(\gamma)} \cdot g(\gamma)\right)^2$$

$$= f_0(1) - 4\gamma f_0(\gamma) = -4\gamma f_0(\gamma) < 0 .$$

Suppose now that, for some $n \geq 1$, $f_n(\gamma) < f_{n-1}(\gamma), \forall \gamma \in [0,1]$. Then

$$F_{n-1}(\gamma, f_n(\gamma)) = 0 = F_n(\gamma, f_{n+1}(\gamma))$$

$$= f_n\left(\gamma^2 + f_{n+1}(\gamma)g(\gamma)\right) - f_{n+1}(\gamma)\gamma\left(1+\gamma^2 + f_{n+1}(\gamma)g(\gamma)\right)^2$$

$$< f_{n-1}\left(\gamma^2 + f_{n+1}(\gamma)g(\gamma)\right) - f_{n+1}(\gamma)\gamma\left(1+\gamma^2 + f_{n+1}(\gamma)g(\gamma)\right)^2$$

$$= F_{n-1}(\gamma, f_{n+1}(\gamma)) .$$

Inasmuch as F_{n-1} is decreasing with respect to the second argument,

$$F_{n-1}(\gamma, f_n(\gamma)) < F_{n-1}(\gamma, f_{n+1}(\gamma)) \Longrightarrow f_{n+1}(\gamma) < f_n(\gamma) .$$

By induction, the claim (3.44) is proved. So, $\inf_n f_n = f_\infty$.

2° Taking limits in (3.41) and $f_n(1)=0$ yields $F_\infty(\gamma, f_\infty(\gamma)) = 0 = f_\infty(1)$, which means that f_∞ satisfies the system (3.38). To prove uniqueness, let x be a solution and define the generator $\mathbf{g} : (p, q) \mapsto (p_+, q_+)$ as follows:

$$p_+ := p^2 + q\,\frac{1+p+p^2}{1+p} , \quad q_+ := x(p)p(1+p_+)^2 .$$

If $p = \gamma$ & $q = x(p)$, then

$$p_+ = p^2 + q\,\frac{1+p+p^2}{1+p} = p^2 + x(p)\,\frac{1+p+p^2}{1+p} = \gamma^2 + x(\gamma)\,\frac{1+\gamma+\gamma^2}{1+\gamma}$$
$$= \gamma^2 + s\,\frac{1+\gamma+\gamma^2}{1+\gamma} = \gamma_+$$

and so

$$q_+ = x(p)p\,(1+p_+)^2 = x(\gamma)\gamma\,(1+\gamma_+)^2 = x\left(\gamma^2 + x(\gamma)\,\frac{1+\gamma+\gamma^2}{1+\gamma}\right)$$
$$= x\left(p^2 + x(p)\,\frac{1+p+p^2}{1+p}\right) = x\left(p^2 + q\,\frac{1+p+p^2}{1+p}\right) = x(p_+)\,.$$

By induction, $p_0 = \gamma_0$ & $q_0 = x(p_0) \Longrightarrow \underset{n}{\&}\big(p_n = \gamma_n$ & $q_n = x(p_n)\big)$. This means that the generator $\mathbf{g}$ coincides with (3.36). Consequently, $x = f_\infty$.

3° Due to the proved uniqueness of the solution of the system (3.38), it suffices to show that the function $\gamma \mapsto (1+\gamma)^2\big(1-\sqrt{\gamma}\big)^2$ satisfies it. Indeed, if $x(\gamma) = (1+\gamma)^2\big(1-\sqrt{\gamma}\big)^2$, then

$$x\big(\gamma^2 + x(\gamma)\,g(\gamma)\big) = \big(1+\gamma^2 + x(\gamma)g(\gamma)\big)^2\Big(1-\sqrt{\gamma^2 + x(\gamma)g(\gamma)}\Big)^2.$$

So, for this $x(\gamma)$, the functional equation (3.38) is satisfied if

$$x(\gamma)\gamma = \Big(1-\sqrt{\gamma^2 + x(\gamma)g(\gamma)}\Big)^2 = 1 - 2\sqrt{\gamma^2 + x(\gamma)g(\gamma)} + \gamma^2 + x(\gamma)g(\gamma)$$

or, equivalently, if

$$2\sqrt{\gamma^2 + x(\gamma)g(\gamma)} = 1 + \gamma^2 + \frac{x(\gamma)}{1+\gamma}\,.$$

In turn, this is true

$$\Longleftrightarrow 4\big(\gamma^2 + x(\gamma)g(\gamma)\big) = \big(1+\gamma^2\big)^2 + 2\frac{1+\gamma^2}{1+\gamma}\,x(\gamma) + \frac{x(\gamma)^2}{(1+\gamma)^2}$$

$$\Longleftrightarrow 4(1+\gamma)\Big(\gamma^2(1+\gamma) + x(\gamma)\,\big(1+\gamma+\gamma^2\big)\Big) =$$
$$\big(1+\gamma^2\big)^2(1+\gamma)^2 + 2\,\big(1+\gamma^2\big)\,(1+\gamma)\,x(\gamma) + x(\gamma)^2$$

$$\Longleftrightarrow x(\gamma)^2 - 2(1+\gamma)^3 x(\gamma) + (1+\gamma)^2\big(1-\gamma^2\big)^2 = 0\,.$$

Solving this equation for $x(\gamma)$ yields

$$\begin{aligned} x(\gamma) &= (1+\gamma)^3 - \sqrt{(1+\gamma)^6 - (1+\gamma)^2\,(1-\gamma^2)^2} \\ &= (1+\gamma)^3 - 2(1+\gamma)\sqrt{(1+\gamma)^4 - (1-\gamma^2)^2} \\ &= (1+\gamma)\Big((1+\gamma)^2 - 2\sqrt{\gamma\,(1+\gamma)^2}\Big) \\ &= (1+\gamma)^2(1+\gamma-\sqrt{\gamma}) = (1+\gamma)^2(1-\sqrt{\gamma})^2\,, \end{aligned}$$

which is our hypothesis.

4° By (3.36), $I(\gamma\,,s)=0 \Longrightarrow s=f_\infty(\gamma) \Longrightarrow I(\gamma_+\,,s_+)=f_\infty(\gamma_+)-s_+$

$$=f_\infty\left(\gamma^2+s\frac{1+\gamma+\gamma^2}{1+\gamma}\right)-s\gamma\left(1+\gamma^2+s\frac{1+\gamma+\gamma^2}{1+\gamma}\right)^2$$

$$=f_\infty\left(\gamma^2+f_\infty(\gamma)\frac{1+\gamma+\gamma^2}{1+\gamma}\right)-f_\infty(\gamma)\gamma\left(1+\gamma^2+f_\infty(\gamma)\frac{1+\gamma+\gamma^2}{1+\gamma}\right)^2=0\,,$$

since f_∞ satisfies (3.38) by 2°. Thus, $I(\gamma\,,s)=0 \Longrightarrow I(\gamma_+\,,s_+)=0$. By induction, $I(\gamma_0\,,s_0)=0 \Longrightarrow \underset{n}{\&}\, I(\gamma_n\,,s_n)=0$.

5° The first equality is a direct consequence of (3.35). As to t_n, note that $s_0 \le f_\infty(\gamma_0) \Longrightarrow \gamma_\infty = s_\infty = 0$ (by 1°) and $t_{n+1}+t_n$

$$\begin{aligned} t_{n+1}+t_n &= \sum_{k=0}^{n}\delta_k + \sum_{k=0}^{n-1}\delta_k = \delta_0 + \sum_{k=0}^{n-1}(\delta_{k+1}+\delta_k) \\ &= \delta_0 + \sum_{k=0}^{n-1}\delta_k\,(1+\gamma_k+\gamma_k^2) \\ &= \frac{s_0}{c(1+\gamma_0)^2} + \sum_{k=0}^{n-1}\frac{s_k}{c\beta_{k+1}(1+\gamma_k)}\,(1+\gamma_k+\gamma_k^2)\;. \end{aligned}$$

On the other hand, by (3.32),

$$\frac{1-\gamma_{k+1}}{c\beta_{k+1}} = \frac{1-\gamma_k^2-s_k\dfrac{1+\gamma_k+\gamma_k^2}{1+\gamma_k}}{c\beta_k(1+\gamma_k)} = \frac{1-\gamma_k}{c\beta_k} - \frac{s_k\,(1+\gamma_k+\gamma_k^2)}{c\beta_{k+1}(1+\gamma_k)}\,,$$

so that

$$\frac{s_k\,(1+\gamma_k+\gamma_k^2)}{c\beta_{k+1}(1+\gamma_k)} = \frac{1-\gamma_k}{c\beta_k} - \frac{1-\gamma_{k+1}}{c\beta_{k+1}}\,.$$

TABLE 3.1: Application of Proposition 3.15

Number of iterations	*Error*
10	0.144171558692232
20	0.143849501717994
30	0.143849153097550
40	0.143849152719163
50	0.143849152718752

Therefore,

$$2t_n + \frac{s_n}{c\beta_{n+1}(1+\gamma_n)} = \frac{s_0}{c(1+\gamma_0)^2} + \sum_{k=0}^{n-1}\left(\frac{1-\gamma_k}{c\beta_k} - \frac{1-\gamma_{k+1}}{c\beta_{k+1}}\right)$$
$$= \frac{s_0}{c(1+\gamma_0)^2} + \frac{1-\gamma_0}{c\beta_0} - \frac{1-\gamma_n}{c\beta_n}$$

and

$$t_n = \frac{1}{2}\left(\frac{(1-\gamma_0^2)(1+\gamma_0)+s_0}{c(1+\gamma_0)^2} - \frac{(1-\gamma_n^2)(1+\gamma_n)+s_n}{c\beta_n(1+\gamma_n)^2}\right).$$

□

As stated by the proposition, f_∞ is the solution of the system (3.38). This fact indicates a possibility of using the recursion defining f_∞ for numerical solution of that system. The practicality of this approach to the solution of such systems depends on the rate of convergence, about which the proposition says nothing. In the absence of a theoretical estimation of the rate, some impression can be obtained from a numerical experiment. The results of such an experiment are presented in the following table. Each function f_n was represented in computer memory as a piecewise linear function defined by the vector of its values on a 16-point Tchebyshev mesh on the segment $[0, 1]$. The equation (3.37) was solved by the secant method. The values of Error shown in the table are maximum absolute values of the difference between the final f_n and the exact solution f_∞ given by the proposition. Figure 3.1 shows the initial approximation f_0 (the solid line), the final one (the dashed line), and the exact solution (the dotted line).

We are in a position now to state the convergence theorem for the method (3.4), which is obtained by summing up the analysis that has been carried out in Lemma 3.11, Lemma 3.12, and Proposition 3.14.

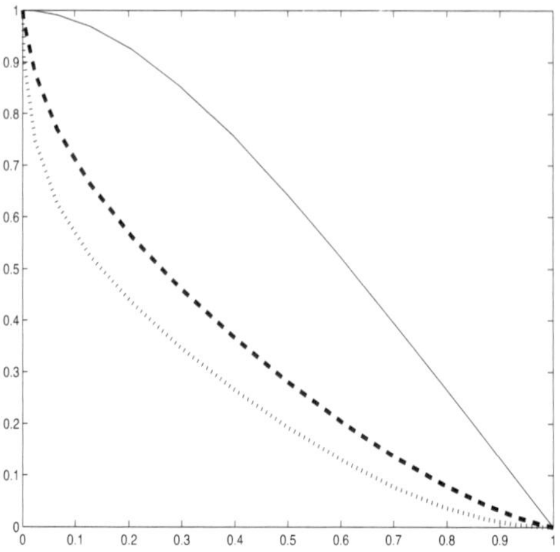

FIGURE 3.1: Application of Proposition 3.15, plots

3.4 Convergence theorem

Theorem 3.16. *Let the selected dd* $[x_1, x_2 \,|\, \mathbf{f}]$ *of* $\mathbf{f}$ *be* ω*-regularly continuous on* D *and let* f_∞ *be the function of Proposition 3.14. If the starters* x_0, γ_0, *and* δ_0 *are such that*

$$\|\mathbf{I} - [x_1, x_0 \,|\, \mathbf{f}]\| \le \gamma_0 \quad \& \quad \|\mathbf{f}(x_0)\| \le \delta_0 \quad \& \quad f_\infty(0, \gamma_0, \delta_0) \ge 1, \tag{3.45}$$

then

1° *if* $\underset{n}{\&}\, t_{n+1} + t_n \le \bar{a} := \omega^{-1}(1 - \underline{h}) - \|x_0 - x_{-1}\|$ *and* $\bar{a} \ge a$, *then*

$$t_\infty = \frac{1}{2}\left(a - \omega^{-1}(a - \delta_0) - (1 - \gamma_0) + \frac{1}{\beta_\infty}\right);$$

if $\exists n$ *with* $t_{n+1} + t_n \le a < t_{n+2} + t_{n+1}$, *then*

$$t_\infty \le \frac{1}{2}\left(t_{n+1} + t_n + \omega^{-1}\left(\frac{1 - \gamma_n}{\beta_n}\right)\right);$$

2° *the sequence* $(x_n, \mathbf{A}_n)$ *generated by the method* (3.4) *from the starter* $(x_0, \mathbf{I})$ *remains in the ball* $B\big((x_0, \mathbf{I}), (t_\infty, \beta_\infty - 1)\big)$ *and converges to a limit* $(x_\infty, \mathbf{A}_\infty)$;

3° *this limit solves for* $(x, \mathbf{X}) \in \mathbb{X} \times \mathcal{L}(\mathbb{Y}, \mathbb{X})$ *the system*

$$\mathbf{f}(x) = 0 \quad \& \quad \mathbf{X}[x, x \,|\, \mathbf{f}] = \mathbf{I}\,. \tag{3.46}$$

4° x_∞ *is the only solution of the equation* $\mathbf{f}(x) = 0$ *in the ball* $B(x_0, R)$, *where*

$$R := \omega^{-1}(1 - \gamma_0) - \delta_0 - t_\infty ;$$

5° *for all* $n = 0, 1, \ldots$,

$$\begin{aligned} &\Delta_n := \|x_\infty - x_n\| \le t_\infty - t_n , \\ &\|\mathbf{A}_\infty - \mathbf{A}_n\| \le \beta_\infty - \beta_n , \\ &\frac{\Delta_{n+1}}{\Delta_n} \le \gamma_n + \beta_n\Big(\omega\big((a - t_n - t_\infty)^+ + \delta_n + \Delta_n\big) - \omega\big((a - t_n - t_\infty)^+\big)\Big); \end{aligned} \tag{3.47}$$

6° *these bounds are exact in the sense that they are attained for the quadratic polynomial*

$$q(x) := x^2 + 2c_1 x - c_2 , \ c_1 > 0 , \ c_2 > 0 ,$$

As γ_n and $\delta_n + \Delta_n$ tend to zero, the bound for the ratio Δ_{n+1}/Δ_n in (3.47) implies that x_n converges to x_∞ superlinearly. However, this bound contains more information about the rate of convergence than just the asymptotics $\Delta_{n+1} = o(\Delta_n)$. It allows us to estimate the errors throughout the whole process, starting from the first iterations.

In the Lipschitz case ($\omega(t) = ct$) when Proposition 3.15 applies, the theorem takes on the more precise form of

Corollary 3.17. *Let the dd* $[x_1, x_2 \,|\, \mathbf{f}]$ *be Lipschitz continuous on* D:

$$\big\|[x_1, x_2 \,|\, \mathbf{f}] - [u_1, u_2 \,|\, \mathbf{f}]\big\| \le c\big(\|x_1 - u_1\| + \|x_2 - u_2\|\big) , \ \forall\, x_1, x_2, u_1, u_2 \in D.$$

If the starters x_0, γ_0, δ_0 *satisfy the condition*

$$\big\|\mathbf{I} - [x_1, x_0 \,|\, \mathbf{f}]\big\| \le \gamma_0 < 1 \ \& \ \big\|\mathbf{f}(x_0)\big\| \le \delta_0 < c^{-1}(1 - \sqrt{\gamma_0})^2 ,$$

then

1° *the sequence* $(\beta_n, \gamma_n, \delta_n)$ *generated by the generator* (3.32) *from* $(1, \gamma_0, \delta_0)$ *converges to* $\big(1/(c\sqrt{I_0}\,), 0, 0\big)$, *where*

$$I_0 := \delta_0^2 - 2\delta_0 \frac{1 + \gamma_0}{c} + \left(\frac{1 - \gamma_0}{c}\right)^2 .$$

Moreover, for all $n = 1, 2, \ldots$,

$$\beta_n = \frac{\sqrt{4\gamma_n \delta_n^2 + I_0(1 - \gamma_n)^2} - \delta_n(1 + \gamma_n)}{c(I_0 - \delta_n^2)} ,$$

$$t_n := \sum_0^{n-1} \delta_k = \frac{1}{2}\left(\frac{1 - \gamma_0}{c} + \delta_0 - \frac{1 - \gamma_n}{c\beta_n} - \delta_n\right);$$

2° *the sequence* $(x_n\,, \mathbf{A}_n)$ *generated by the method* (3.4) *from the starter*

$(x_0, \mathbf{I})$ remains in the ball $B\big((x_0, \mathbf{I}), (t_\infty, \beta_\infty - 1)\big)$ and converges to a solution $(x_\infty, \mathbf{A}_\infty)$ of the system (3.46);

3° *x_∞ is the only solution of the equation $\mathbf{f}(x) = 0$ in the ball $B(x_0, R)$, where*

$$R := \frac{1-\gamma_0}{c} - \delta_0 - t_\infty\,;$$

4° *for all $n = 1, 2, \dots$,*

$$\Delta_n \le \frac{1}{2}\left(\frac{1-\gamma_n}{c\beta_n} + \delta_n - \sqrt{I_0}\right),$$

$$\|\mathbf{A}_\infty - \mathbf{A}_n\| \le \frac{1}{c\sqrt{I_0}} - \frac{\sqrt{4\gamma_n\delta_n^2 + I_0(1-\gamma_n)^2} - \delta_n(1+\gamma_n)}{c(I_0 - \delta_n^2)},$$

$$\frac{\Delta_{n+1}}{\Delta_n} \le \gamma_n + c\beta_n(\delta_n + \Delta_n);$$

5° *all these bounds are exact.*

3.5 Research project

When studying an iterative method that does not involve evaluation of derivatives of the operator like the method (3.4), it is unnatural to assume its differentiability. I have come to believe that the differentiability of the operator is irrelevant concerning convergence properties of derivative-free iterative methods. I dare to conjecture that, given the sequence of iterations generated by such a method from some starter, it is possible to construct a *smooth* operator for which this method generates the same iterations. The problem I suggest for a research project is to prove or disprove this conjecture.

Chapter 4

Broyden's method

Let $\mathbb{X}$, $\mathbb{H}$, and $\mathcal{L}(\mathbb{H},\mathbb{X})$ be a Banach space, a Hilbert one with the inner product $\langle \cdot , \cdot \rangle$, and the Banach space of linear bounded operators acting from $\mathbb{H}$ into $\mathbb{X}$. Broyden's method [4] (in its inversion-free form), given a starting pair $(x_0 , \mathbf{A}_0) \in \mathbb{X} \times \mathcal{L}(\mathbb{H},\mathbb{X})$, generates the sequence of iterations $(x_n, \mathbf{A}_n)$ according to the following rule:

$$x_+ := x - \mathbf{A}\mathbf{f}(x) \ , \ \mathbf{A}_+ := \mathbf{A} - \frac{\mathbf{A}\mathbf{f}(x_+)}{\langle \mathbf{A}^*\mathbf{A}\mathbf{f}(x) , \mathbf{f}(x_+) - \mathbf{f}(x)\rangle} \langle \mathbf{A}^*\mathbf{A}\mathbf{f}(x) , \cdot \rangle \ , \tag{4.1}$$

where $\mathbf{A}^*$ is the adjoint of $\mathbf{A}$.

4.1 Motivation

The method (4.1) was designed for operator equations in finite dimensions. Broyden in [4] motivates his choice $x_+ - x = -\mathbf{A}\mathbf{f}(x)$ of the parameter v in the family of rank 1 updates satisfying the secant equation,

$$\mathbf{A}_+^{-1} := \mathbf{A}^{-1} + \frac{\mathbf{f}(x_+)}{\langle v , x_+ - x\rangle} \langle v , \cdot \rangle \ , \tag{4.2}$$

by lack of information about the rate of change of $\mathbf{f}$ in directions different from $u := x_+ - x$. This consideration led him to choose v in (4.2) so that $\mathbf{A}_+^{-1} v = \mathbf{A}^{-1} v \ , \ \forall v \perp u$. Used in (4.2), this condition implies $\langle v , \cdot \rangle = 0 \ , \ \forall v \perp u$, which can be true only if v is proportional to u. Application of the Sherman–Morrison formula (Lemma 1.3) to Broyden's update

$$\mathbf{A}_+^{-1} := \mathbf{A}^{-1} + \frac{\mathbf{f}(x_+)}{\langle u , u\rangle} \langle u , \cdot \rangle \tag{4.3}$$

translates it into the update in (4.1).

Later it was found that u as v in (4.2) minimizes the Frobenius norm of the matrix $\mathbf{A}_+^{-1} - \mathbf{A}^{-1}$ subject to the secant equation $\mathbf{A}_+\big(\mathbf{f}(x_+) - \mathbf{f}(x)\big) = x_+ - x$. Another observation that can motivate Broyden's choice is that it gives to the denominator in (4.2) the maximum value among all unit v: $\max_v\big\{ \langle v , u\rangle \ \big| \ \|v\| = 1 \big\} = \|u\|$.

4.2 Majorant generator and convergence lemma

As seen from (4.1), $\mathbf{A}_+\big(\mathbf{f}(x_+)-\mathbf{f}(x)\big)=x_+-x$ and $\mathbf{A}_+$ is invertible provided $\mathbf{A}$ is invertible and $x_+\neq x$ (see (4.3)). So, invertibility of $\mathbf{A}_0$ implies invertibility of all $\mathbf{A}_n$ generated by Broyden's method and the equalities

$$\underset{n\geq 1}{\&}\ \mathbf{A}_n=[x_n,x_{n-1}\,|\,\mathbf{f}]^{-1}. \tag{4.4}$$

Moreover, if $\mathbf{A}_0$ is invertible, then $\mathbf{A}$ and $\mathbf{f}$ in (4.1) can be replaced by their normalizations $\mathbf{A}\mathbf{A}_0^{-1}$ and $\mathbf{A}_0\mathbf{f}$ without affecting either the set of solutions of the equation $\mathbf{f}(x)=0$ or the method. To save indexation, let us assume (with a minor loss of generality) that $\mathbb{X}=\mathbb{H}$ (that is , $\mathbf{f}$ acts on $\mathbb{H}$) and $\mathbf{A}$ and $\mathbf{f}$ are already normalized:

$$\mathbf{A}_0=\mathbf{I}\,. \tag{4.5}$$

This $\mathbf{A}_0$ determines

$$\mathbf{A}_1=\mathbf{I}-\frac{\mathbf{f}(x_0)}{\big\langle\mathbf{f}(x_0)\,,\,\mathbf{f}(x_0-\mathbf{f}(x_0))-\mathbf{f}(x_0)\big\rangle}\langle\mathbf{f}(x_0)\,,\cdot\rangle\ .$$

An iteration $(x\,,\mathbf{A})$ of Broyden's method induces the quadruple of reals

$$\bar t:=\|x-x_0\|\ ,\ \bar\beta:=\|\mathbf{A}\|\ ,\ \bar\delta:=\|x_+-x\|=\big\|\mathbf{A}\mathbf{f}(x)\big\|\ ,\ \bar\sigma:=\big\|\mathbf{f}(x_+)\big\|\,.$$

Lemma 4.3 below relates the next quadruple $\big(\bar t_+,\bar\beta_+,\bar\delta_+,\bar\sigma_+\big)$ with $\big(\bar t,\bar\beta,\bar\delta,\bar\sigma\big)$. Its proof involves the norm of the operator $\mathbf{I}+a\langle b\,,\cdot\rangle$, where a and b are any two vectors of $\mathbb{H}$. So, we have first to get an exact expression of this norm in terms of a and b.

Lemma 4.1. *Let a and b be two vectors of a Hilbert space $\mathbb{H}$.*

$$\|\mathbf{I}+a\langle b\,,\cdot\rangle\|^2=1+\langle a,b\rangle+0.5\|a\|\,\|b\|\left(\|a\|\,\|b\|+\sqrt{\|a\|^2\|b\|^2+4\big(1+\langle a,b\rangle\big)}\right).$$

Proof.

$$\begin{aligned}\nu:=\|\mathbf{I}+a\langle b\,,\cdot\rangle\|^2&=\max_{\|x\|\leq 1}\|x+a\langle b\,,x\rangle\|^2\\&=\max_{\|x\|\leq 1}\big(\|x\|^2+2\langle a,x\rangle\langle b\,,x\rangle+\|a\|^2\langle b\,,x\rangle^2\big)\\&=\max_{(\alpha,\beta,x)\in A}\big(\|x\|^2+2\alpha\beta+\beta^2\|a\|^2\big)\ ,\end{aligned}$$

where

$$A:=\Big\{(\alpha,\beta,x)\ \Big|\ \langle a\,,x\rangle=\alpha\ \&\ \langle b\,,x\rangle=\beta\ \&\ \|x\|\leq 1\Big\}.$$

Let

$$B(\alpha,\beta) := \Big\{x \;\Big|\; (\alpha,\beta,x)\in A\Big\} = \Big\{x \;\Big|\; \langle a\,,x\rangle = \alpha \;\;\&\;\; \langle b\,,x\rangle = \beta \;\;\&\;\; \|x\|\le 1\Big\}.$$

By Corollary 1.7, $B(\alpha,\beta)\neq\emptyset \iff \|\alpha b-\beta a\|^2 \le \|a\|^2\|b\|^2 - \langle a\,,b\rangle^2 =: det$. Hence,

$$C := \Big\{(\alpha,\beta) \;\Big|\; B(\alpha,\beta)\neq\emptyset\Big\} = \Big\{(\alpha,\beta) \;\Big|\; \|\alpha b-\beta a\|^2 \le det\Big\}.$$

By the lemma on sections (Lemma 1.4),

$$\begin{aligned}\nu &= \max_{(\alpha,\beta)\in C}\;\max_{x\in B(\alpha,\beta)} \left(\|x\|^2 + 2\alpha\beta + \beta^2\|a\|^2\right)\\ &= \max_{(\alpha,\beta)\in C}\left(2\alpha\beta+\beta^2\|a\|^2 + \max_{x\in B(\alpha,\beta)}\|x\|^2\right).\end{aligned} \tag{4.6}$$

If a and b are linearly independent and $(\alpha,\beta)\in C$, then, by Corollary 1.7,

$$B(\alpha,\beta) = \Big\{\alpha' a' + \beta' b' + y \;\Big|\; \langle a', y\rangle = \langle b', y\rangle = 0 \;\;\&\;\; \|y\|^2 \le 1-\alpha'^2-\beta'^2\Big\},$$

where

$$a' := \frac{a}{\|a\|}\,,\quad \alpha' := \frac{\alpha}{\|a\|}\,,\quad b' := \frac{\|a\|^2 b - \langle a\,,b\rangle a}{\|\|a\|^2 b - \langle a\,,b\rangle a\|}\,,\quad \beta' := \frac{\beta\|a\|^2 - \alpha\langle a\,,b\rangle}{\|\|a\|^2 b - \langle a\,,b\rangle a\|}\,,$$

and $\langle b', a'\rangle = 0$. So, $\|x\|^2 = \alpha'^2 + \beta'^2 + \|y\|^2$, the interior maximum in (4.6)

$$= \max_y\Big\{\alpha'^2+\beta'^2+\|y\|^2 \;\Big|\; \langle a', y'\rangle = \langle b', y\rangle = 0 \;\;\&\;\; \|y\|^2 \le 1-\alpha'^2-\beta'^2\Big\} = 1\,,$$

and $\nu - 1 = \max_{(\alpha,\beta)\in C}\left(2\alpha\beta+\beta^2\|a\|^2\right)$. To evaluate the last maximum, we use the lemma on sections once more. Let

$$A(\beta) := \Big\{\alpha \;\Big|\; (\alpha,\beta)\in C\Big\} = \Big\{\alpha \;\Big|\; \alpha^2\|b\|^2 + 2\alpha\beta\langle a\,,b\rangle + \beta^2\|a\|^2 - det \le 0\Big\}$$

$$= \begin{cases} \left\{\alpha \;\middle|\; \dfrac{\beta\langle a\,,b\rangle - \sqrt{\delta}}{\|b\|^2} \le \alpha \le \dfrac{\beta\langle a\,,b\rangle + \sqrt{\delta}}{\|b\|^2}\right\}, \\ \qquad\qquad \text{if } \delta := \beta^2\langle a\,,b\rangle^2 - \|b\|^2\left(\beta^2\|a\|^2 - det\right) \ge 0 \\ \emptyset\,, \text{ otherwise.}\end{cases}$$

As linear independence of a and b implies $det>0$, it follows that $B{:=}\{\beta \mid A(\beta)\neq\emptyset\} = \{\beta \mid \beta^2 \le \|b\|^2\}$ and $\nu-1$

$$\begin{aligned}&= \max_{\beta\in B}\;\max_{\alpha\in A(\beta)}\left(2\alpha\beta+\beta^2\|a\|^2\right)\\ &= \max_{\beta^2\le\|b\|^2}\left(\|a\|^2\beta^2 + 2\max_\alpha\left\{\beta\alpha \;\middle|\; \frac{\beta\langle a\,,b\rangle - \sqrt{\delta}}{\|b\|^2} \le \alpha \le \frac{\beta\langle a\,,b\rangle + \sqrt{\delta}}{\|b\|^2}\right\}\right).\end{aligned}$$

If $0 \le \beta \le \|b\|$, then the interior maximum $= \beta\left(\beta\langle a\,,b\rangle + \sqrt{\delta}\right)\big/\,\|b\|^2$. Otherwise, it $= \beta\left(\beta\langle a\,,b\rangle - \sqrt{\delta}\right)\big/\,\|b\|^2$. Hence, $\nu - 1 = \max\{m_1\,,\,m_2\}$, where

$$m_1 := \max_{0\le\beta\le\|b\|}\left(\|a\|^2\beta^2 + 2\beta\frac{\beta\langle a\,,b\rangle + \sqrt{\delta}}{\|b\|^2}\right)$$

$$= \max_{0\le\beta\le\|b\|}\left(\left(\|a\|^2\|b\|^2 + 2\langle a\,,b\rangle\right)\frac{\beta^2}{\|b\|^2} + 2\frac{\beta}{\|b\|}\sqrt{det\left(1 - \frac{\beta^2}{\|b\|^2}\right)}\right)$$

and

$$m_2 := \max_{-\|b\|\le\beta\le 0}\left(\|a\|^2\beta^2 + 2\beta\frac{\beta\langle a\,,b\rangle - \sqrt{\delta}}{\|b\|^2}\right)$$

$$= \max_{0\le t\le\|b\|}\left(\left(\|a\|^2\|b\|^2 + 2\langle a\,,b\rangle\right)\frac{t^2}{\|b\|^2} + 2\frac{t}{\|b\|}\sqrt{det\left(1 - \frac{t^2}{\|b\|^2}\right)}\right) = m_1.$$

Hence, $\nu - 1$

$$= \max_{0\le\beta\le\|b\|^2}\left(\left(\|a\|^2\|b\|^2 + 2\langle a\,,b\rangle\right)\frac{\beta^2}{\|b\|^2} + 2\frac{\beta}{\|b\|}\sqrt{det\left(1 - \frac{\beta^2}{\|b\|^2}\right)}\right)$$

$$= \max_{0\le\theta\le\pi/2}\left(\sin^2\theta\left(\|a\|^2\|b\|^2 + 2\langle a\,,b\rangle\right) + 2\sin\theta\cos\theta\sqrt{det}\right)$$

$$= \max_{0\le\phi\le\pi}\left(\frac{1-\cos\phi}{2}\left(\|a\|^2\|b\|^2 + 2\langle a\,,b\rangle\right) + \sin\phi\sqrt{det}\right)$$

$$= \frac{\|a\|^2\|b\|^2}{2} + \langle a\,,b\rangle + \frac{1}{2}\max_{0\le\phi\le\pi}\left(2\sin\phi\,\sqrt{det} - \cos\phi\,\left(\|a\|^2\|b\|^2 + 2\langle a\,,b\rangle\right)\right).$$

Now we define $\psi \in [-\pi/2\,,\,\pi/2]$ by setting

$$\cos\psi := \frac{2\sqrt{det}}{\sqrt{4det + \left(\|a\|^2\|b\|^2 + 2\langle a\,,b\rangle\right)^2}}\,,$$

$$\sin\psi := \frac{\|a\|^2\|b\|^2 + 2\langle a\,,b\rangle}{\sqrt{4det + \left(\|a\|^2\|b\|^2 + 2\langle a\,,b\rangle\right)^2}}\,.$$

This yields $\nu - 1 - 0.5\|a\|^2\|b\|^2 - \langle a\,,b\rangle$

$$= \frac{1}{2}\sqrt{4det + \left(\|a\|^2\|b\|^2 + 2\langle a\,,b\rangle\right)^2}\max_{0\le\phi\le\pi}\left(\sin\phi\cos\psi - \cos\phi\sin\psi\right)$$

$$= \frac{1}{2}\sqrt{\|a\|^2\|b\|^2\left(\|a\|^2\|b\|^2 + 4(1 + \langle a\,,b\rangle)\right)}\max_{0\le\phi\le\pi}\sin(\phi - \psi).$$

When ϕ scans the segment $[0,\pi]$, the difference $\phi-\psi$ scans the segment $[-\psi\,,\pi-\psi]$ containing $\pi/2$. So, $\max\limits_{0\le\phi\le\pi}\sin(\phi-\psi)=1$ and

$$\nu = 1+\langle a\,,b\rangle + 0.5\|a\|\,\|b\|\left(\|a\|\,\|b\| + \sqrt{\|a\|^2\|b\|^2+4(1+\langle a\,,b\rangle)}\right).$$

If $b=\lambda a$, $\beta=\lambda\alpha$, and $\alpha^2\le\|a\|^2$, then, by Corollary 1.6,

$$B(\alpha,\beta)=\left\{\frac{\alpha}{\|a\|^2}a+y\;\middle|\;\langle a\,,y\rangle=0\;\;\&\;\;\|y\|^2\le 1-\frac{\alpha}{\|a\|^2}\right\}$$

and so

$$\max_{x\in B(\alpha,\beta)}\|x\|^2=\max_y\left\{\frac{\alpha}{\|a\|^2}a+y\;\middle|\;\langle a\,,y\rangle=0\;\;\&\;\;\|y\|^2\le 1-\frac{\alpha}{\|a\|^2}\right\}=1\,.$$

Hence (see (4.6)),

$$\begin{aligned}\nu &= 1+\max_{\alpha^2\le\|a\|^2}\left(\lambda(\lambda\|a\|^2+2)\right)\alpha^2\\ &= \begin{cases}\left(1+\lambda\|a\|^2\right)^2\,, & \text{if } \lambda\ge 0 \bigvee \lambda\le -2/\|a\|^2\,,\\ 1\,, \text{ if } -2/\|a\|^2\le\lambda\le 0\,.\end{cases}\end{aligned}$$

On the other hand, $b=\lambda\,a \Longrightarrow \langle a\,,b\rangle=\lambda\|a\|^2\;\;\&\;\;\|b\|=|\lambda|\cdot\|a\| \Longrightarrow$

$$\begin{aligned}&1+\langle a\,,b\rangle+0.5\|a\|\,\|b\|\left(\|a\|\,\|b\|+\sqrt{\|a\|^2\|b\|^2+4(1+\langle a\,,b\rangle)}\right)\\ &= 1+\lambda\|a\|^2+0.5|\lambda|\,\|a\|^2\left(|\lambda|\,\|a\|^2+\sqrt{\lambda^2\|a\|^4+4(1+\lambda\|a\|)}\right)\\ &= 1+\lambda\|a\|^2+0.5\lambda^2\|a\|^4+0.5|\lambda|\|a\|^2|2+\lambda\|a\|^2|\\ &= 1+\lambda\|a\|^2+0.5\lambda^2\|a\|^4+0.5\|a\|^2\cdot\begin{cases}\lambda\left(2+\lambda\|a\|^2\right)\,, \\ \qquad\text{if } \lambda\ge 0\bigvee\lambda\le -2/\|a\|^2\,,\\ -\lambda\left(2+\lambda\|a\|^2\right)\,,\\ \qquad\text{if } -2/\|a\|^2\le\lambda\le 0\,,\end{cases}\\ &= \begin{cases}\left(1+\lambda\|a\|^2\right)^2\,, & \text{if } \lambda\ge 0\bigvee\lambda\le -2/\|a\|^2\,,\\ 1\,, \text{ if } -2/\|a\|^2\le\lambda\le 0\,.\end{cases}\end{aligned}$$

Thus, again

$$\nu = 1+\langle a\,,b\rangle + 0.5\|a\|\,\|b\|\left(\|a\|\,\|b\| + \sqrt{\|a\|^2\|b\|^2+4(1+\langle a\,,b\rangle)}\right).$$

□

Corollary 4.2. $\|\mathbf{I}-a\|a\|^{-2}\langle a\,,\cdot\rangle\|=1$.

Lemma 4.3. 1° $\bar{t}_+ := \|x_+ - x_0\| \le \bar{t} + \bar{\delta}$.

$$2^\circ\ \bar{\delta}_+ := \|x_{++} - x_+\| \le \bar{\delta}\frac{\bar{\beta}\bar{\sigma}}{\left(\bar{\delta} - \bar{\beta}\bar{\sigma}\right)^+} .$$

$$3^\circ\ \bar{\beta}_+ := \|\mathbf{A}_+\| \le \bar{\beta}\left(\frac{\bar{\beta}\bar{\sigma}}{\left(\bar{\delta} - \bar{\beta}\bar{\sigma}\right)^+} + 1\right) .$$

4° *If the selected dd* $[x_1, x_2 \,|\, \mathbf{f}]$ *of* $\mathbf{f}$ *is regularly continuous, then*

$$\bar{\sigma}_+ := \|\mathbf{f}(x_{++})\| \le \bar{\delta}_+\Big[\omega\Big(\min\Big\{\left(\bar{a} - 2\bar{t}_+ - \bar{\delta}_+\right)^+, \left(\bar{a} - 2\bar{t} - \bar{\delta}\right)^+\Big\} + \bar{\delta} + \bar{\delta}_+\Big) - \omega\Big(\min\Big\{\left(\bar{a} - 2\bar{t}_+ - \bar{\delta}_+\right)^+, \left(\bar{a} - 2\bar{t} - \bar{\delta}\right)^+\Big\}\Big)\Big],$$

where $\bar{a} := \omega^{-1}\big(\|[x_1, x_0 \,|\, \mathbf{f}]\| - \underline{h}\big) - \|x_1 - x_0\|$.

5° *All four bounds are exact: they are attained for the quadratic polynomial*

$$p(x) := x^2 + 2c_1x - c_2 \ , \ c_1 > 0 \ , \ c_2 > 0 .$$

Proof. 1° is an obvious consequence of the triangle inequality.

2°

$$\begin{aligned}
\bar{\delta}_+ &= \|\mathbf{A}_+\mathbf{f}(x_+)\| \\
&= \left\|\left(\mathbf{A} - \frac{\mathbf{Af}(x_+)}{\langle \mathbf{Af}(x_+), x_+ - x\rangle + \|x_+ - x\|^2}\langle \mathbf{A}^*(x_+ - x), \cdot\rangle\right)\mathbf{f}(x_+)\right\| \\
&= \|\mathbf{Af}(x_+)\|\left|1 - \frac{\langle \mathbf{Af}(x_+), x_+ - x\rangle}{\langle \mathbf{Af}(x_+), x_+ - x\rangle + \|x_+ - x\|^2}\right| \\
&= \|\mathbf{Af}(x_+)\|\frac{\bar{\delta}^2}{\left|\langle \mathbf{Af}(x_+), x_+ - x\rangle + \bar{\delta}^2\right|} = \frac{\bar{\delta}\|\mathbf{Af}(x_+)\|}{\left|\left\langle \mathbf{Af}(x_+), \dfrac{x_+ - x}{\bar{\delta}}\right\rangle + \bar{\delta}\right|} .
\end{aligned}$$

As $\left|\langle a, b\rangle + \alpha\right| \ge \max\{0, \langle a, b\rangle + \alpha\} \ge \max\{0, \alpha - \|a\|\cdot\|b\|\} = \left(\alpha - \|a\|\cdot\|b\|\right)^+$, it follows that

$$\bar{\delta}_+ \le \bar{\delta}\frac{\|\mathbf{Af}(x_+)\|}{\left(\bar{\delta} - \|\mathbf{Af}(x_+)\|\right)^+} \le \bar{\delta}\frac{\|\mathbf{A}\|\cdot\|\mathbf{f}(x_+)\|}{\left(\bar{\delta} - \|\mathbf{A}\|\cdot\|\mathbf{f}(x_+)\|\right)^+} = \bar{\delta}\frac{\bar{\beta}\bar{\sigma}}{\left(\bar{\delta} - \bar{\beta}\bar{\sigma}\right)^+} .$$

3° By (4.1),

$$\begin{aligned}
\|\mathbf{A}_+\| &= \left\|\mathbf{A} - \frac{\mathbf{Af}(x_+)}{\langle \mathbf{Af}(x_+), x_+ - x\rangle + \|x_+ - x\|^2}\langle \mathbf{A}^*(x_+ - x), \cdot\rangle\right\| \\
&\le \|\mathbf{A}\|\cdot\left\|\mathbf{I} - \frac{\mathbf{f}(x_+)}{\langle \mathbf{Af}(x_+), x_+ - x\rangle + \|x_+ - x\|^2}\langle \mathbf{A}^*(x_+ - x), \cdot\rangle\right\| . \quad (4.7)
\end{aligned}$$

By Lemma 4.1,

$$\|\mathbf{I}-a\langle b\,,\cdot\rangle\|^2 = 1-\langle a\,,b\rangle+0.5\|a\|\cdot\|b\|\left(\|a\|\cdot\|b\|+\sqrt{\|a\|^2\|b\|^2+4(1-\langle a\,,b\rangle)}\right),$$

where $\langle a\,,b\rangle \geq -\|a\|\cdot\|b\|$. Therefore, $\|\mathbf{I}-a\langle b\,,\cdot\rangle\|^2$

$$\leq 1+\|a\|\cdot\|b\|+0.5\|a\|\cdot\|b\|\left(\|a\|\cdot\|b\|+\sqrt{\|a\|^2\|b\|^2+4(1+\|a\|\cdot\|b\|)}\right)$$

$$= 1+\|a\|\cdot\|b\|+0.5\|a\|\cdot\|b\|\left(\|a\|\cdot\|b\|+\sqrt{(\|a\|\cdot\|b\|+2)^2}\right)$$

$$= 1+\|a\|\cdot\|b\|+\|a\|\cdot\|b\|\big(\|a\|\cdot\|b\|+1\big) = \big(\|a\|\cdot\|b\|+1\big)^2$$

and $\big\|\mathbf{I}-a\langle b\,,\cdot\rangle\big\| \leq \|a\|\cdot\|b\|+1$. Hence, the last norm in (4.7)

$$\leq \frac{\big\|\mathbf{f}(x_+)\big\|\cdot\big\|\mathbf{A}^*(x_+-x)\big\|}{\big|\bar{\delta}^{\,2}+\big\langle\mathbf{Af}(x_+)\,,\,x_+-x\big\rangle\big|}+1 \leq \frac{\bar{\delta}\bar{\sigma}\|\mathbf{A}^*\|}{\big|\bar{\delta}^{\,2}+\big\langle\mathbf{Af}(x_+)\,,\,x_+-x\big\rangle\big|}+1$$
$$= \frac{\bar{\sigma}\|\mathbf{A}^*\|}{\left|\bar{\delta}+\left\langle\mathbf{Af}(x_+)\,,\,\dfrac{x_+-x}{\bar{\delta}}\right\rangle\right|}+1\,.$$

As we have seen above, the denominator $\geq \big(\bar{\delta}-\bar{\beta}\bar{\sigma}\big)^+$. Besides, one of the basic facts of linear functional analysis is the equality of the norms $\|\mathbf{A}^*\|$ and $\|\mathbf{A}\|$. Therefore, that norm $\leq \bar{\beta}\bar{\sigma}/\big(\bar{\delta}-\bar{\beta}\bar{\sigma}\big)^+ + 1$ and

$$\bar{\beta}_+ \leq \bar{\beta}\left(\frac{\bar{\beta}\bar{\sigma}}{\big(\bar{\delta}-\bar{\beta}\bar{\sigma}\big)^+}+1\right).$$

4° According to (4.3), $\big\|\mathbf{A}_{++}^{-1}-\mathbf{A}_+^{-1}\big\| = \big\|\mathbf{f}(x_{++})\big\|/\|x_{++}-x_+\|$, so that

$$\bar{\sigma}_+ = \big\|\mathbf{f}(x_{++})\big\| = \bar{\delta}_+\big\|\mathbf{A}_{++}^{-1}-\mathbf{A}_+^{-1}\big\| = \bar{\delta}_+\big\|[x_{++},x_+\,|\,\mathbf{f}]-[x_+,x\,|\,\mathbf{f}]\big\|. \tag{4.8}$$

As in (3.25), the last norm

$$\leq \omega\Big(\min\big\{\big(\bar{a}-2\bar{t}_+-\bar{\delta}_+\big)^+,\big(\bar{a}-2\bar{t}-\bar{\delta}\big)^+\big\}+\bar{\delta}_++\bar{\delta}\Big)-$$
$$\omega\Big(\min\big\{\big(\bar{a}-2\bar{t}_+-\bar{\delta}_+\big)^+,\big(\bar{a}-2\bar{t}-\bar{\delta}\big)^+\big\}\Big).$$

So,

$$\bar{\sigma}_+ \leq \bar{\delta}_+\Big[\omega\Big(\min\big\{\big(\bar{a}-2\bar{t}_+-\bar{\delta}_+\big)^+,\big(\bar{a}-2\bar{t}-\bar{\delta}\big)^+\big\}+\bar{\delta}_++\bar{\delta}\Big)-$$
$$\omega\Big(\min\big\{\big(\bar{a}-2\bar{t}_+-\bar{\delta}_+\big)^+,\big(\bar{a}-2\bar{t}-\bar{\delta}\big)^+\big\}\Big)\Big].$$

5° When $\mathbb{H} = \mathbb{R}$, the divided difference operator is just a real multiplier (the divided difference) $\big(f(x_+) - f(x)\big)/(x_+ - x)$ and so

$$A_+ = A - \frac{Af(x_+)}{Af(x)\big(Af(x_+) - Af(x)\big)} \cdot A^2 f(x) = A - \frac{Af(x_+)}{f(x_+) - f(x)}$$
$$= \frac{-Af(x)}{f(x_+) - f(x)} = \frac{x_+ - x}{f(x_+) - f(x)} = [x_+, x \,|\, f\,]^{-1},$$

i.e., for scalar functions Broyden's method coincides with the secant one (0.3). Having noted this, we observe that the mapping $\mathbf{U} : \mathbb{R}^2 \to \mathbb{R}^2$, $\mathbf{U}(x, A) = (x_+ , A_+)$ (the iteration of Broyden's method) maps the set

$$M := \Big\{(x\,, A) \;\Big|\; x^* < x < x_0 \;\;\&\;\; A > 0\Big\},$$

where $x^* := \sqrt{c_1^2 + c_2} - c_1$ (the zero of the polynomial p), into itself: $\mathbf{U}(M) \subset M$. Indeed,

$$(x\,, A) \in M \Longrightarrow p(x) > 0 \;\;\&\;\; A > 0 \Longrightarrow x_+ := x - Ap(x) < x < x_0\,.$$

In turn, due to the convexity and monotonicity of p,

$$x_+ < x \Longrightarrow p(x_+) > p(x) + A(x_+ - x) = 0 = p(x^*) \Longrightarrow x_+ > x^*.$$

Besides,

$$A_+ = \frac{x_+ - x}{p(x_+) - p(x)} = \frac{1}{x_+ + x + 2c_1} > \frac{1}{2(x_0 + c_1)} > 0\,.$$

Thus, $x^* < x < x_0 \;\;\&\;\; A > 0 \Longrightarrow x^* < x_+ < x_0 \;\;\&\;\; A_+ > 0\,.$

Suppose now that $\bar t = |x - x_0|$, $\bar\beta = |A|$, $\bar\delta = |x_+ - x|$, $\bar\sigma = \big|p(x_+)\big|$. Then

$$\bar t_+ := |x_+ - x_0| = x_0 - x_+ = (x_0 - x) + (x - x_+)$$
$$= |x - x_0| + |x_+ - x| = \bar t + \bar\delta\,,$$

$$\bar\beta_+ := |A_+| = A_+ = \frac{x_+ - x}{p(x_+) - p(x)} = \frac{-Ap(x)}{p(x_+) - p(x)} = \frac{A^2 p(x)}{Ap(x) - Ap(x_+)}$$
$$= \frac{\big|A(x_+ - x)\big|}{|x_+ - x| - \big|Ap(x_+)\big|} = \frac{\bar\beta\bar\sigma}{\bar\delta - \bar\beta\bar\sigma}\,,$$

$$\bar\delta_+ = \big|A_+ p(x_+)\big| = A_+ p(x_+) = \frac{x_+ - x}{p(x_+) - p(x)}\, p(x_+) = -Ap(x)\frac{p(x_+)}{p(x_+) - p(x)}$$
$$= |x_+ - x|\frac{Ap(x_+)}{Ap(x) - Ap(x_+)} = |x_+ - x|\frac{\big|Ap(x_+)\big|}{|x_+ - x| - \big|Ap(x_+)\big|} = \bar\delta\frac{\bar\beta\bar\sigma}{\bar\delta - \bar\beta\bar\sigma}\,,$$

and, by (4.8),

$$\begin{aligned}\bar\sigma_+ &= \bar\delta_+\big|[x_{++},x_+\,|\,p]-[x_+,x\,|\,p]\big| = \bar\delta_+\big|(x_{++}+x_++2c_1)-(x_++x+2c_1)\big|\\ &= \bar\delta_+\big|(x_{++}-x_+)+(x_+-x)\big| = \bar\delta_+\big(|x_{++}-x_+|+|x_+-x|\big) = \bar\delta_+\big(\bar\delta_++\bar\delta\big)\\ &= \bar\delta_+\Big[\omega\Big(\min\{(\bar a-2\bar t_+-\bar\delta_+)^+,(\bar a-2\bar t-\bar\delta)^+\}+\bar\delta_++\bar\delta\Big)-\\ &\qquad\qquad \omega\Big(\min\{(\bar a-2\bar t_+-\bar\delta_+)^+,(\bar a-2\bar t-\bar\delta)^+\}\Big)\Big]\,,\end{aligned}$$

since the regularity modulus ω of the dd $[x_1,x_2\,|\,p]$ is the function $t\mapsto t$. □

The lemma suggests the following majorant generator:

$$\gamma := \frac{\beta\sigma/\delta}{1-\beta\sigma/\delta}\,,\ t_+ := t+\delta\,,\ \beta_+ := \beta(1+\gamma)\,,\ \delta_+ := \gamma\delta\,, \tag{4.9}$$

$$\begin{aligned}\sigma_+ := \delta_+\Big[\omega\Big(\min\{(a-2t_+-\delta_+)^+,(a-2t-\delta)^+\}+\delta+\delta_+\Big)-\\ \omega\Big(\min\{(a-2t_+-\delta_+)^+,(a-2t-\delta)^+\}\Big)\Big],\end{aligned}$$

where a is the best lower bound for $\bar a$ available. As

$$a-2t_+-\delta_+ = a-2t-\delta-(\delta+\delta_+) < a-2t-\delta,$$

we have

$$\begin{aligned}\sigma_+ &:= \delta_+\Big[\omega\Big((a-2t_+-\delta_+)^++\delta+\delta_+\Big)-\omega\Big((a-2t_+-\delta_+)^+\Big)\Big]\\ &= \delta_+e(a-2t-\delta\,,\delta+\delta_+)\,,\end{aligned} \tag{4.10}$$

where e is the function (2.8):

$$\begin{aligned}e(u\,,t) &:= \omega(((\alpha-t)^++t)-\omega(((\alpha-t)^+)\\ &= \begin{cases}\omega(\alpha)-\omega(\alpha-t)\,, \text{ if } 0\le t\le\alpha\\ \omega(t)\,,\text{if } t\ge\alpha\,.\end{cases}\end{aligned} \tag{4.11}$$

Lemma 4.4. *Let the selected dd* $[x_1,x_2\,|\,\mathbf{f}]$ *of* $\mathbf{f}$ *be* ω*-regularly continuous on* D. *If*

$$\bar q_0 := (\bar\beta_0\,,\bar\delta_0\,,\bar\sigma_0)\prec q_0 \ \ \&\ \ \frac{\bar\sigma_0}{\bar\delta_0}\le\frac{\sigma_0}{\delta_0}\ \ \&\ \ \beta_\infty<\infty\,,$$

then

1° *the sequence* $t_n := \sum_0^{n-1}\delta_k$ *(equivalently, the series* $\sum\delta_k$*) converges:* $t_\infty<\infty$*;*

2° $\delta_\infty = \sigma_\infty = 0$;

3° $\underset{n}{\&}\left(\bar{q}_n \prec q_n \ \ \& \ \ \dfrac{\bar{\beta}_{n+1}\bar{\sigma}_{n+1}}{\bar{\delta}_{n+1}} \le \dfrac{\beta_{n+1}\sigma_{n+1}}{\delta_{n+1}}\right)$;

4° *the sequence* $(x_n\,,\mathbf{A}_n)$ *generated by the method* (4.1) *from the starter* $(x_0\,,\mathbf{I})$ *converges to a limit* $(x_\infty\,,\mathbf{A}_\infty)$;

5° *this limit solves the system* $\mathbf{f}(x) = 0 \ \ \& \ \ \mathbf{X}[x,x\,|\,\mathbf{f}] = \mathbf{I}$ *for* $(x\,,\mathbf{X})$;

6° x_∞ *is the only solution of the equation* $\mathbf{f}(x) = 0$ *in the ball* $B(x_0\,,r)$, *where*

$$r := \omega^{-1}\left(1 - \frac{\sigma_0}{\delta_0} - \omega\big((a-t_\infty)^+ + \delta_0 + t_\infty\big) + \omega\big((a-t_\infty)^+\big)\right).$$

7° *for all* $n = 0,1,\dots$

$$\big\|\mathbf{f}(x_{n+1})\big\| \le \sigma_n \ \ \& \ \ \Delta_n := \|x_\infty - x_n\| \le t_\infty - t_n \ \ \& \ \ \frac{\Delta_{n+1}}{\Delta_n} \le \beta_n\omega(\Delta_{n-1}).$$

Proof. 1° If $\beta_\infty < \infty$, then $\gamma_n := \beta_{n+1}/\beta_n - 1 \to 0$. Then there exists an m such that $\underset{n\ge m}{\&}\ \gamma_n \le \gamma_m < 1$ and so (see (4.9))

$$\sum_{m+1}^{\infty}\delta_n = \sum_{m+1}^{\infty}\gamma_{n-1}\delta_{n-1} = \sum_{m}^{\infty}\gamma_n\delta_n \le \gamma_m\sum_{m}^{\infty}\delta_n = \gamma_m(t_\infty - t_m).$$

Therefore, $t_\infty = t_{m+1} + \sum_{m+1}^{\infty}\delta_n \le t_{m+1} + \gamma_m(t_\infty - t_m)$. Solving this inequality for t_∞ yields

$$t_\infty \le \frac{t_{m+1} - \gamma_m t_m}{1-\gamma_m}.$$

2° $t_\infty < \infty \Longrightarrow \delta_n = t_{n+1} - t_n \to 0$. Besides, $\beta_n\sigma_n/\delta_n = \gamma_n/(1+\gamma_n) \to 0$, so that

$$\beta_\infty < \infty \ \ \& \ \ \frac{\beta_n\sigma_n}{\delta_n} \to 0 \Longrightarrow \frac{\sigma_n}{\delta_n} \to 0 \Longrightarrow \sigma_n \to 0\,.$$

3° As the generator (4.9), (4.10) is monotone in the same sense as the generators (2.21) and (3.18), $\bar{q} \prec q \Longrightarrow \mathbf{g}(\bar{q}) \prec \mathbf{g}(q)$. On the other hand, Lemma 4.3 shows that $\bar{q}_+ \prec \mathbf{g}(\bar{q})$. Hence, $\bar{q}_+ \prec \mathbf{g}(q) = q_+$. By induction, $\bar{q}_0 \prec q_0 \Longrightarrow \underset{n}{\&}\,\bar{q}_n \prec q_n$. In particular, $\bar{t}_n \le t_n\,, \bar{\beta}_n \le \beta_n\,, \bar{\delta}_n \le \delta_n$. Then, by (4.9), (4.10), $\bar{\beta}_{n+1}\bar{\sigma}_{n+1}/\bar{\delta}_{n+1}$

$$= \bar{\beta}_{n+1}\Big[\omega\Big(\big(a - 2\bar{t}_{n+1} - \bar{\delta}_{n+1}\big)^+ + \bar{\delta}_n + \bar{\delta}_{n+1}\Big) - \omega\Big(\big(a - 2\bar{t}_{n+1} - \bar{\delta}_{n+1}\big)^+\Big)\Big]$$

$$\le \beta_{n+1}\Big[\omega\Big(\big(a - 2\bar{t}_{n+1} - \bar{\delta}_{n+1}\big)^+ + \delta_n + \delta_{n+1}\Big) - \omega\Big(\big(a - 2\bar{t}_{n+1} - \bar{\delta}_{n+1}\big)^+\Big)\Big].$$

As $2\bar{t}_{n+1} + \bar{\delta}_{n+1} \le 2t_{n+1} + \delta_{n+1}$ and because of the concavity of ω,

$$\omega\Big(\big(a - 2\bar{t}_{n+1} - \bar{\delta}_{n+1}\big)^+ + \delta_n + \delta_{n+1}\Big) - \omega\Big(\big(a - 2\bar{t}_{n+1} - \bar{\delta}_{n+1}\big)^+\Big)$$

$$\le \omega\Big(\big(a - 2t_{n+1} - \delta_{n+1}\big)^+ + \delta_n + \delta_{n+1}\Big) - \omega\Big(\big(a - 2t_{n+1} - \delta_{n+1}\big)^+\Big).$$

It follows that $\bar{\beta}_{n+1}\bar{\sigma}_{n+1}/\bar{\delta}_{n+1}$

$$\leq \beta_{n+1}\Big[\omega\Big(\big(a-2t_{n+1}-\delta_{n+1}\big)^{+}+\delta_n+\delta_{n+1}\Big)-\omega\Big(\big(a-2t_{n+1}-\delta_{n+1}\big)^{+}\Big)\Big]$$

$$= \frac{\beta_{n+1}\sigma_{n+1}}{\delta_{n+1}}$$

by (4.10).

4°–5° By 3°, $\underset{n}{\&}\,\bar{\delta}_n \leq \delta_n$ and so

$$\|x_{n+m}-x_n\| \leq \sum_{k=n}^{n+m-1}\|x_{k+1}-x_k\| = \sum_{k=n}^{n+m-1}\bar{\delta}_k \leq \sum_{k=n}^{n+m-1}\delta_k < \sum_{k=n}^{\infty}\delta_k = t_\infty - t_n\,.$$

This shows that x_n is a Cauchy sequence and so converges: $\exists \lim x_n =: x_\infty$. Setting $n=0$ results in $x_m \in B(x_0, t_\infty)$, while forcing m to ∞ yields $\|x_\infty - x_n\| \leq t_\infty - t_n$. Moreover, $\bar{\sigma}_n = \big\|\mathbf{f}(x_{n+1})\big\| \leq \sigma_n \to 0$ by 2°, so that $\mathbf{f}(x_\infty)=0$.

Now consider the operators $\mathbf{A}_n$. In view of (4.1), $\|\mathbf{A}_{m+n}-\mathbf{A}_n\|$

$$\leq \sum_{k=n}^{m+n-1}\|\mathbf{A}_{k+1}-\mathbf{A}_k\|$$

$$= \sum_{k=n}^{m+n-1}\left\|\frac{\mathbf{A}_k\mathbf{f}(x_{k+1})}{\big\langle \mathbf{A}_n\mathbf{f}(x_{k+1})\,,x_{k+1}-x_k\big\rangle + \|x_{k+1}-x_k\|^2}\big\langle \mathbf{A}_n^*(x_{k+1}-x_k)\,,\cdot\big\rangle\right\|.$$

It is an easy exercise to see that the norm of the operator $x \mapsto a\langle b\,,x\rangle$ is equal to $\|a\|\cdot\|b\|$. So,

$$\|\mathbf{A}_{m+n}-\mathbf{A}_n\| \leq \sum_{k=n}^{m+n-1}\frac{\big\|\mathbf{A}_k\mathbf{f}(x_{k+1})\big\|\cdot\big\|\mathbf{A}_k^*\,(x_{k+1}-x_k)\big\|}{\big|\big\langle \mathbf{A}_k\mathbf{f}(x_{k+1})\,,x_{k+1}-x_k\big\rangle+\bar{\delta}_k^2\big|}$$

$$\leq \sum_{k=n}^{m+n-1}\frac{\|\mathbf{A}_k\|\big\|\mathbf{f}(x_{k+1})\big\|\cdot\|\mathbf{A}_k^*\|\bar{\delta}_k}{\big|\big\langle \mathbf{A}_k\mathbf{f}(x_{k+1})\,,x_{k+1}-x_k\big\rangle+\bar{\delta}_k^2\big|}$$

$$= \sum_{k=n}^{m+n-1}\frac{\bar{\beta}_k^2\bar{\sigma}_k}{\left|\left\langle \mathbf{A}_k\mathbf{f}(x_{k+1})\,,\dfrac{x_{k+1}-x_k}{\bar{\delta}_k}\right\rangle+\bar{\delta}_k\right|}\,.$$

As in Lemma 4.3, the denominator $\geq \bar{\delta}_k - \bar{\beta}_k\bar{\sigma}_k$. Therefore,

$$\|\mathbf{A}_{m+n}-\mathbf{A}_n\| \leq \sum_{k=n}^{m+n-1}\bar{\beta}_k\frac{\bar{\beta}_k\bar{\sigma}_k}{\bar{\delta}_k-\bar{\beta}_k\bar{\sigma}_k} = \sum_{k=n}^{m+n-1}\bar{\beta}_k\frac{\bar{\beta}_k\bar{\sigma}_k/\bar{\delta}_k}{1-\bar{\beta}_k\bar{\sigma}_k/\bar{\delta}_k}\,.$$

According to 3°, $\bar{\beta}_k\bar{\sigma}_k/\bar{\delta}_k \le \beta_k\sigma_k/\delta_k$ and so

$$\frac{\bar{\beta}_k\bar{\sigma}_k/\bar{\delta}_k}{1-\bar{\beta}_k\bar{\sigma}_k/\bar{\delta}_k} \le \frac{\beta_k\sigma_k/\delta_k}{1-\beta_k\sigma_k/\delta_k}\,.$$

Consequently,

$$\begin{aligned}\|\mathbf{A}_{m+n}-\mathbf{A}_n\| &\le \sum_{k=n}^{m+n-1}\beta_k\frac{\beta_k\sigma_k/\delta_k}{1-\beta_k\sigma_k/\delta_k} = \sum_{k=n}^{m+n-1}\beta_k\gamma_k = \sum_{k=n}^{m+n-1}(\beta_{k+1}-\beta_k)\\ &= \beta_{m+n}-\beta_n < \beta_\infty - \beta_n\,.\end{aligned}$$

It follows that $\mathbf{A}_n$ is a Cauchy sequence in the Banach space $\mathcal{L}(\mathbb{H})$ of linear operators acting on $\mathbb{H}$ and so converges to a limit $\mathbf{A}_\infty$.

$$n=1 \Longrightarrow \|\mathbf{A}_{m+1}-\mathbf{A}_1\| \le \beta_{m+1}-\beta_1 = \beta_{m+1}-\delta_0/(\delta_0-\sigma_0)$$

and $m\to\infty \Longrightarrow \|\mathbf{A}_\infty-\mathbf{A}_n\| \le \beta_\infty-\beta_n$. Taking limits in $\mathbf{A}_n[x_n,x_{n-1}\,|\,\mathbf{f}]=\mathbf{I}$ results in $\mathbf{A}_\infty[x_\infty,x_\infty\,|\,\mathbf{f}]=\mathbf{I}$. Thus,

$$\mathbf{f}(x_\infty)=0 \;\;\&\;\; \mathbf{A}_\infty[x_\infty,x_\infty\,|\,\mathbf{f}]=\mathbf{I}\,.$$

6° Let x_* be another solution of the equation $\mathbf{f}(x)=0$. Then

$$0=\mathbf{f}(x_*)-\mathbf{f}(x_\infty)=[x_*,x_\infty\,|\,\mathbf{f}](x_*-x_\infty),$$

which shows that the dd $[x_*,x_\infty\,|\,\mathbf{f}]$ is not invertible and so $\big\|\mathbf{I}-[x_*,x_\infty\,|\,\mathbf{f}]\big\|\ge 1$ (otherwise, $\sum_{k=0}^{\infty}\big(\mathbf{I}-[x_*,x_\infty\,|\,\mathbf{f}]\big)^k = [x_*,x_\infty\,|\,\mathbf{f}]^{-1}$). On the other hand, $\big\|\mathbf{I}-[x_*,x_\infty\,|\,\mathbf{f}]\big\|$

$$\le \big\|\mathbf{I}-[x_1,x_0\,|\,\mathbf{f}]\big\| + \big\|[x_1,x_0\,|\,\mathbf{f}]-[x_0,x_\infty\,|\,\mathbf{f}]\big\| + \big\|[x_0,x_\infty\,|\,\mathbf{f}]-[x_*,x_\infty\,|\,\mathbf{f}]\big\|\,. \tag{4.12}$$

By (4.3) and exploiting the assumption $\bar{\sigma}_0/\bar{\delta}_0 \le \sigma_0/\delta_0$, we infer that the first norm $=\|\mathbf{I}-\mathbf{A}^{-1}\| = \|\mathbf{f}(x_1)\|/\|x_1-x_0\| = \bar{\sigma}_0/\bar{\delta}_0 \le \sigma_0/\delta_0$. By Lemma 3.8, the second norm

$$\begin{aligned}&\le \omega\Big(\min\{\omega^{-1}(\|[x_1,x_0\,|\,\mathbf{f}]\|-\underline{h}),\omega^{-1}(\|[x_0,x_\infty\,|\,\mathbf{f}]\|-\underline{h})\}+\\ &\qquad\qquad \|x_1-x_0\|+\|x_0-x_\infty\|\Big)-\\ &\qquad \omega\Big(\min\{\omega^{-1}(\|[x_1,x_0\,|\,\mathbf{f}]\|-\underline{h}),\omega^{-1}(\|[x_0,x_\infty\,|\,\mathbf{f}]\|-\underline{h})\}\Big).\end{aligned}$$

According to (3.14),

$$\begin{aligned}\omega^{-1}(\|[x_0,x_\infty\,|\,\mathbf{f}]\|-\underline{h}) &\ge \Big(\omega^{-1}(\|[x_1,x_0\,|\,\mathbf{f}]\|-\underline{h})-\|x_0-x_1\|-\|x_\infty-x_0\|\Big)^+\\ &= (\bar{a}-\bar{t}_\infty)^+.\end{aligned}$$

Hence,

$$\begin{aligned}\big\|[x_1,x_0\,|\,\mathbf{f}]-[x_0,x_\infty\,|\,\mathbf{f}]\big\| &\le \omega\big((\bar a-\bar t_\infty)^+ +\bar\delta_0+\bar t_\infty\big)-\omega\big((\bar a-\bar t_\infty)^+\big)\\ &\le \omega\big((a-t_\infty)^+ +\delta_0+t_\infty\big)-\omega\big((a-t_\infty)^+\big).\end{aligned}$$

Similarly, the third norm in (4.12)

$$\begin{aligned}&\le \omega\Big(\min\{\omega^{-1}(\big\|[x_0,x_\infty\,|\,\mathbf{f}]\big\|-\underline{h}),\omega^{-1}(\big\|[x_*,x_\infty\,|\,\mathbf{f}]\big\|-\underline{h})\}+\|x_*-x_0\|\Big)-\\ &\qquad\qquad \omega\Big(\min\{\omega^{-1}(\big\|[x_0,x_\infty\,|\,\mathbf{f}]\big\|-\underline{h}),\omega^{-1}(\big\|[x_*,x_\infty\,|\,\mathbf{f}]\big\|-\underline{h})\}\Big)\\ &\le \omega\Big(\big(\omega^{-1}(\big\|[x_0,x_\infty\,|\,\mathbf{f}]\big\|-\underline{h})-\|x_*-x_0\|\big)^+ +\|x_*-x_0\|\Big)-\\ &\qquad\qquad \Big(\big(\omega^{-1}(\big\|[x_0,x_\infty\,|\,\mathbf{f}]\big\|-\underline{h})-\|x_*-x_0\|\big)^+\Big)\\ &\le \omega\big(\|x_*-x_0\|\big).\end{aligned}$$

Thus,

$$1\le \frac{\sigma_0}{\delta_0}+\omega\big((a-t_\infty)^+ +\delta_0+t_\infty\big)-\omega\big((a-t_\infty)^+\big)+\omega\big(\|x_*-x_0\|\big)$$

and

$$\|x_*-x_0\|\ge \omega^{-1}\Big(1-\frac{\sigma_0}{\delta_0}-\omega\big((a-t_\infty)^+ +\delta_0+t_\infty\big)+\omega\big((a-t_\infty)^+\big)\Big)\,.$$

7° As

$$x_{n+1}-x_\infty=x_n-x_\infty-\mathbf{A}_n\big(\mathbf{f}(x_n)-\mathbf{f}(x_\infty)\big)=x_n-x_\infty-\mathbf{A}_n[x_n,x_\infty\,|\,\mathbf{f}](x_n-x_\infty),$$

by the secant equation, and because of (4.4), we have

$$\begin{aligned}x_{n+1}-x_\infty &= \Big(\mathbf{I}-\mathbf{A}_n[x_n,x_\infty\,|\,\mathbf{f}]\Big)(x_n-x_\infty)\\ &= \Big(\mathbf{A}_n[x_n,x_{n-1}\,|\,\mathbf{f}]-\mathbf{A}_n[x_n,x_\infty\,|\,\mathbf{f}]\Big)(x_n-x_\infty),\end{aligned}$$

whence $\Delta_{n+1}\le\Delta_n\|\mathbf{A}_n\|\big\|[x_n,x_{n-1}\,|\,\mathbf{f}]-[x_n,x_\infty\,|\,\mathbf{f}]\big\|$. By Lemma 3.8 and (3.14), the last norm

$$\begin{aligned}&\le\omega\Big(\min\{\omega^{-1}(\big\|[x_n,x_{n-1}\,|\,\mathbf{f}]\big\|-\underline{h}),\omega^{-1}(\big\|[x_n,x_\infty\,|\,\mathbf{f}]\big\|-\underline{h})\}+\|x_{n-1}-x_\infty\|\Big)-\\ &\qquad\qquad \omega\Big(\min\{\omega^{-1}(\big\|[x_n,x_{n-1}\,|\,\mathbf{f}]\big\|-\underline{h}),\omega^{-1}(\big\|[x_n,x_\infty\,|\,\mathbf{f}]\big\|-\underline{h})\}\Big)\\ &\le \omega\Big(\big(\omega^{-1}(\big\|[x_n,x_{n-1}\,|\,\mathbf{f}]\big\|-\underline{h})-\|x_{n-1}-x_\infty\|\big)^+ +\|x_{n-1}-x_\infty\|\Big)-\\ &\qquad\qquad \omega\Big(\big(\omega^{-1}(\big\|[x_n,x_{n-1}\,|\,\mathbf{f}]\big\|-\underline{h})-\|x_{n-1}-x_\infty\|\big)^+\Big).\end{aligned}$$

So, Δ_{n+1}

$$\leq \bar{\beta}_n \Delta_n \Big[\omega\Big(\big(\omega^{-1}(\|[x_n, x_{n-1} \,|\, \mathbf{f}]\| - \underline{h}) - \Delta_{n-1}\big)^+ + \Delta_{n-1}\Big) - \omega\Big(\big(\omega^{-1}(\|[x_n, x_{n-1} \,|\, \mathbf{f}]\| - \underline{h}) - \Delta_{n-1}\big)^+\Big)\Big]$$

$$\leq \beta_n \Delta_n \omega(\Delta_{n-1}).$$

Therefore, $\Delta_{n+1}/\Delta_n \leq \beta_n \omega(\Delta_{n-1}) \to 0$. □

4.3 Convergence theorem

The lemma poses the question: precisely which starters $q_0 = (0\,,1\,,\delta_0\,,\sigma_0)$ cause β_∞ to be finite? In other words, what is the convergence domain of the generator (4.9), (4.10)? To answer the question, it is expedient to change t and σ for $u := a - 2t - \delta$ and $\gamma := (\beta\sigma/\delta)/(1 - \beta\sigma/\delta)$. After the change, the generator becomes

$$\beta_+ := \beta(1+\gamma)\,,\ \delta_+ := \gamma\delta\,,\ u_+ := u - \delta(1+\gamma)\,,\ \gamma_+ := \frac{\beta_+ e\big(u\,,\delta(1+\gamma)\big)}{1 - \beta_+ e\big(u\,,\delta(1+\gamma)\big)}\,, \tag{4.13}$$

where (recall (2.8))

$$e(u\,,t) := \omega\big((u-t)^+ + t\big) - \omega\big((u-t)^+\big) = \begin{cases} \omega(u) - \omega(u-t)\,, & \text{if } 0 \leq t \leq u\,,\\ \omega(t)\,, & \text{if } t \geq u\,. \end{cases}$$

One can speak about convergence of the sequence $(\beta_n\,,\delta_n\,,u_n\,,\gamma_n)$ generated by this generator only if $\underset{n}{\&} 0 \leq \gamma_n < 1$, since $\gamma \geq 1 \Longrightarrow \beta_+ \geq 2\beta \Longrightarrow \beta_\infty = \infty$.

Proposition 4.5. *Suppose that* $\underset{n}{\&} 0 \leq \gamma_n < 1$. *Then*

1° $\beta_\infty < \infty \Longleftrightarrow f_\infty(\delta_0\,,u_0\,,\gamma_0) \geq 1$, *where* $f_0(\delta, u, \gamma) := \beta_\infty/(1+\gamma))$ *and* $f_{n+1}(\delta\,,u\,,\gamma)$ *is the (unique) solution for* β *of the equation*

$$f_n\left(\gamma\delta\,, u - \delta(1+\gamma)\,, \frac{\beta(1+\gamma)e\big(u\,,\delta(1+\gamma)\big)}{1 - \beta(1+\gamma)e\big(u\,,\delta(1+\gamma)\big)}\right) = \beta(1+\gamma)\,.$$

2° *The function* f_∞ *is the only solution of the system*

$$x\left(\gamma\delta\,, u - \delta(1+\gamma)\,, \frac{x(\delta, u, \gamma)(1+\gamma)\, e\big(u, \delta(1+\gamma)\big)}{1 - x(\delta, u, \gamma)(1+\gamma)\, e\big(u, \delta(1+\gamma)\big)}\right) = x(\delta\,,u\,,\gamma)(1+\gamma)\,,$$

$$x(0, u, 0) = \beta_\infty. \tag{4.14}$$

Proof. 1° As seen from (4.13), $\underset{n}{\&}\, \beta_n < \beta_\infty$ and

$$\beta_{n+1} < \beta_\infty \Longleftrightarrow \beta_n(1+\gamma_n) < \beta_\infty \Longleftrightarrow \beta_n < \frac{\beta_\infty}{1+\gamma_n} =: f_0(\delta_n\,, u_n\,, \gamma_n)\,.$$

Suppose that, for some $k \geq 0$,

$$\beta_{n+1} < \beta_\infty \Longleftrightarrow \beta_{n-k} < f_k(\delta_{n-k}\,, u_{n-k}\,, \gamma_{n-k})\,, \tag{4.15}$$

where f_k is not increasing with respect to the first argument, not decreasing with respect to the second, decreasing with respect to the third, and $f_k(0, u, 0) = \beta_\infty\,, \forall\, u \geq 0$. Let, for brevity, $m := n - k - 1$ and rewrite (using (4.13)) the inequality on the right in (4.15) as

$$\beta_m(1+\gamma_m) < f_k\left(\gamma_m\delta_m\,, u_m - \delta_m(1+\gamma_m)\,, \frac{\beta_m(1+\gamma_m)\, e\big(u_m\,, \delta_m(1+\gamma_m)\big)}{1-\beta_m(1+\gamma_m)\, e\big(u_m\,, \delta_m(1+\gamma_m)\big)}\right),$$

or, equivalently, as

$$F_k\big(\beta_m\,, \delta_m\,, u_m\,, \gamma_m\big) > 0\,, \tag{4.16}$$

where

$$F_k\big(\beta\,, \delta\,, u\,, \gamma\big) := f_k\left(\gamma\delta\,, u - \delta(1+\gamma)\,, \frac{\beta(1+\gamma)\, e\big(u\,, \delta(1+\gamma)\big)}{1-\beta(1+\gamma)\, e\big(u\,, \delta(1+\gamma)\big)}\right) - \beta(1+\gamma)\,. \tag{4.17}$$

As f_k is decreasing in the third argument by the induction hypothesis, the function $\beta \mapsto F_k(\beta\,, \delta_m\,, u_m\,, \gamma_m)$ is decreasing in $[1\,, \beta_\infty)$ from

$$F_k(1\,, \delta_m\,, u_m\,, \gamma_m) > F_k(\beta_m\,, \delta_m\,, u_m\,, \gamma_m) > 0$$

(by (4.16)) to $F_k(\beta_\infty\,, \delta_m\,, u_m\,, \gamma_m)$

$$\begin{aligned}
&= f_k\left(\gamma_m\delta_m\,, u_m - \delta_m(1+\gamma_m)\,, \frac{\beta_\infty(1+\gamma_m)e\big(u_m\,, \delta_m(1+\gamma_m)\big)}{1-\beta_\infty(1+\gamma_m)e\big(u_m\,, \delta_m(1+\gamma_m)\big)}\right) - \\
&\qquad \beta_\infty(1+\gamma_m) \\
&< f_k\left(\gamma_m\delta_m\,, u_m - \delta_m(1+\gamma_m)\,, \frac{\beta_m(1+\gamma_m)e\big(u_m\,, \delta_m(1+\gamma_m)\big)}{1-\beta_m(1+\gamma_m)e\big(u_m\,, \delta_m(1+\gamma_m)\big)}\right) - \\
&\qquad \beta_\infty(1+\gamma_m) \\
&= f_k\big(\delta_{m+1}\,, u_{m+1}\,, v_{m+1}\big) - \beta_\infty(1+\gamma_m)\,.
\end{aligned}$$

Since f_k is decreasing in the third argument and not increasing in the first,

$$f_k\big(\delta_{m+1}\,, u_{m+1}\,, \gamma_{m+1}\big) < f_k(0\,, u_{m+1}\,, 0) = \beta_\infty$$

and

$$F_k(\beta_\infty\,,\delta_m\,,u_m\,,\gamma_m) < f_k(0\,,u_{m+1}\,,0) - \beta_\infty(1+\gamma_m) = \beta_\infty - \beta_\infty(1+\gamma_m) < 0.$$

It follows that the equation $F_k(\beta\,,\delta_m\,,u_m\,,\gamma_m) = 0$ is uniquely solvable for $\beta \in [1\,,\beta_\infty)$. Denote the solution $f_{k+1}(\delta_m\,,u_m\,,\gamma_m)$:

$$F_k\big(f_{k+1}(\delta\,,u\,,\gamma)\,,\delta\,,u\,,\gamma\big) = 0\,. \tag{4.18}$$

In particular,

$$F_k\big(f_{k+1}(\delta_m\,,u_m\,,\gamma_m)\,,\delta_m\,,u_m\,,\gamma_m\big) = 0\,.$$

Comparison with (4.16) shows that $\beta_m < f_{k+1}(\delta_m\,,u_m\,,\gamma_m)$. The function f_{k+1} is decreasing with respect to the third argument. Indeed, as seen from (4.17), F_k is also decreasing in γ, so that $\gamma < \gamma'$

$$\Longrightarrow F_k\big(f_{k+1}(\delta\,,u\,,\gamma)\,,\delta\,,u\,,\gamma\big) = 0 = F_k\big(f_{k+1}(\delta\,,u\,,\gamma')\,,\delta\,,u\,,\gamma'\big) < F_k\big(f_{k+1}(\delta\,,u\,,\gamma')\,,\delta\,,u\,,\gamma\big)$$

$$\Longrightarrow f_{k+1}(\delta\,,u\,,\gamma) > f_{k+1}(\delta\,,u\,,\gamma')\,.$$

Similarly, f_{k+1} is not decreasing in u. Namely, as e is not increasing with respect to the first argument,

$$u < u' \Longrightarrow e\big(u\,,\delta(1+\gamma)\big) \ge e\big(u'\,,\delta(1+\gamma)\big)$$

$$\Longrightarrow f_{k+1}(\delta\,,u'\,,\gamma)(1+\gamma)e\big(u\,,\delta(1+\gamma)\big) \ge f_{k+1}(\delta\,,u'\,,\gamma)(1+\gamma)e\big(u'\,,\delta(1+\gamma)\big)\,.$$

Then

$$f_k\left(\gamma\delta\,,u-\delta(1+\gamma)\,,\frac{f_{k+1}(\delta\,,u'\,,\gamma)(1+\gamma)e\big(u\,,\delta(1+\gamma)\big)}{1-f_{k+1}(\delta\,,u'\,,\gamma)(1+\gamma)e\big(u\,,\delta(1+\gamma)\big)}\right) \le f_k\left(\gamma\delta\,,u'-\delta(1+\gamma)\,,\frac{f_{k+1}(\delta\,,u'\,,\gamma)(1+\gamma)e\big(u'\,,\delta(1+\gamma)\big)}{1-f_{k+1}(\delta\,,u'\,,\gamma)(1+\gamma)e\big(u'\,,\delta(1+\gamma)\big)}\right)$$

and $F_k\big(f_{k+1}(\delta\,,u\,,\gamma)\,,\delta\,,u\,,\gamma\big) = 0 = F_k\big(f_{k+1}(\delta\,,u'\,,\gamma)\,,\delta\,,u'\,,\gamma\big)$

$$= f_k\left(\gamma\delta\,,u'-\delta(1+\gamma)\,,\frac{f_{k+1}(\delta\,,u'\,,\gamma)(1+\gamma)e\big(u'\,,\delta(1+\gamma)\big)}{1-f_{k+1}(\delta\,,u'\,,\gamma)(1+\gamma)e\big(u'\,,\delta(1+\gamma)\big)}\right) - f_{k+1}(\delta\,,u'\,,\gamma)(1+\gamma)$$

$$\ge f_k\left(\gamma\delta\,,u-\delta(1+\gamma)\,,\frac{f_{k+1}(\delta\,,u'\,,\gamma)(1+\gamma)e\big(u\,,\delta(1+\gamma)\big)}{1-f_{k+1}(\delta\,,u'\,,\gamma)(1+\gamma)e\big(u\,,\delta(1+\gamma)\big)}\right) - f_{k+1}(\delta\,,u'\,,\gamma)(1+\gamma)$$

$$= F_k\big(f_{k+1}(\delta\,,u'\,,\gamma)\,,\delta\,,u\,,\gamma\big)\,.$$

Inasmuch as F_k is decreasing in the first argument,

$$F_k\big(f_{k+1}(\delta\,,u\,,\gamma)\,,\delta\,,u\,,\gamma\big) \geq F_k\big(f_{k+1}(\delta\,,u'\,,\gamma)\,,\delta\,,u\,,\gamma\big) \Longrightarrow$$
$$f_{k+1}(\delta\,,u'\,,\gamma) \geq f_{k+1}(\delta\,,u\,,\gamma)\,.$$

Likewise, as e is increasing in its second argument,

$$\delta < \delta \Longrightarrow e\big(u\,,\delta(1+\gamma)\big) \leq e\big(u\,,\delta\,'(1+\gamma)\big)$$
$$\Longrightarrow \frac{f_{k+1}(\delta\,'\,,u\,,\gamma)(1+\gamma)e\big(u\,,\delta(1+\gamma)\big)}{1-f_{k+1}(\delta\,'\,,u\,,\gamma)(1+\gamma)e\big(u\,,\delta(1+\gamma)\big)}$$
$$\leq \frac{f_{k+1}(\delta\,'\,,u\,,\gamma)(1+\gamma)e\big(u\,,\delta\,'(1+\gamma)\big)}{1-f_{k+1}(\delta\,'\,,u\,,\gamma)(1+\gamma)e\big(u\,,\delta\,'(1+\gamma)\big)}$$

and so $F_k\big(f_{k+1}(\delta\,,u\,,\gamma)\,,\delta\,,u\,,\gamma\big) = 0 = F_k\big(f_{k+1}(\delta\,'\,,u\,,\gamma)\,,\delta\,'\,,u\,,\gamma\big)$

$$= f_k\left(\gamma\delta\,'\,,u-\delta\,'(1+\gamma)\,,\frac{f_{k+1}(\delta\,'\,,u\,,\gamma)(1+\gamma)e\big(u\,,\delta\,'(1+\gamma)\big)}{1-f_{k+1}(\delta\,'\,,u\,,\gamma)(1+\gamma)e\big(u\,,\delta\,'(1+\gamma)\big)}\right) -$$
$$f_{k+1}(\delta\,'\,,u\,,\gamma)(1+\gamma)$$
$$\leq f_k\left(\gamma\delta\,,u-\delta(1+\gamma)\,,\frac{f_{k+1}(\delta\,'\,,u\,,\gamma)(1+\gamma)e\big(u\,,\delta(1+\gamma)\big)}{1-f_{k+1}(\delta\,,u'\,,\gamma)(1+\gamma)e\big(u\,,\delta(1+\gamma)\big)}\right) -$$
$$f_{k+1}(\delta\,'\,,u\,,\gamma)(1+\gamma)$$
$$= F_k\big(f_{k+1}(\delta\,'\,,u\,,\gamma)\,,\delta\,,u\,,\gamma\big)\,.$$

Since F_k is decreasing in the first argument,

$$F_k\big(f_{k+1}(\delta\,,u\,,\gamma)\,,\delta\,,u\,,\gamma\big) \leq F_k\big(f_{k+1}(\delta\,,u'\,,\gamma)\,,\delta\,,u\,,\gamma\big) \Longrightarrow$$
$$f_{k+1}(\delta\,'\,,u\,,\gamma) \leq f_{k+1}(\delta\,,u\,,\gamma)\,.$$

Besides, (4.18) implies

$$0 = F_k\big(f_{k+1}(\delta_n\,,u_n\,,\gamma_n)\,,\delta_n\,,u_n\,,\gamma_n\big)$$
$$= f_k\big(\gamma_n\delta_n\,,u_n-\delta_n(1+\gamma_n)\,,f_{k+1}(\delta_n\,,u_n\,,\gamma_n)(1+\gamma_n)\,e\big(u_n\,,\delta_n(1+\gamma_n)\big)\big) -$$
$$f_{k+1}(\delta_n\,,u_n\,,\gamma_n)(1+\gamma_n)\,,$$

whence, by forcing n to infinity,

$$0 = f_k(0\,,u_\infty\,,0) - f_{k+1}(0\,,u_\infty\,,0) = \beta_\infty - f_{k+1}(0\,,u_\infty\,,0)\,,$$

i.e., $f_{k+1}(0, u_\infty, 0) = \beta_\infty$. Thus, (4.15) implies

$$\beta_{n+1} < \beta_\infty \Longleftrightarrow \beta_{n-k-1} < f_{k+1}(\delta_{n-k-1}, u_{n-k-1}, \gamma_{n-k-1}).$$

By induction, $\beta_{n+1} < \beta_\infty \Longleftrightarrow 1 = \beta_0 < f_n(\delta_0, u_0, \gamma_0)$ and $\underset{n}{\&} \beta_n < \beta_\infty \Longleftrightarrow 1 \le \inf_n f_n(\delta_0, u_0, \gamma_0)$. The sequence f_n is pointwise decreasing:

$$\underset{n}{\&} f_{n+1}(\delta, u, \gamma) < f_n(\delta, u, \gamma). \tag{4.19}$$

This is verified inductively. First, let us see that $f_1(\delta, u, \gamma) < f_0(\delta, u, \gamma)$ or, since F_0 is decreasing with respect to the first argument, that

$$0 = F_0\big(f_1(\delta, u, \gamma), \delta, u, \gamma\big) > F_0\big(f_0(\delta, u, \gamma), \delta, u, \gamma\big).$$

Indeed, by (4.17), $F_0\big(f_0(\delta, u, \gamma), \delta, u, \gamma\big)$

$$= f_0\left(\gamma\delta, u - \delta(1+\gamma), \frac{f_0(\delta, u, \gamma)(1+\gamma)\, e\big(u, \delta(1+\gamma)\big)}{1 - f_0(\delta, u, \gamma)(1+\gamma)\, e\big(u, \delta(1+\gamma)\big)}\right) - f_0(\delta, u, \gamma)(1+\gamma)$$

$$= f_0\left(\gamma\delta, u - \delta(1+\gamma), \frac{\beta_\infty\, e\big(u, \delta(1+\gamma)\big)}{1 - \beta_\infty\, e\big(u, \delta(1+\gamma)\big)}\right) - \beta_\infty$$

$$= \frac{\beta_\infty}{1 + \dfrac{\beta_\infty\, e\big(u, \delta(1+\gamma)\big)}{1 - \beta_\infty\, e\big(u, \delta(1+\gamma)\big)}} - \beta_\infty$$

$$= \beta_\infty\big(1 - \beta_\infty e\big(u, \delta(1+\gamma)\big)\big) - \beta_\infty = -\beta_\infty^2\, e\big(u, \delta(1+\gamma)\big) < 0.$$

Suppose now that $f_n(\delta, u, \gamma) < f_{n-1}(\delta, u, \gamma)$ for some $n \ge 1$. Then by (4.18)

$$F_{n-1}\big(f_n(\delta, u, \gamma), \delta, u, \gamma\big) = 0 = F_n\big(f_{n+1}(\delta, u, \gamma), \delta, u, \gamma\big)$$

$$= f_n\left(\gamma\delta, u - \delta(1+\gamma), \frac{f_{n+1}(\delta, u, \gamma)(1+\gamma)\, e\big(u, \delta(1+\gamma)\big)}{1 - f_{n+1}(\delta, u, \gamma)(1+\gamma)\, e\big(u, \delta(1+\gamma)\big)}\right) - f_{n+1}(\delta, u, \gamma)(1+\gamma)$$

$$< f_{n-1}\left(\gamma\delta, u - \delta(1+\gamma), \frac{f_{n+1}(\delta, u, \gamma)(1+\gamma)\, e\big(u, \delta(1+\gamma)\big)}{1 - f_{n+1}(\delta, u, \gamma)(1+\gamma)\, e\big(u, \delta(1+\gamma)\big)}\right) - f_{n+1}(\delta, u, \gamma)(1+\gamma)$$

$$= F_{n-1}\big(f_{n+1}(\delta, u, \gamma), \delta, u, \gamma\big)$$

and, because F_{n-1} is decreasing in the first argument,

$$F_{n-1}\big(f_n(\delta,u,\gamma),\beta,u,\gamma\big) < F_{n-1}\big(f_{n+1}(\delta,u,\gamma),\delta,u,\gamma\big) \Longrightarrow$$
$$f_n(\delta,u,\gamma) > f_{n+1}(\delta,u,\gamma)\,.$$

By induction, the claim (4.19) is proved. Now $\inf_n f_n = f_\infty$ and so

$$\beta_\infty < \infty \Longleftrightarrow f_\infty(\delta_0\,,u_0\,,\gamma_0) \geq 1.$$

2° Taking limits in (4.18) and in $f_k(0\,,u\,,0) = \beta_\infty$ yields

$$F_\infty\big(f_\infty(\delta\,,u\,,\gamma)\,,\beta\,,u\,,\gamma\big) = 0 \ \ \& \ \ f_\infty(0\,,u\,,0) = \beta_\infty\,,$$

that is, f_∞ solves the system (4.14). To prove uniqueness, let x be a solution and consider the generator $\mathbf{g} : (b\,,d\,,p\,,q) \mapsto (b_+\,,d_+\,,p_+\,,q_+)$ defined as follows:

$$b_+ := x(d\,,p\,,q)(1+q)\ ,\ d_+ := dq\ ,\ p_+ := p - d(1+q)\ ,$$

$$q_+ := \frac{x(d\,,p\,,q)(1+q)\,e\big(p\,,d(1+q)\big)}{1 - x(d\,,p\,,q)(1+q)\,e\big(p\,,d(1+q)\big)}\,.$$

If $(b\,,d\,,p\,,q) = (\beta\,,\delta\,,u\,,\gamma)$ & $\beta = x(\delta\,,u\,,\gamma)$, then

$$b_+ = x(d\,,p\,,q)(1+q) = x(\delta\,,u\,,\gamma)(1+\gamma) = \beta(1+\gamma) = \beta_+\,,$$

$$d_+ = dq = \gamma\delta = \delta_+\,,$$

$$p_+ = p - d(1+q) = u - \delta(1+\gamma) = u_+\,,$$

$$q_+ = \frac{x(d\,,p\,,q)(1+q)\,e\big(p\,,d(1+q)\big)}{1 - x(d\,,p\,,q)(1+q)\,e\big(p\,,d(1+q)\big)} = \frac{x(\delta\,,u\,,\gamma)(1+\gamma)\,e\big(u\,,\delta(1+\gamma)\big)}{1 - x(\delta\,,u\,,\gamma)(1+\gamma)\,e\big(u\,,\delta(1+\gamma)\big)}$$

$$= \frac{\beta(1+\gamma)\,e\big(u\,,\delta(1+\gamma)\big)}{1 - \beta(1+\gamma)\,e\big(u\,,\delta(1+\gamma)\big)} = \gamma_+\,,$$

and, because x is a solution, $\beta_+ = \beta(1+\gamma)$

$$= x(\delta\,,u\,,\gamma)(1+\gamma)$$

$$= x\left(\gamma\delta\,,u - \delta(1+\gamma)\,,\frac{x(\delta\,,u\,,\gamma)(1+\gamma)\,e\big(u\,,\delta(1+\gamma)\big))}{1 - x(\delta\,,u\,,\gamma)(1+\gamma)\,e\big(u\,,\delta(1+\gamma)\big))}\right)$$

$$= x\left(\gamma\delta\,,u - \delta(1+\gamma)\,,\frac{\beta(1+\gamma)\,e\big(u\,,\delta(1+\gamma)\big)}{1 - \beta(1+\gamma)\,e\big(u\,,\delta(1+\gamma)\big)}\right)$$

$$= x(\delta_+\,,u_+\,,\gamma_+)\,.$$

By induction, $(b_0\,,d_0\,,p_0\,,q_0)=(\beta_0\,,\delta_0\,,u_0\,,\gamma_0)\ \ \&\ \ \beta_0=x(\delta_0\,,u_0\,,\gamma_0)$ implies

$$\underset{n}{\&}\Big((b_n\,,d_n\,,p_n\,,q_n)=(\beta_n\,,\delta_n\,,u_n\,,\gamma_n)\ \ \&\ \ \beta_n=x(\delta_n\,,u_n\,,\gamma_n)\Big)\,.$$

This means that the generator $\mathbf{g}$ coincides with (4.13). Consequently, $x=f_\infty\,.$ □

To get the convergence theorem for Broyden's method, it remains now to replace the condition $\beta_\infty<\infty$ in Lemma 4.4 by its equivalent in terms of the variables $\delta\,,u\,,\gamma$ established by the last proposition.

Theorem 4.6. *Let the selected dd* $[x_1,x_2\,|\,\mathbf{f}]$ *of* $\mathbf{f}$ *be* ω*-regularly continuous on* D*. If the starters* $x_0\,,\delta_0\,,u_0\,,\gamma_0$ *are such that*

$$\|\mathbf{f}(x_0)\|\le\delta_0\,,$$

$$\bar a:=\omega^{-1}\big(\|[x_0-f(x_0),x_0\,|\,\mathbf{f}]\|-\underline{h}\big)-\|\mathbf{f}(x_0)\|\le u_0\,,$$

$$\|\mathbf{f}(x_0-\mathbf{f}(x_0))\|\le\frac{\gamma_0}{1+\gamma_0}\|f(x_0)\|\,,$$

and

$$f_\infty(\delta_0\,,u_0\,,\gamma_0)\ge 1\,,$$

where f_∞ *is the function of Proposition 4.5, then*

1° *the sequence* $t_n:=\sum_0^{n-1}\delta_k$ *(equivalently, the series* $\sum\delta_k$*) converges:*

$$t_\infty<\infty;$$

2° $\gamma_\infty=\delta_\infty=0$*;*

3° *the sequence* $(x_n\,,\mathbf{A}_n)$ *generated by the method* (4.1) *from the starter* $(x_0\,,\mathbf{I})$ *converges to a limit* $(x_\infty\,,\mathbf{A}_\infty)$*;*

4° *this limit solves the system* $\mathbf{f}(x)=0\ \ \&\ \ \mathbf{X}[x,x\,|\,\mathbf{f}]=\mathbf{I}$ *for* $(x\,,\mathbf{X})$*;*

5° x_∞ *is the only solution of the equation* $\mathbf{f}(x)=0$ *in the ball* $B(x_0\,,r)$*, where*

$$r:=\omega^{-1}\bigg(1-\frac{\gamma_0}{\delta_0(\delta_0-\gamma_0)}-\omega\big((a-t_\infty)^++\delta_0+t_\infty\big)+\omega\big((a-t_\infty)^+\big)\bigg)$$

and a *is any lower bound for* $\bar a$*;*

6° *for all* $n=0,1,\dots$

$$\|\mathbf{f}(x_{n+1})\|\le\frac{\beta_n\gamma_n}{\delta_n-\beta_n\gamma_n}$$

$$\Delta_n:=\|x_\infty-x_n\|\le t_\infty-t_n$$

$$\frac{\Delta_{n+1}}{\Delta_n}\le\beta_n\omega(\Delta_{n-1}).\qquad(4.20)$$

For linear ω ($\omega(t) = ct$, Lipschitz continuity of dd) the generator (4.13) simplifies into

$$\beta_+ := \beta(1+\gamma)\ ,\ \gamma_+ := \frac{c\beta_+\delta(1+\gamma)}{1-c\beta_+\delta(1+\gamma)}\ ,\ \delta_+ := \gamma\delta\,. \tag{4.21}$$

Fixed points of this generator fill up the nonnegative half $\{(\beta\,,0\,,0) \mid \beta \geq 0\}$ of the β-axis. It can be simplified still further by changing δ for

$$s := c\beta_+\delta(1+\gamma)\,.$$

After this change of variables, the generator turns into

$$\beta_+ := \beta(1+\gamma)\ ,\ \gamma_+ := \frac{s}{1-s}\ ,\ s_+ := \frac{s\gamma}{(1+\gamma)(1-s)^2}\,. \tag{4.22}$$

Note that β does not appear in the second and third equations. Convergence of the sequence $(\beta_n\,,\gamma_n\,,s_n)$ generated by the generator (4.22) (equivalent to (4.21)) from the starter $(1\,,\gamma_0\,,s_0)$ occurs simultaneously with convergence to $(0\,,0)$ of the sequence $(\gamma_n\,,s_n)$, generated by the truncated generator

$$\gamma_+ := \frac{s}{1-s}\ ,\ s_+ := \frac{s\gamma}{(1+\gamma)(1-s)^2}\,, \tag{4.23}$$

defined by the last two equations in (4.22), from $(\gamma_0\,,s_0)$. This generator simplifies even further if we change γ for $u := \gamma/(1+\gamma)$. After the change, it reduces to

$$u_+ := s\ ,\ s_+ := \frac{su}{(1-s)^2}\,. \tag{4.24}$$

Clearly, $(\gamma_n\,,s_n) \to (0\,,0)$ if and only if the sequence $(u_n\,,s_n)$, generated by the generator (4.24), converges to $(0\,,0)$. So, it is preferable to analyze the simpler generator (4.24). The latter has three fixed points

$$(0\,,0)\ ,\ \left(\frac{3-\sqrt{5}}{2}\,,\frac{3-\sqrt{5}}{2}\right)\ ,\ \left(\frac{3+\sqrt{5}}{2}\,,\frac{3+\sqrt{5}}{2}\right).$$

If $s \geq 0.5\left(3-\sqrt{5}\right) =: a$, then

$$s_+ := \frac{us}{(1-s)^2} \geq u \inf_{s\geq a} \frac{s}{(1-s)^2} = u\frac{a}{(1-a)^2} = u \Longrightarrow u_{++} = s_+ \geq u\,.$$

So, the condition $\underset{n}{\&}\, s_n < a$ is necessary for convergence of $(u_n\,,s_n)$ to $(0\,,0)$. The attraction basin of the origin (and the attraction basin of the fixed point $(\beta_\infty\,,0\,,0)$ of the generators (4.21) and (4.22)) is determined by

Proposition 4.7.

1° $u_\infty = s_\infty = 0 \Longleftrightarrow s_0 \le f_\infty(u_0)$, *where*

$$f_0(u) := 1 + \frac{u}{2a} - \sqrt{\left(1+\frac{u}{2a}\right)^2 - 1}\ , \quad a := \frac{3-\sqrt{5}}{2}\ ,$$

and $f_{n+1}(u)$ *is the (unique) solution for* s *of the equation*

$$f_n(s) = \frac{us}{(1-s)^2}\ .$$

2° f_∞ *is the only solution of the system*

$$x(x(u)) = \frac{u\,x(u)}{\big(1-x(u)\big)^2} \quad \& \quad x(a) = a\ . \tag{4.25}$$

3° $s_0 < f_\infty(u_0) \Longrightarrow \underset{n}{\&}\, s_n < f_\infty(u_n)$ *and* $s_0 = f_\infty(u_0) \Longrightarrow \underset{n}{\&}\, s_n = f_\infty(u_n)$.

Proof. 1° Let $\overset{n}{\underset{k=0}{\&}}\ 0 \le s_k < a$. Then, by (4.24), $s_{n+1} < a$

$$\Longleftrightarrow \frac{u_n s_n}{(1-s_n)^2} < a \Longleftrightarrow \frac{u_n}{a}\, s_n < 1 - 2s_n + s_n^2$$

$$\Longleftrightarrow s_n^2 - 2\left(1+\frac{u_n}{2a}\right) s_n + 1 > 0$$

$$\Longleftrightarrow s_n < 1 + \frac{u_n}{2a} - \sqrt{\left(1+\frac{u_n}{2a}\right)^2 - 1}\ \bigvee\ s_n < 1 + \frac{u_n}{2a} + \sqrt{\left(1+\frac{u_n}{2a}\right)^2 - 1}\ .$$

The second possibility contradicts the supposition $s_n < a$. So, if $s_n < a$, then

$$s_{n+1} < a \Longleftrightarrow s_n < 1 + \frac{u_n}{2a} - \sqrt{\left(1+\frac{u_n}{2a}\right)^2 - 1} =: f_0(u_n)\ .$$

As seen from this definition, f_0 is decreasing in $(0\,,\infty)$ and $f_0(a) = a$. Suppose that, for some $k \ge 1$,

$$s_{n+1} < a \Longleftrightarrow s_{n-k} < f_k(u_{n-k})\ , \tag{4.26}$$

where f_k is positive and decreasing in $(0\,,\infty)$ and $f_k(a) = a$. Using (4.24), rewrite the inequality on the right in (4.26) as

$$F_k(u_{n-k-1}, s_{n-k-1}) = f_k(s_{n-k-1}) - \frac{u_{n-k-1}s_{n-k-1}}{\big(1-s_{n-k-1}\big)^2} > 0\ , \tag{4.27}$$

where

$$F_k(u\,,s) := f_k(s) - \frac{us}{(1-s)^2}\ , \quad u \ge 0\ ,\ \ 0 \le s < 1\ .$$

As f_k is decreasing by the induction hypothesis, the function $s \mapsto F_k(u\,,s)$ is decreasing in $(0\,,1)$ from $F_k(u\,,0) = f_k(0) > f_k(a) = a > 0$ to $F_k(u\,,1) =$

$f_k(1) - \infty = -\infty$. So, the equation $F_k(u\,,s) = 0$ is uniquely solvable for $s \in (0\,,\infty)$. Denote the solution $f_{k+1}(u)$:

$$F_k\big(u\,,f_{k+1}(u)\big) = 0\;,\;\;\forall\,u \geq 0\,. \tag{4.28}$$

In particular,

$$F_k\big(u_{n-k-1}\,,f_{k+1}(u_{n-k-1})\big) = 0\,.$$

Since F_k is decreasing with respect to the second argument, comparison with (4.27) shows that $s_{n-k-1} < f_{k+1}(u_{n-k-1})$. The function f_{k+1} is decreasing in $(0\,,\infty)$. Indeed, inasmuch as F_k is decreasing in each of its two arguments,

$$\begin{aligned} 0 < u < u' &\Longrightarrow F_k\big(u\,,f_{k+1}(u)\big) = 0 = F_k\big(u'\,,f_{k+1}(u')\big) < F_k\big(u\,,f_{k+1}(u')\big) \\ &\Longrightarrow f_{k+1}(u') < f_{k+1}(u)\,. \end{aligned}$$

Besides, as $f_k(a) = a$ by the induction hypothesis,

$$F_k(a\,,a) = f_k(a) - \left(\frac{a}{1-a}\right)^2 = a - a = 0\,.$$

On the other hand, (4.28) implies

$$0 = F_k\big(a\,,f_{k+1}(a)\big) = f_k\big(f_{k+1}(a)\big) - \frac{a f_{k+1}(a)}{\big(1 - f_{k+1}(a)\big)^2}\,.$$

Since the equation $F_k(a\,,s) = 0$ has only one s-solution in $(0\,,1)$, it follows that $f_{k+1}(a) = a$. Thus, (4.26) implies

$$s_{n+1} < a \Longleftrightarrow s_{n-k-1} < f_{k+1}(u_{n-k-1})\,.$$

By induction, $s_{n+1} < a \Longleftrightarrow s_0 < f_n(u_0)$ and $\underset{n}{\&}\, s_n < a \Longleftrightarrow s_0 \leq \inf_n f_n(u_0)$. The sequence f_n is pointwise decreasing in $(0\,,a)$:

$$\underset{n}{\&}\, f_{n+1}(s) < f_n(s)\;,\;\;\forall\,s \in (0\,,a)\,. \tag{4.29}$$

This is verified inductively. First we have to show that $0 < s < a \Longrightarrow f_1(s) < f_0(s)$ or, as F_0 is decreasing in the second argument, that

$$0 < s < a \Longrightarrow 0 = F_0\big(s\,,f_1(s)\big) > F_0\big(s\,,f_0(s)\big) = f_0\big(f_0(s)\big) - \frac{s f_0(s)}{\big(1 - f_0(s)\big)^2}\,.$$

By definition of f_0, $s f_0(s)/\big(1 - f_0(s)\big)^2 = a$ and so, for $s \in (0\,,a)$,

$$\begin{aligned} f_0\big(f_0(s)\big) < \frac{s f_0(s)}{\big(1 - f_0(s)\big)^2} &\Longleftrightarrow f_0\big(f_0(s)\big) < a \\ &\Longleftrightarrow f_0(s) > f_0^{-1}(a) = a \Longleftrightarrow s < f_0^{-1}(a) = a\,, \end{aligned}$$

which is true. Suppose next that $f_n(s) \le f_{n-1}(s)$ for some $n \ge 1$ and all $s \in (0\,, a)$. Then,

$$\begin{aligned}
F_{n-1}\big(s\,, f_n(s)\big) = 0 = F_n\big(s\,, f_{n+1}(s)\big) &= f_n\big(f_{n+1}(s)\big) - \frac{s\,f_{n+1}(s)}{\big(1-f_{n+1}(s)\big)^2} \\
&< f_{n-1}\big(f_{n+1}(s)\big) - \frac{s\,f_{n+1}(s)}{\big(1-f_{n+1}(s)\big)^2} \\
&= F_{n-1}\big(s\,, f_{n+1}(s)\big)\,.
\end{aligned}$$

Inasmuch as F_{n-1} is decreasing with respect to the second argument,

$$F_{n-1}\big(s\,, f_n(s)\big) < F_{n-1}\big(s\,, f_{n+1}(s)\big) \Longrightarrow f_n(s) > f_{n+1}(s)\,.$$

By induction, (4.29) is proved. Then $\inf_n f_n = \lim f_n =: f_\infty$ and

$$\mathop{\&}_n s_n < a \Longleftrightarrow s_0 \le f_\infty(u_0)\,.$$

2° Taking limits in (4.28) and $f_n(a) = a$, we obtain $F_\infty\big(u\,, f_\infty(u)\big) = 0$ and $f_\infty(a) = a$, that is, f_∞ solves the system (4.25). To see its uniqueness, let the function x be a solution and consider the generator $\mathbf{g} : (p\,, q) \mapsto (p_+\,, q_+)$ defined as follows:

$$p_+ := q\ ,\ q_+ := \frac{p\,x(p)}{(1-x(p)))^2}\,.$$

Then $(p\,, q) = (u\,, s)$ & $q = x(p)$ implies $p_+ := q = x(p) = x(u) = s = u_+$ and so

$$q_+ := \frac{p\,x(p)}{\big(1-x(p)\big)^2} = \frac{ps}{(1-s)^2} = \frac{us}{(1-s)^2} = s_+\,.$$

At the same time (because x is a solution of (4.25)),

$$q_+ := \frac{p\,x(p)}{\big(1-x(p)\big)^2} = \frac{ux(u)}{\big(1-x(u)\big)^2} = x\big(x(u)\big) = x\big(x(p)\big) = x(q) = x(p_+)\,.$$

Thus,

$$(p\,, q) = (u\,, s)\ \ \&\ \ q = x(p) \Longrightarrow (p_+\,, q_+) = (u_+\,, s_+)\ \ \&\ \ q_+ = x(p_+)\,.$$

By induction,

$$(p_0\,, q_0) = (u_0\,, s_0)\ \ \&\ \ q_0 = x(p_0) \Longrightarrow \mathop{\&}_n (p_n\,, q_n) = (u_n\,, s_n)\ \ \&\ \ q_n = x(p_n)\,.$$

This means that the generator $\mathbf{g}$ coincides with (4.24). Consequently, $x = f_\infty$.

3° By (4.24) and in view of 2°,

$$s < f_\infty(u) \Longrightarrow s_+ = \frac{us}{(1-s)^2} < \frac{uf_\infty(u)}{\big(1-f_\infty(u)\big)^2} = f_\infty\big(f_\infty(u)\big)\,.$$

As follows from 1°, f_∞ is not increasing, so that $s < f_\infty(u) \Longrightarrow f_\infty(s) \geq f_\infty\big(f_\infty(u)\big)$. Therefore, $s < f_\infty(u) \Longrightarrow s_+ < f_\infty(s) = f_\infty(u_+)$. By induction, $s_0 < f_\infty(u_0) \Longrightarrow \underset{n}{\&}\, s_n < f_\infty(u_n)$. Likewise, $s_0 = f_\infty(u_0) \Longrightarrow \underset{n}{\&}\, s_n = f_\infty(u_n)$. □

Corollary 4.8. *The sequence* $(\beta_n\,, \gamma_n\,, s_n)$, *generated by the generator* (4.22) *from the starter* $(1\,, \gamma_0\,, s_0)$, *converges* $\big($*to* $(\beta_\infty\,, 0\,, 0)\big)$ *if and only if*

$$s_0 \leq f_\infty\left(\frac{\gamma_0}{1+\gamma_0}\right),$$

where f_∞ *is as defined in the proposition.*

In contrast with the system (3.38), this proposition offers no expression in the finite number of elementary functions for the solution of the system (4.25). So, testing numerically the convergence of the iterative procedure for solving this system described in the proposition, we terminate the iterations when the maximum absolute value of the difference $f_{n+1} - f_n$ (called Error) becomes less than 10^{-12}. The values of Error obtained after 10, 20, and 30 iterations are presented in the table below. Unlike the similar table in Chapter 3, it

TABLE 4.1: Application of Proposition 4.7

Number of iterations	*Error*
10	8.68e-06
20	1.75e-09
30	3.54e-13

shows that sometimes (that is, for some systems of the type (3.38) and (4.25)) such a procedure can prove to be practical. Figure 4.1 demonstrates plots of the starter f_0 (the solid line) and the final approximation (the dashed line) obtained after 30 iterations.

Corollary 4.9. *Let the selected dd* $[x_1, x_2\,|\,\mathbf{f}]$ *of* $\mathbf{f}$ *be Lipschitz continuous. The sequence* $(\beta_n\,, \gamma_n\,, s_n)$ *generated by the generator* (4.22) *from the starter* $(1\,, \gamma_0\,, s_0)$, *converges* $\big($*to* $(\beta_\infty\,, 0\,, 0)\big)$ *if and only if*

$$s_0 \leq f_\infty\left(\frac{\gamma_0}{1+\gamma_0}\right),$$

where f_∞ *is as defined in the proposition.*

This result allows us to state the following special case of Theorem 4.6 for operators with Lipschitz continuous dd's.

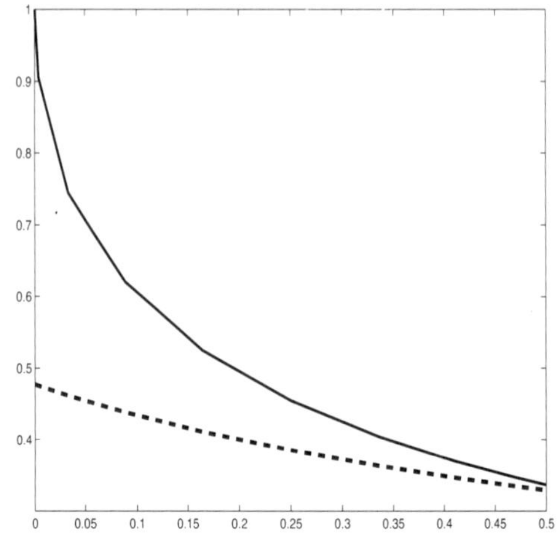

FIGURE 4.1: Application of Proposition 4.7, plots

Corollary 4.10. *Let the selected dd* $[x_1, x_2 \,|\, \mathbf{f}]$ *of* $\mathbf{f}$ *be Lipschitz continuous. If*

$$\|\mathbf{f}(x_0)\| \le s_0 \;\;\&\;\; \|\mathbf{f}(x_0 - \mathbf{f}(x_0))\| \le \frac{\gamma_0}{1+\gamma_0}\|\mathbf{f}(x_0)\|$$

and

$$s_0 \le f_\infty\left(\frac{\gamma_0}{1+\gamma_0}\right),$$

where f_∞ *is the function of Proposition 4.7, then*

1° *the sequence* $t_n := \sum_0^{n-1} \delta_k$, *(equivalently, the series* $\sum \delta_k$*) converges:* $t_\infty < \infty$;

2° $\gamma_\infty = s_\infty = 0$;

3° *the sequence* $(x_n\,, \mathbf{A}_n)$ *generated by the method* (4.1) *from the starter* $(x_0\,, \mathbf{I})$ *converges to a limit* $(x_\infty\,, \mathbf{A}_\infty)$;

4° *this limit solves the system* $\mathbf{f}(x) = 0 \;\;\&\;\; \mathbf{X}[x, x \,|\, \mathbf{f}] = \mathbf{I}$ *for* $(x\,, \mathbf{X})$;

5° x_∞ *is the only solution of the equation* $\mathbf{f}(x) = 0$ *in the ball* $B(x_0\,, r)$, *where*

$$r := c^{-1}\left(1 - \frac{s_0}{1+\gamma_0}\right) - \frac{\gamma_0}{s_0}(1+\gamma_0) - t_\infty\,;$$

6° *for all* $n = 0, 1, \ldots$

$$\|\mathbf{f}(x_{n+1})\| \le \gamma_n \;\;\&\;\; \Delta_n := \|x_\infty - x_n\| \le t_\infty - t_n \;\;\&\;\; \frac{\Delta_{n+1}}{\Delta_n} \le c\beta_n \Delta_{n-1}.$$

4.4 Rate of convergence

The equation (4.20) shows that the sequence x_n converges (if and when) to x_∞ superlinearly. It should be stressed that this result applies also to nonsmooth operators. By (4.20),

$$\Delta_{n+1} \leq \Delta_n \beta_n \omega(\Delta_{n-1}) < \Delta_n \omega_0(\Delta_{n-1}) ,$$

where $\omega_0 := \beta_\infty \omega$. Consider the related difference equation:

$$u_+ := u\,\omega_0(v) \ , \ v_+ := u \ . \tag{4.30}$$

It is easy to see that this generator produces a majorant sequence u_n for Δ_n:

$$\Delta_0 \leq u_0 \Longrightarrow \underset{n}{\&}\, \Delta_n \leq u_n \ .$$

This sequence can be described by a one-dimensional difference equation of the type $u_{n+1} = f(u_n)$. If such f exists, then, by (4.30), $v = f(u)$ and $v_+ = u = f(u_+) = f\big(u\,\omega_0(v)\big) = f\big(u\,\omega_0(f(u))\big)$, so that f must satisfy the functional equation

$$x\big(u\,\omega_0(x(u))\big) = u \ . \tag{4.31}$$

Conversely, a solution f of this equation determines the one-dimensional difference equation $u_{n+1} = f(u_n)$ that, given u_0, generates the sequence u_n of estimates for errors Δ_n : $\underset{n}{\&}\, \Delta_n \leq u_n$. By Theorem 4.6, $\Delta_0 \leq t_\infty$, so that t_∞ is a natural candidate for u_0. Thus, if one knows a solution of the functional equation (4.31), he is able to get a priori estimates for Δ_n before running the infinite-dimensional process (4.1). The equation (4.31) can be solved numerically by one of the iterative methods applicable to nondifferentiable operators, Broyden's method included.

In the case of linear ω $(\omega(t) = ct)$, (4.20) implies

$$\Delta_{n+1} \leq \alpha \Delta_n \Delta_{n-1} \ , \ \alpha := c\beta_\infty$$

or, equivalently, $\alpha\Delta_{n+1} \leq \alpha\Delta_n \cdot \alpha\Delta_{n-1}$. The corresponding difference equation is $u_{n+1} := u_n\, u_{n-1}$. The variable change $v_n := \ln u_n$ yields $v_{n+1} = v_n + v_{n-1}$, the familiar Fibonacci difference equation [1]. Its solution is $v_n = Fib_{n-1} v_1 + Fib_n v_0$, where Fib_n is the n-th Fibonacci number:

$$Fib_n = \frac{1}{\sqrt{5}}\left(\frac{1+\sqrt{5}}{2}\right)^n - \frac{1}{\sqrt{5}}\left(\frac{1-\sqrt{5}}{2}\right)^n .$$

It follows that

$$\begin{aligned}\alpha\Delta_n \le u_n &= \exp(v_n) = \exp\big(Fib_{n-1}v_1 + Fib_n v_0\big) \\ &= \exp\big(Fib_{n-1}\ln u_1 + Fib_n \ln u_0\big) = \exp\left(\ln\left(u_1^{Fib_{n-1}} u_0^{Fib_n}\right)\right) \\ &= u_1^{Fib_{n-1}} u_0^{Fib_n}.\end{aligned}$$

4.5 Evaluation of the function f_∞ of Proposition 4.5

The appearance of the function f_∞ in Theorem 4.6 raises the question: how to evaluate it? Proposition 4.5 points out two possible ways to approach the problem of numerical solution of functional equations of the kind of (4.14): (i) the iterative process used in its definition of f_∞ and (ii) solving the similar system induced by the majorant generator. Both seem impractical. The first because the convergence of iterations can be prohibitively slow (as in the example of Section 3.3), especially when the starter is close to the boundary of the convergence domain of the generator. The second because efficient methods for numerical solution of functional equations of this class are (as far as I know) still absent. (We will make an attempt to drop a bit into the void in Chapter 7). Hence the need for another, more practical, approach to evaluation of f_∞. In this section, we are trying to look at f_∞ from yet another angle. Whether its practicality is different from the previous two is an open question.

We illustrate the idea on the model generator

$$t_+ := t + \delta \ , \quad d_+ := \delta \ , \quad \delta_+ := \delta\left(\frac{\omega(a - 2t + d)}{\omega(a - 2t - \delta)} - 1\right) \tag{4.32}$$

we have used in convergence analysis of the secant method (0.3) in [18]. We try to find a scalar function f whose dd $[x, x' \,|\, f\,]$ is ω-regularly continuous and for which the secant method generates from the starter $(0\,, d_0, \delta_0)$ the same sequence $(x_n\,,\, x_n - x_{n-1}\,,\, x_{n+1} - x_n)$ as (t_n, d_n, δ_n) generated by the generator (4.32). If $(t, d, \delta) = (x\,,\, x - x_-\,,\, x_+ - x)$, then $t_+ := t + \delta = x + x_+ - x = x_+$, $d_+ := \delta = x_+ - x$, and

$$\delta_+ := \delta\left(\frac{\omega(a - 2t + d)}{\omega(a - 2t - \delta)} - 1\right) = (x_+ - x)\left(\frac{\omega(a - x - x_-)}{\omega(a - x_+ - x)} - 1\right).$$

So, $(t, d, \delta) = (x\,,\, x - x_-\,,\, x_+ - x)$ implies $(t_+, d_+, \delta_+) = (x_+\,,\, x_+ - x\,,\, x_{++} - x_+)$ if and only if

$$(x_+ - x)\left(\frac{\omega(a - x - x_-)}{\omega(a - x_+ - x)} - 1\right) = -\frac{f(x_+)}{[x_+, x \,|\, f\,]}\ .$$

Since $[x_+, x \mid f] = \big(f(x_+) - f(x)\big)/(x_+ - x)$, this equation can be rewritten as

$$\frac{\omega(a - x - x_-)}{\omega(a - x - x_+)} - 1 = -\frac{f(x_+)}{f(x_+) - f(x)}$$

or, equivalently,

$$f(x_+) = f(x)\left(1 - \frac{\omega(a - x - x_+)}{\omega(a - x - x_-)}\right) . \tag{4.33}$$

If f meets this condition, then, by induction,

$$t_0 = x_0 = 0 \;\;\&\;\; d_0 = x_0 - x_{-1} = -x_{-1} \;\;\& \tag{4.34}$$

$$\delta_0 = -\frac{f(x_0)}{[x_0, x_{-1} \mid f]} = -\frac{f(0)}{[0, -d_0 \mid f]}$$

$$\Longrightarrow \mathop{\&}_n \big(t_n = x_n \;\;\&\;\; d_n = x_n - x_{n-1} \;\;\&\;\; \delta_n = x_{n+1} - x_n\big)$$

$$\Longrightarrow \mathop{\&}_n \delta_n = -\frac{f(x_n)}{[x_n, x_n - d_n \mid f]} = -\frac{f(t_n)}{[t_n\,, t_n - d_n \mid f]} = -\frac{f(t_n)}{[t_n\,, t_{n-1} \mid f]} . \tag{4.35}$$

We are going to show now that the function

$$f(t) := f(t \mid t_0, d_0, \delta_0) := \delta\,\omega(a - 2t + d) \tag{4.36}$$

(which depends on the paremeters $t_0, d_0, \delta_0)$) solves (4.33) for f on the sequence (t_n, d_n, δ_n):

$$\mathop{\&}_n f(t_n) = \delta_n \omega(a - 2t_n + d_n) \Longrightarrow \mathop{\&}_n f(t_{n+1}) = f(t_n)\left(1 - \frac{\omega(a - 2t_n - \delta_n)}{\omega(a - 2t_n + d_n)}\right) . \tag{4.37}$$

Indeed, $\mathop{\&}_n f(t_n) = \delta_n \omega(a - 2t_n + d_n) \Longrightarrow$

$$\begin{aligned} f(t_{n+1}) &= \delta_{n+1}\omega(a - 2t_{n+1} + d_{n+1}) \\ &= \delta_n\left(\frac{\omega(a - 2t_n + d_n)}{\omega(a - 2t_n - \delta_n)} - 1\right)\omega(a - 2t_n - 2\delta_n + \delta_n) \\ &= \delta_n\big[\omega(a - 2t_n + d_n) - \omega(a - 2t_n - \delta_n)\big] \\ &= f(t_n)\left(1 - \frac{\omega(a - 2t_n - \delta_n)}{\omega(a - 2t_n + d_n)}\right) . \end{aligned}$$

Thus, f in (4.35) is the function (4.36). By (4.37), the sequence $f(t_n)$ is decreasing and so f is invertible: $\mathop{\&}_n t_n = f^{-1}(\delta_n \omega(a - 2t_n + d_n) \mid t_0, d_0, \delta_0)$. This sequence is increasing and (if defined) remains $< t_\infty = f^{-1}(0 \mid t_0, d_0, \delta_0) =$

$\underset{n}{\&} f^{-1}(0 \,|\, t_n, d_n, \delta_n)$. Thus, $f^{-1}(0 \,|\, t, d, \delta)$ is an invariant of the generator (4.32).

If $\underset{n}{\&} 2t_n + \delta_n < a$, then the generator (4.32) generates infinite sequences (t_n, d_n, δ_n), $\exists t_\infty \le 0.5a$, and $d_\infty = \delta_\infty = 0$, so that $t_\infty = f^{-1}(0 \,|\, t_0, d_0, \delta_0) \le 0.5a$. Conversely, if $f^{-1}(0 \,|\, t_0, d_0, \delta_0) \le 0.5a$, then

$$\underset{n}{\&} t_n = f^{-1}(\delta_n \omega(a - 2t_n + d_n) \,|\, t_0, d_0, \delta_0) < f^{-1}(0 \,|\, t_0, d_0, \delta_0) \le 0.5a$$

and $2t_n + \delta_n = t_n + t_{n+1} < a$. Thus, $\underset{n}{\&} 2t_n + \delta_n < a \iff f^{-1}(0 \,|\, t_0, d_0, \delta_0) \le 0.5a$.

For linear ω (the case of Lipschitz continuity of the dd), the generator (4.32) simplifies into

$$t_+ := t + \delta \ , \quad d_+ := \delta \ , \quad \delta_+ := \frac{\delta(d + \delta)}{a - 2t - \delta} \tag{4.38}$$

and the equation (4.33) becomes

$$f(x_+) = f(x) \frac{x_+ - x_-}{a - x - x_-} \ . \tag{4.39}$$

This equation is solved by the quadratic polynomial $f(x) := x^2 - ax + c$. Indeed, for it $[x_+, x \,|\, f\,] = x_+ + x - a$,

$$x_+ - x = -\frac{f(x)}{[x, x_- \,|\, f\,]} = -\frac{f(x)}{x + x_- - a} = \frac{f(x)}{a - 2x + d} \ ,$$

and $x_+ - x_- = d + f(x)/(a - 2x + d)$, so that (4.39) can be rewritten as

$$\frac{f(x)}{a - 2x + d}\left(d + \frac{f(x)}{a - 2x + d}\right) = f\left(x + \frac{f(x)}{a - 2x + d}\right)$$

$$= \left(x + \frac{f(x)}{a - 2x + d}\right)^2 - a\left(x + \frac{f(x)}{a - 2x + d}\right) + c$$

$$= \frac{1}{(a - 2x + d)^2}\Big(\big(f(x) + x(a - 2x + d)\big)^2 -$$

$$a\big(f(x) + x(a - 2x + d)\big)(a - 2x + d) + c(a - 2x + d)^2\Big)$$

or, equivalently,

$$\begin{aligned} f(x)\big[f(x) + d(a - 2x + d)\big] &= \big[f(x) + x(a - 2x + d)\big]^2 \\ &\quad - a\big[f(x) + x(a - 2x + d)\big](a - 2x + d) + c(a - 2x + d)^2. \end{aligned}$$

Expanding the polynomials in x on both sides, we see that they are identical:

$$x^4 - 2(a + d)x^3 + \big(2c + ad + (a + d)^2\big)\, x^2 - (2c + ad)(a + d)x + c(c + d(a + d)) \ .$$

Now $f(t_n) = \delta_n(a - 2t_n + d_n)$

$$\iff t_n^2 - at_n + c = \delta_n(a - 2t_n + d_n)$$

$$\iff I(t_n, d_n, \delta_n) := (a - 2t_n)^2 - 4\delta_n(a - 2t_n + d_n) = a^2 - 4c \ ,$$

i.e., I is an invariant of (4.38).

4.6 Comparative analysis of iterative methods

Superlinear convergence alone (and, generally, rate or order of convergence), frequently used in the literature as a quality index of an iterative method, is an unsatisfactory measure of its merits. First of all, because of its asymptotic character. It reflects the behavior of a method only on the final iterations in the vicinity of a solution and says nothing about the initial ones. A similar shortcoming is inherent in the popular Ostrovsky efficiency index [43], which uses order of convergence. Second, the rate of convergence is not the only valuable quality of iterative methods. Another one is the size of the convergence domain. It is an even more important consideration than the rate of convergence when one contemplates the choice of a starter x_0 to get convergence in the first place. The third property that should be taken into account is the computational cost of an iteration.

Kantorovich's majorization principle used in this book offers more solid ground for comparison of various methods. Convergence theorems based on this principle (like Theorems 2.15, 3.16, 4.6 and their corollaries) provide the convergence domain of the majorant generator and an upper bound u_n for the current error $\Delta_n := \|x_n - x_\infty\|$: $\underset{n}{\&} \Delta_n \le u_n$. The convergence domain of the generator gives a sufficient condition for convergence of iterations $(x_n, \mathbf{A}_n)$, while the error bounds u_n allow us to estimate the rate of convergence not only asymptotically, but beginning with the initial iterations.

To be more specific, consider a situation when analyses of two iterative methods

$$(x_+, \mathbf{A}_+) := \mathbf{F}_i(x, \mathbf{A}) \ , \ i = 1, 2 \ ,$$

have produced two different convergence domains $(x_0, \mathbf{A}_0) \in S_i$ and two sequences $u_n^{(i)}$ of bounds for the respective errors $\Delta_n^{(i)}$. If, for example, $S_1 \supset S_2$, then it is reasonable to start the iterations with $(x_0, \mathbf{A}_0) \in S_1$. In this case, there is little sense in comparing the rate of convergence of two methods until the iterations enter S_2. From this moment on, we would prefer the first method, if $\underset{n}{\&} u_n^{(1)} \le u_n^{(2)}$. However, this is an ideal situation that generally cannot be expected. More realistically, suppose that $\overset{m}{\underset{n=0}{\&}} u_n^{(1)} \le u_n^{(2)}$ and

$\underset{n=m+1}{\overset{\infty}{\&}} u_n^{(2)} < u_n^{(1)}$. Then, the obvious decision is to use $\mathbf{F}_1$ for the iterations $0, 1, \ldots, m$ and then switch to $\mathbf{F}_2$. In general, $\underset{n}{\&} \Delta_n \le \min\left\{u_n^{(1)}, u_n^{(2)}\right\}$ and one should use $\mathbf{F}_1$, if $u_n^{(1)} \le u_n^{(2)}$, and $\mathbf{F}_2$, otherwise. Thus, a combination of methods can prove to be more efficient than each of the combined methods.

The point made above can be illustrated by the comparison of Ulm's method (3.4) with Broyden's method (4.1). Their analyses have been carried out under the same assumption, regular continuity of the selected dd $[x_1, x_2 \,|\, \mathbf{f}]$, and with the same parameter $\bar{a} = \omega^{-1}\big(\big\|[x_1, x_0 \,|\, \mathbf{f}]\big\| - \underline{h}\big) - \|x_1 - x_0\|$. So, a in (3.18) and (4.9) is also the same. Moreover, we can safely assume that $a = 1$, since the variables t and δ can always be replaced by at and $a\delta$, respectively, and the function $t \mapsto \omega(at)$ is as a general representative of the class $\mathcal{N}$ as ω. To make the comparison more transparent, we restrict ourselves to the simplification of regular continuity of dd's, Lipschitz continuity. Under Lipschitz continuity (3.5) (with $c = 1$ without loss of generality), Corollary 3.17 guarantees convergence of Ulm's iterations $(x_n, \mathbf{A}_n)$ provided

$$\big\|\mathbf{I} - [x_1, x_0 \,|\, \mathbf{f}]\big\| \le \gamma_0 < 1 \;\;\&\;\; \big\|\mathbf{f}(x_0)\big\| \le \delta_0 < \left(1 - \sqrt{\gamma_0}\right)^2 . \tag{4.40}$$

By the secant equation and in view of (4.5),

$$\begin{aligned}\mathbf{f}(x_1) &= \mathbf{f}(x_0) + [x_1, x_0 \,|\, \mathbf{f}\,](x_1 - x_0) = \mathbf{f}(x_0) - [x_1, x_0 \,|\, \mathbf{f}\,]\mathbf{f}(x_0)\\ &= \big(\mathbf{I} - [x_1, x_0 \,|\, \mathbf{f}\,]\big)\mathbf{f}(x_0)\end{aligned}$$

and so $\big\|\mathbf{f}(x_1)\big\| \le \big\|\mathbf{I} - [x_1, x_0 \,|\, \mathbf{f}]\big\| \cdot \big\|\mathbf{f}(x_0)\big\|$. Therefore, condition (4.40) implies

$$\big\|\mathbf{f}(x_1)\big\| \le \gamma_0 \left(1 - \sqrt{\gamma_0}\right)^2 , \forall\, \gamma_0 \in (0\,, 1)\,.$$

At the same time, Corollary 4.10 says that Broyden's iterations $(x_n, \mathbf{A}_n)$, generated from the same starters $x_0, x_{-1}, \mathbf{I}$, converge if

$$\big\|\mathbf{f}(x_1)\big\| \le \frac{\gamma_0}{1 + \gamma_0}\big\|\mathbf{f}(x_0)\big\| \;\;\&\;\; \big\|\mathbf{f}(x_0)\big\| \le s_0 \le f_\infty\left(\frac{\gamma_0}{1 + \gamma_0}\right),$$

where f_∞ is the function of Proposition 4.7 represented by its approximation obtained in Section 4.3. It follows that they converge if

$$\big\|\mathbf{f}(x_1)\big\| \le \frac{\gamma_0}{1 + \gamma_0} f_\infty\left(\frac{\gamma_0}{1 + \gamma_0}\right).$$

The graphs of the functions $\gamma \mapsto \gamma\left(1 - \sqrt{\gamma}\right)^2$ and $\gamma \mapsto \gamma/(1+\gamma) f_\infty(\gamma/(1+\gamma))$ are depicted in Figure 4.2 by the solid and the dotted lines, respectively. These graphs suggest that the starter x_0 ensures convergence more often, if it meets the conditions of Corollary 4.10. This impression is corroborated by my (limited) numerical experimentation with both methods. One should remember, though, that Ulm's method is applicable to equations in Banach spaces, while Broyden's method makes sense only in Hilbert ones.

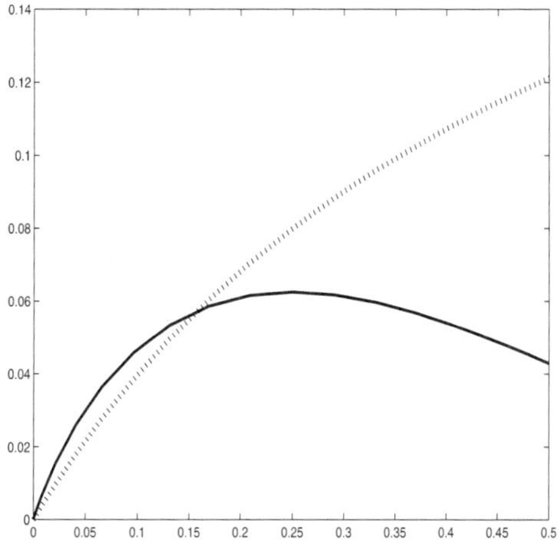

FIGURE 4.2: Comparison of two methods

Next, let us compare the error bounds for both methods, as stated by Corollaries 3.17 (with $c = 1$ and $\beta_0 = 1$)

$$u_n^{(U)} = \frac{1}{2}\left(\frac{1-\gamma_n^{(U)}}{\beta_n^{(U)}} + \delta_n^{(U)} - \sqrt{I_0}\right),$$

$$I_0 := \left(\delta_0^{(U)}\right)^2 - 2\delta_0^{(U)}\left(1+\gamma_0^{(U)}\right) + \left(1-\gamma_0^{(U)}\right)^2, \tag{4.41}$$

and 4.10

$$u_n^{(B)} = t_\infty^{(B)} - t_n^{(B)}, \quad t_n^{(B)} := \sum_{k=0}^{n-1} \delta_k^{(B)}. \tag{4.42}$$

First, we run the generator (4.21) starting from the starter $(1, \gamma_0, \delta_0)$ with $\delta_0 \le \left(1-\sqrt{\gamma_0}\right)^2$ (this condition ensures convergence of both sequences $\left(\beta_n^{(U)}, \gamma_n^{(U)}, \delta_n^{(U)}\right)$ and $\left(\beta_n^{(B)}, \gamma_n^{(B)}, \delta_n^{(B)}\right)$) to get $t_\infty^{(B)}$ as a function of γ_0. The result is presented in the table below. With $t_\infty^{(B)}$ in hand, we run the

TABLE 4.2: $t_\infty^{(B)}$ as a function of γ_0

γ_0	0.1	0.2	0.3	0.5	0.7	0.8	0.9
$t_\infty^{(B)}$	6.09e-1	4.33e-1	3.06e-1	1.40e-1	4.70e-2	2.04e-2	5.03e-3

TABLE 4.3: Comparison of two methods

Iter.	$\gamma_0 = 0.2$		$\gamma_0 = 0.5$		$\gamma_0 = 0.8$	
	$u_n^{(U)}$	$u_n^{(B)}$	$u_n^{(U)}$	$u_n^{(B)}$	$u_n^{(U)}$	$u_n^{(B)}$
0	5.53e-1	4.33e-1	2.93e-1	1.40e-1	1.05e-1	2.04e-2
1	2.47e-1	1.27e-1	2.07e-1	5.43e-2	9.44e-2	9.27e-3
2	1.74e-1	6.60e-2	1.43e-1	1.14e-2	7.84e-2	3.40e-4
3	1.20e-1	1.80e-2	9.77e-2	1.15e-3	5.97e-2	5.88e-6
4	8.18e-2	3.32e-3	6.67e-2	2.71e-5	4.26e-2	3.82e-9
5	5.58e-2	2.02e-4	4.44e-2	6.59e-8	2.94e-2	2.05e-11
6	3.81e-2	2.43e-6	3.10e-2	3.77e-12	2.02e-2	2.04e-11
7	2.60e-2	1.79e-9	2.12e-2	0.	1.38e-2	2.04e-11

generators (3.35) and (4.22) in parallel for four values of γ_0, getting the bounds (4.41) and (4.42). They are shown in the next table. It makes plain the superiority of Broyden's method in Hilbert spaces.

4.7 Applications

4.7.1 Complementarity problem.

Let $\mathbb{H}, C$, and $\mathbf{g} : \mathbb{H} \to \mathbb{H}$ be a Hilbert space, a closed cone in it, and a (generally nonsmooth) operator acting on $\mathbb{H}$. The problem of finding a solution of the system

$$x \in C \ \ \& \ \ \mathbf{g}(x) \in C^* \ \ \& \ \ \langle x\,, \mathbf{g}(x) \rangle = 0\ , \tag{4.43}$$

where C^* is the dual cone, $C^* := \big\{ y \in \mathbb{H} \,\big|\, \langle x\,, y \rangle \geq 0\,, \forall\, x \in C \big\}$, is called [26] the complementarity problem (or CP$(C, \mathbf{g})$). Its special case

$$x \in \mathbb{E}^n \ \ \& \ \ \mathop{\&}_{i=1}^{n} (x_i \geq 0 \ \ \& \ \ g_i(x) \geq 0 \ \ \& \ \ x_i g_i(x) = 0) \tag{4.44}$$

is the subject of the vast amount of literature (see [9] ,[26]). In particular, Mangasarian in [38] proved a theorem establishing equivalence between the problem (4.44) and an operator equation. His result can be stated as

Proposition 4.11. [38]

> *Let* C *and* $\varphi : \mathbb{R} \to \mathbb{R}$ *be the standard positive cone of* $\mathbb{E}^n$ *and any strictly increasing function with* $\varphi(0) = 0$*. Define the operator* $\mathbf{f} : \mathbb{E}^n \to \mathbb{E}^n$ *by setting*
> $$\mathbf{f}(x)_i := \varphi(|x_i - g_i(x)|) - \varphi(x_i) - \varphi(g_i(x)),\ i = 1, \ldots, n\,.$$
> *A vector* $x = (x_1, \ldots, x_n)$ *solves CP(*$C, \mathbf{g}$*) if and only if* $\mathbf{f}(x) = 0$*.*

Mangasarian's theorem can be extended to any separable Hilbert space $\mathbb{H}$ using the fact that it has an orthonormal basis $\{e_k\}_1^\infty$ (see, for example, [32, Ch. IV] or [8, Corollary 2.1.8]), so that each $x \in \mathbb{H}$ has the unique representation $x = \sum_1^\infty \langle x\,, e_k\rangle e_k$. Taking for C the standard positive cone

$$C := \left\{ x \in \mathbb{H} \;\middle|\; \mathop{\&}_k \langle x\,, e_k\rangle \geq 0 \right\},$$

we obtain $C^* = C$ and $\mathbf{g}(x) \in C^* \iff \mathop{\&}_k \langle \mathbf{g}(x)\,, e_k\rangle \geq 0$. With this choice of C, CP($C, \mathbf{g}$) becomes: find $x \in \mathbb{H}$ such that

$$\mathop{\&}_k \Big(\langle x\,, e_k\rangle \geq 0 \;\;\&\;\; \langle \mathbf{g}(x)\,, e_k\rangle \geq 0 \;\;\&\;\; \langle x\,, e_k\rangle\langle \mathbf{g}(x)\,, e_k\rangle = 0 \Big)\,. \tag{4.45}$$

Proposition 4.12. *Let* $\mathbb{H}$ *and* $\{e_k\}_1^\infty$ *be a separable Hilbert space and an orthonormal basis in it. Let* φ *be any strictly increasing function on* $\mathbb{R}$ *with* $\varphi(0) = 0$*. Define* $\mathbf{f} : \mathbb{H} \to \mathbb{H}$ *by setting coordinates* $\langle \mathbf{f}(x)\,,\, e_k\rangle$ *of* $\mathbf{f}(x)$ *to*

$$\varphi\big(|\langle \mathbf{g}(x) - x\,,\, e_k\rangle|\big) - \varphi\big(\langle \mathbf{g}(x)\,,\, e_k\rangle\big) - \varphi\big(\langle x\,,\, e_k\rangle\big)\,.$$

Then $x \in \mathbb{H}$ *solves* (4.45) *if and only if* $\mathbf{f}(x) = 0$.

Proof. Fix a $k \in \mathbb{N}$ and let for short $a := \langle x\,, e_k\rangle$ and $b := \langle \mathbf{g}(x)\,, e_k\rangle$. The claim reduces to

$$a \geq 0 \;\;\&\;\; b \geq 0 \;\;\&\;\; ab = 0 \iff \varphi(|a - b|) = \varphi(a) + \varphi(b)\,. \tag{4.46}$$

Because of the symmetry $a \leftrightarrow b$, it suffices to consider the case $a \geq b$. If the left side of (4.46) is true, then $a \geq b = 0$ and so the right side is true also. Conversely, suppose that $\varphi(a - b) = \varphi(a) + \varphi(b)$. The following situations are conceivable:

(*i*) $a \geq b > 0$, (*ii*) $a \geq b = 0$, (*iii*) $a > 0 > b$, (*iv*) $a = 0 > b$, (*v*) $0 > a \geq b$.

We must show that only (ii) is possible if $\varphi(a-b) = \varphi(a) + \varphi(b)$. Indeed, as φ is increasing and $\varphi(0) = 0$,

$$(i) \Longrightarrow \varphi(a) \geq \varphi(b) > 0 \Longrightarrow \varphi(a) > \varphi(a-b) = \varphi(a) + \varphi(b) > \varphi(a),$$

$$(iii) \Longrightarrow \varphi(a) > 0 > \varphi(b) \Longrightarrow \varphi(a) < \varphi(a-b) = \varphi(a) + \varphi(b) < \varphi(a),$$

$$(iv) \Longrightarrow \varphi(a) = 0 > \varphi(b) \Longrightarrow 0 = \varphi(a) < \varphi(|b|) = \varphi(a-b) = \varphi(a) + \varphi(b) = \varphi(b) < 0,$$

$$(v) \Longrightarrow 0 > \varphi(a) \geq \varphi(b) \Longrightarrow 0 \leq \varphi(a-b) = \varphi(a) + \varphi(b) < 0.$$

Thus, in each of these four cases, $\varphi(a-b) \neq \varphi(a) + \varphi(b)$, while (ii) implies $a \geq 0$ & $b \geq 0$ & $ab = 0$. □

Inasmuch as the Hilbert space $L_2[0,1]$ of square integrable functions on $[0,1]$ is separable [32],[8] and the function $t \mapsto t$ can be taken for φ, we have

Corollary 4.13. *A function* $x \in L_2[0,1]$ *solves the system*

$$\mathop{\&}_{0\leq t\leq 1}\Big(x(t) \geq 0 \;\;\&\;\; \mathbf{g}(x)(t) \geq 0 \;\;\&\;\; x(t)\mathbf{g}(x)(t) = 0\Big) \tag{4.47}$$

if and only if $\mathbf{f}(x)(t) := \min\{x(t)\,, \mathbf{g}(x)(t)\} = 0$.

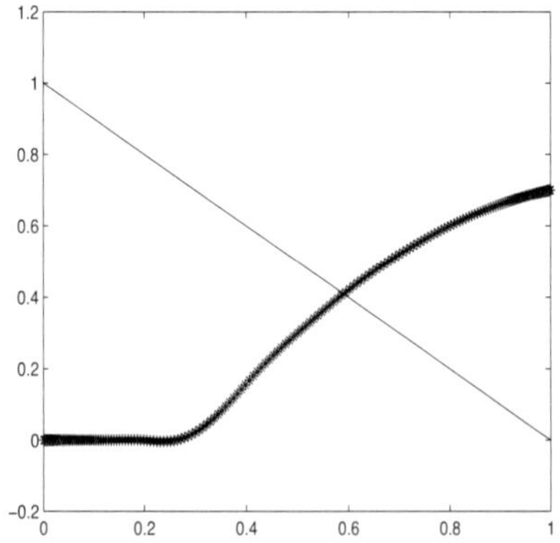

FIGURE 4.3: Complementarity problem

We apply the method (4.1) to the problem (4.47) with $\mathbf{g}(x)(t) := x(t) - (t-c)(2-t)$, where c is some constant within $(0,1)$. For this $\mathbf{g}$, the problem (4.47) has only the solution $x(t) = (t-c)^+(2-t)$ (agree?). Starting from $x_0(t) := 1-t$, $x_{-1}(t) := 0.9x_0(t) + 0.001$, and

$$(\mathbf{A}_0 h)(t) := \frac{x_0(t) - x_{-1}(t)}{\mathbf{f}(x_0)(t) - \mathbf{f}(x_{-1})(t)} h(t)\,, \tag{4.48}$$

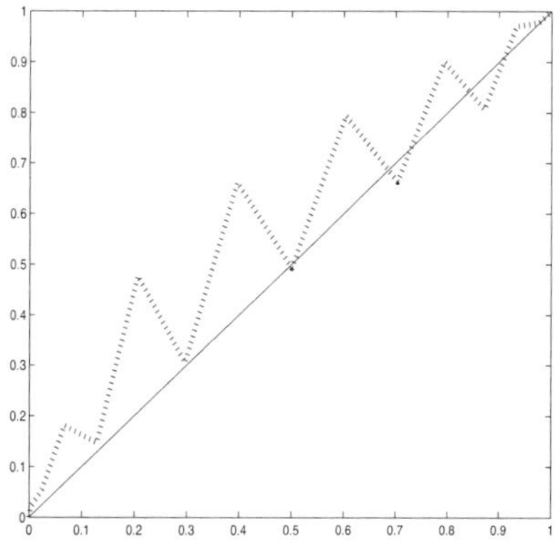

FIGURE 4.4: Functional equation

the method attains the requested accuracy of approximation (measured by the max-norm of the function $\mathbf{f}(x)(t)$) of 10^{-12} after eigth iterations. Figure 4.3 shows the initial (solid line) and the final (asterisked line) approximations for $c = 0.3$.

4.7.2 Functional equation

We have applied Broyden's method to the functional equation (4.31) with $\omega_0(t) := \sqrt{t}$. Starting from $x_0(t) := t$, $x_{-1}(t) := 0.9x_0(t)$ and $\mathbf{A}_0$ as in (4.48), the method has reduced the max-norm of the function $\mathbf{f}(x)(t) := x\big(t\omega_0(x(t))\big) - t$ to 10^{-12} after 40 iterations. Figure 4.4 shows the initial (solid line) and the final (dashed line) approximations.

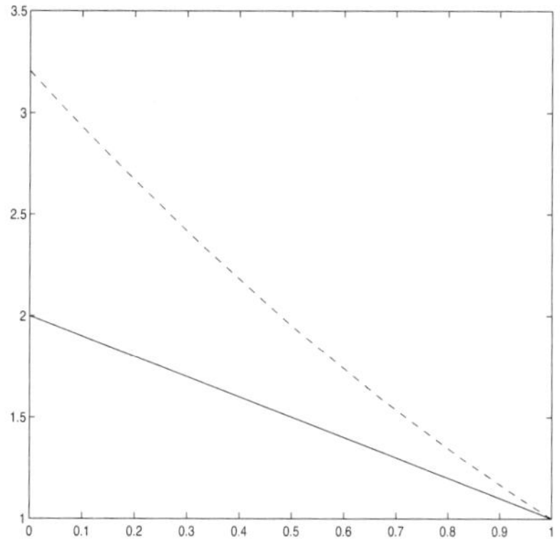

FIGURE 4.5: Integral equation

4.7.3 Integral equation

Pimbley [46] considered the following integral equation arising in a model of statistical mechanics:

$$\mathbf{f}(x)(t) := x(t) - \lambda \int_t^1 x(s-t)x(s)\,ds - 1 = 0, \quad 0 \le t \le 1, \quad x \in \mathbb{C}[0,1]\,. \quad (4.49)$$

He found that it has two positive decreasing twice differentiable solutions for each $\lambda \in (0, 0.5)$, one for $\lambda = 0.5$, and none for $\lambda > 0.5$, and has investigated their properties. However, he did not try to solve this equation numerically.

We apply the method (4.1) to the equation (4.49) with $\lambda = 0.5$ starting from $x_0(t) := x_{-1}(t) := 2 - t$ (as seen from (4.49) the condition $x(1) = 1$ is necessary for any solution) and $\mathbf{A}_0 := \mathbf{I}$. The successive approximations $x_n(t)$ were represented by cubic splines created on the Tchebyshev mesh of 16 points on $[0,1]$. The quality of an approximation x_n is measured by the max-norm of the corresponding function (4.49) shown in Table 4.4 under the heading Error. The final approximation x_{15} is tabulated in columns 3 and 4 of the table. The plots of the initial approximation (the solid line) and the final one (the dashed line) are shown in Figure 4.5.

TABLE 4.4: Integral equation

Iteration	*Error*	t	$x_{15}(t)$
0	1.6597e-1	0.0000e-0	3.0410e+0
1	1.1076e-1	3.1439e-3	3.0329e+0
2	4.6124e-2	2.8058e-2	2.9698e+0
3	2.2092e-2	7.6638e-2	2.8485e+0
4	7.2765e-3	1.4645e-1	2.6781e+0
5	9.3770e-4	2.3398e-1	2.4714e+0
6	6.1311e-5	3.3486e-1	2.2428e+0
7	6.6470e-6	4.4402e-1	2.0071e+0
8	4.8785e-6	5.5598e-1	1.7782e+0
9	3.3993e-6	6.6514e-1	1.5677e+0
10	3.3742e-7	7.6602e-1	1.3843e+0
11	2.3144e-9	8.5355e-1	1.2338e+0
12	1.0981e-9	9.2336e-1	1.1196e+0
13	4.7155e-10	9.7194e-1	1.0431e+0
14	4.7808e-12	9.9686e-1	1.0048e+0
15	3.8636e-14	1.0000e+0	1.0000e+0

4.8 Research projects

The equivalence between the complementarity problem $\mathrm{CP}(C, \mathbf{g})$ and an operator equation is established by Proposition 4.12 only for the standard positive cone. This eqvivalence should be extended for general positive cones. Another interesting research direction is to find an operator equation equivalent to a problem more general than the complementarity one: find a vector $x^* \in X \subset \mathbb{H}$ such that

$$\langle \mathbf{g}(x^*) \, , x - x^* \rangle \geq 0 \ , \ \forall \, x \in X \ . \tag{4.50}$$

This problem is known [26] as *variational inequality.*

For a scalar function f on $(0, \infty)$, Broyden's method reduces to the system of difference equations

$$x_+ := x - Af(x) \ , \ A_+ := -\frac{Af(x)}{f(x_+) - f(x)} \, ,$$

or, after the change of variables $y = Af(x)$,

$$x_+ := x + y \ , \ y_+ := y + \frac{f(x)}{f(x_+) - f(x)} \, .$$

Fixed points of this generator are zeroes of f. So, it is very tempting to determine its convergence domain.

The Riccati matrix equation

$$\mathbf{f}(x) := \frac{1}{2} xax + bx + xc + d = 0 \ , \ \forall \, x \in \mathbb{R}^{m \times n} \, ,$$

arises frequently in estimation and control problems and so was a great draw among researchers. However, all publications on the subject that I know use the language and tools of the matrix theory. Why not try to solve it iteratively, by Broyden's method, for example?

Chapter 5

Optimal secant updates of low rank

5.1 Motivation

Broyden's method is only one (though the one most widely known and used) representative of the broad class of iterative methods collectively called secant update methods. The methods of this class, given an initial pair $(x_0, \mathbf{A}_0) \in D \times \mathcal{L}(\mathbb{H})$ with invertible $\mathbf{A}_0$, generate the sequence $(x_n, \mathbf{A}_n) \in D \times \mathcal{L}(\mathbb{H})$ as follows:

$$x_+ := x - \mathbf{A}\mathbf{f}(x) \ , \ \mathbf{A}_+ := \mathbf{A} + \mathbf{B} \ , \tag{5.1}$$

where the update $\mathbf{B}$ is a linear operator of a finite rank (most often 1 or 2) such that $\mathbf{A}_+$ is invertible and satisfies the secant equation

$$\mathbf{A}_+^{-1}(x_+ - x) = \mathbf{f}(x_+) - \mathbf{f}(x) \,. \tag{5.2}$$

As we have seen in Chapter 3, there is a great variety of updates $\mathbf{B}$ satisfying this condition. This fact raises the question: which of them is most preferable? The answer to this question depends on a criterion enabling us to compare any two given updates and to decide which one is better than the other. The criteria that have been used for justifying Broyden's update hardly can serve in that capacity because they are unrelated to the purpose of the iterative method being designed: to locate a solution of the equation

$$\mathbf{f}(x) = 0 \,. \tag{5.3}$$

In this chapter, we use as such a criterion one that directly reflects this purpose: the entropy of the solution's position within a set of its guaranteed existence and uniqueness (see Section 1.4).

The goal of any method intended to solve the equation (5.3) is to reduce the uncertainty in the solution's whereabouts using the information obtained at one iteration. So, it is quite natural to borrow the notion of entropy to measure efficiency of iterative methods. A particular representative of a class of methods for solving operator equations is optimal, if its iteration reduces the uncertainty of the solution's position (measured by the entropy) as much as possible for the methods of this class. This optimality criterion was introduced in [21] and used in [22] to determine the most efficient secant-type methods.

To be able to speak about the entropy of the solution's position, one has to know a bounded subset of the space where a solution to be located exists and is unique. To determine such a set for a given operator is not a trivial matter. This task is dealt with in the next two sections. First, we need to obtain a convergence condition for the modified Newton method. This condition will be used in the existence and uniqueness theorem in Section 5.3.

5.2 Modified Newton method

This termi is usually used in the literature to refer to the following method:

$$x_+ := x - \mathbf{f}'(x_0)^{-1}\mathbf{f}(x) . \tag{5.4}$$

Paying attention to this method, we do not renege on the promise not to consider methods that involve inversion. This method is used solely as a theoretical tool. As in the previous chapter, we replace $\mathbf{f}$ by its normalization $\mathbf{f}'(x_0)^{-1}\mathbf{f}$ and examine the method (5.4) in the form

$$x_+ := x - \mathbf{f}(x) . \tag{5.5}$$

The current iteration $x \in D$ of the method (5.5) induces the pair $(\bar{t}, \bar{\delta})$ of reals

$$\bar{t} := \|x - x_0\| \ , \ \bar{\delta} := \|x_+ - x\| = \|\mathbf{f}(x)\| .$$

Obviously, $\bar{t}_+ := \|x_+ - x_0\| \le \bar{t} + \bar{\delta}$. The following lemma provides a similar bound for $\bar{\delta}_+ := \|x_{++} - x_+\| = \|\mathbf{f}(x_+)\|$.

Lemma 5.1. *Suppose that the (normalized) operator* $\mathbf{f}$ *in* (5.5) *is* ω*-regularly smooth relative to* x_0 *(Definition 2.2). Then*

$$\bar{\delta}_+ \le \Psi(\alpha, \bar{t} + \bar{\delta}) - \Psi(\alpha, \bar{t}) \ ,$$

where $\alpha := \omega^{-1}\big(\|\mathbf{f}'(x_0)\|\big) = \omega^{-1}(1)$ *and* Ψ *is the function* (2.9).

Proof. By the Newton–Leibnitz theorem,

$$\mathbf{f}(x_+) = \mathbf{f}(x) + \int_0^1 \mathbf{f}'\big(x + s(x_+ - x)\big)(x_+ - x)\, ds ,$$

where, by (5.5), $\mathbf{f}(x) = -(x_+ - x) = -\mathbf{f}'(x_0)(x_+ - x)$. So,

$$\mathbf{f}(x_+) = \int_0^1 \Big(\mathbf{f}'\big(x + s(x_+ - x)\big) - \mathbf{f}'(x_0)\Big)(x_+ - x) ds$$

and

$$\bar{\delta}_+ = \|\mathbf{f}(x_+)\| \le \int_0^1 \Big\|\mathbf{f}'\big(x + s(x_+ - x)\big) - \mathbf{f}'(x_0)\Big\| \bar{\delta}\, ds \, .$$

If $\mathbf{f}$ is ω-smooth relative to x_0, as assumed, then the norm in the integrand

$$\le \omega\big(\omega^{-1}(\min\big\{\big\|\mathbf{f}'\big(x + s(x_+ - x)\big)\big\|, \big\|\mathbf{f}'(x_0)\big\|\big\}) + \|x + s(x_+ - x) - x_0\|\big) \\ - \min\big\{\big\|\mathbf{f}'\big(x_n + s(x_+ - x)\big)\big\|\, ,\, \|\mathbf{f}'(x_0)\|\big\}$$
$$\le \omega\left(\min\left\{\omega^{-1}\big(\big\|\mathbf{f}'\big(x + s(x_+ - x)\big)\big\|\big)\, ,\, \alpha\right\} + \bar{t} + s\bar{\delta}\right) \\ - \omega\left(\min\left\{\omega^{-1}\big(\big\|\mathbf{f}'\big(x + s(x_+ - x)\big\|\big)\, ,\, \alpha\right\}\right). \quad (5.6)$$

By Lemma 2.3,

$$\omega^{-1}\big(\big\|\mathbf{f}'\big(x+s(x_+-x)\big)\big\|\big) \ge \big(\omega^{-1}(\|\mathbf{f}'(x_0)\|) - \|x+s(x_+-x)-x_0\|\big)^+ \ge (\alpha - \bar{t} - s\bar{\delta})^+,$$

$$\min\left\{\omega^{-1}\big(\big\|\mathbf{f}'\big(x+s(x_+-x)\big)\big\|\big)\, ,\, \alpha\right\} \ge (\alpha - \bar{t} - s\bar{\delta})^+ \, ,$$

and, taking into account the concavity of ω, the difference (5.6)

$$\le \omega\left(\big(\alpha - \bar{t} - s\bar{\delta}\big)^+ + \bar{t} + s\bar{\delta}\right) - \omega\left(\big(\alpha - \bar{t} - s\bar{\delta}\big)^+\right) \, .$$

Then

$$\bar{\delta}_+ \le \int_0^1 \left[\omega\left(\big(\alpha - \bar{t} - s\bar{\delta}\big)^+ + \bar{t} + s\bar{\delta}\right) - \omega\left(\big(\alpha - \bar{t} - s\bar{\delta}\big)^+\right)\right] \bar{\delta}\, ds$$
$$= \int_{\bar{t}}^{\bar{t}+\bar{\delta}} \big[\omega\big((\alpha - \tau)^+ + \tau\big) - \omega\big((\alpha - \tau)^+\big)\big]\, d\tau = \Psi\big(\alpha, \bar{t} + \bar{\delta}\big) - \Psi(\alpha, \bar{t}) \, .$$

□

The lemma suggests the following majorant generator:

$$t_+ := t + \delta \, , \quad \delta_+ := \Psi(\alpha, t + \delta) - \Psi(\alpha, t) \, . \quad (5.7)$$

For it, $t_{++} - \Psi(\alpha, t_+) = t_+ - \Psi(\alpha, t)$, i.e., the generator has an invariant:

$$\underset{n}{\&}\, t_{n+1} - \Psi(\alpha, t_n) = t_1 - \Psi(\alpha, t_0) \, .$$

In particular, $t_0 = 0 \Longrightarrow \underset{n}{\&}\, t_{n+1} - \Psi(\alpha, t_n) = \delta_0 \Longrightarrow t_{n+1} = t_n - \Phi(t_n)/\Phi'(t_0)$, where

$$\Phi(t) := \Psi(\alpha, t) - t + \delta_0 \, , \quad \forall\, t \ge 0 \, . \quad (5.8)$$

We see that the sequence t_n is obtained also by application of the method (5.5) to the convex function Φ starting from $t_0 = 0$. Therefore, it converges if and only if

$$0 \ge \min_{t \ge 0} \Phi(t) = \Phi(\alpha) = \Psi(\alpha\, , \alpha) - \alpha + \delta_0 = \alpha\, \omega(\alpha) - w(\alpha) - \alpha + \delta_0 = \delta_0 - w(\alpha) \, ,$$

that is, $\delta_0 \le w(\alpha)$. This condition guarantees convergence of the sequence x_n, generated by the method (5.5) from the starter x_0, to a solution of (5.3).

Proposition 5.2. *Suppose that the (normalized) operator* $\mathbf{f}$ *in* (5.5) *is* ω*-regularly smooth on* D *relative to* x_0 . *If*

$$\|\mathbf{f}(x_0)\| \le \delta_0 \le w(\alpha) \ , \tag{5.9}$$

then

1° *the sequence* t_n *generated by the generator* (5.7) *from the initial pair* $(0, \delta_0)$ *converges to* t_* *(the lesser of two zeroes of the function* (5.8)) *and*

$$\underset{n}{\&} \left(\bar{t}_n \le t_n \ \& \ \bar{\delta}_n \le \delta_n\right) \ , \tag{5.10}$$

2° *the sequence* x_n *generated by the method* (5.5) *from* x_0 *remains in the ball* $B(x_0, t_*)$ *and converges to a solution* x_∞ *of* (5.3).

Proof. The first claim is proved above. The proof of the second follows the standard argument (see, for example, the proof of Lemma 4.4, 3°). □

5.3 Existence and uniqueness of solutions (regular smoothness)

The question of solvability of a given operator equation makes sense (and is interesting) in general Banach spaces. Therefore, it is advisable to look for an answer in the Banach space setting. Having this in mind, we assume here that the operator $\mathbf{f}$ acts from a subset D of one Banach space, $\mathbb{X}$, into another, $\mathbb{Y}$.

Let $\mathbf{f}$ be differentiable and its derivative $\mathbf{f}'(x_0)$ at a point $x_0 \in D$ be boundedly invertible. Suppose that the normalized operator $\mathbf{f}'(x_0)^{-1}\mathbf{f}$ (which acts on $\mathbb{X}$) is ω-regularly smooth on D relative to x_0 and denote:

$$\delta_0 := \left\|\mathbf{f}'(x_0)^{-1}\mathbf{f}(x_0)\right\| \ , \ \alpha := \omega^{-1}\left(\|\mathbf{f}'(x_0)\|\right) ,$$

$$p(\alpha\,, t) := t\omega(\alpha) + \Psi(\alpha\,, t) \ , \ q(\alpha\,, t) := t\omega(\alpha) - \Psi(\alpha\,, t).$$

The function $t \mapsto p(\alpha\,, t)$ is increasing in $[0, \infty)$ from 0 to ∞ and so the equation $p(\alpha\,, t) = \delta_0$ has the unique t-solution for any $\delta_0 \ge 0$ (let us refer to it as $p^{-1}(\alpha\,, \delta_0)$). As to $t \mapsto q(\alpha\,, t)$, it is increasing in $[0, \alpha]$ and decreasing in $[\alpha\,, \infty)$. To see it, consider the derivative

$$\frac{\partial q}{\partial t}(\alpha, t) = \omega(\alpha) - \frac{\partial \Psi}{\partial t}(\alpha, t) = \omega(\alpha) - \begin{cases} \omega(\alpha) - \omega(\alpha - t) \ , \ \text{if } \ 0 \le t < \alpha \ , \\ \omega(t) \ , \ \text{if } \ t > \alpha \ , \end{cases}$$
$$= \begin{cases} \omega(\alpha - t) > 0 \ , \ \text{if } \ 0 \le t < \alpha \ , \\ \omega(\alpha) - \omega(t) \le 0 \ , \ \text{if } \ t > \alpha \ . \end{cases}$$

So, either the equation $q(\alpha, t) = \delta_0$ has no t-solutions, if $q(\alpha, \alpha) < \delta_0$, or else it has two $q_-^{-1}(\alpha, \delta_0)$ and $q_+^{-1}(\alpha, \delta_0)$, where $q_-(\alpha, \cdot)$ and $q_+(\alpha, \cdot)$ are the restrictions of $q(\alpha, \cdot)$ to $[0, \alpha]$ and $[\alpha, \infty)$, respectively ($q_-^{-1}(\alpha, \delta_0) = q_+^{-1}(\alpha, \delta_0) = \alpha \iff q(\alpha, \alpha) = \delta_0$). The function $q(\alpha, \cdot)$ is ω-regularly smooth in $[0, \alpha]$. Indeed, if $0 \le t < t' \le \alpha$, then

$$\begin{aligned}
&\omega^{-1}\Big(\min\{|q'(\alpha, t)|, q'(\alpha, t')|\} + \big|q'(\alpha, t) - q'(\alpha, t')\big|\Big) - \\
&\qquad\qquad\qquad \omega^{-1}\Big(\min\{|q'(\alpha, t)|, |q'(\alpha, t')|\}\Big) \\
&= \omega^{-1}\Big(\min\{\omega(\alpha - t), \omega(\alpha - t')\} + \big|\omega(\alpha - t) - \omega(\alpha - t')\big|\Big) - \\
&\qquad\qquad\qquad \omega^{-1}\Big(\min\{\omega(\alpha - t), \omega(\alpha - t')\}\Big) \\
&= \omega^{-1}\Big(\omega(\alpha - t') + \omega(\alpha - t) - \omega(\alpha - t')\Big) - \omega^{-1}(\omega(\alpha - t')) \\
&= (\alpha - t) - (\alpha - t') = t' - t\,.
\end{aligned}$$

Note also that $q(\alpha, \alpha) = w(\alpha)$, where (to recall (2.8)) $w(t) := \int_0^t \omega(s)\,ds$.

Theorem 5.3. 1° *If the equation* $\mathbf{f}(x) = 0$ *has a solution* $x_* \in D$, *then*

$$\|x_* - x_0\| \ge p^{-1}(\alpha, \delta_0)\,.$$

2° *If it has a solution* $x_* \in D$ *and* $q(\alpha, \alpha) \ge \delta_0$, *then*

$$p^{-1}(\alpha, \delta_0) \le \|x_* - x_0\| \le q_-^{-1}(\alpha, \delta_0) \tag{5.11}$$

and this solution is unique in the ball $B\big(x_0, q_+^{-1}(\alpha, \delta_0)\big)$.

3° *If* $q(\alpha, \alpha) \ge \delta_0$, *then the equation* $\mathbf{f}(x) = 0$ *has a solution in the set*

$$D(x_0) := \big\{x \in D \;\big|\; p^{-1}(\alpha, \delta_0) \le \|x - x_0\| \le q_-^{-1}(\alpha, \delta_0)\big\}\,. \tag{5.12}$$

4° *The existence radius* $q_-^{-1}(\alpha, \delta_0)$ *is sharp: it is attained for the* (ω*-regularly smooth in* $[0, \alpha]$) *function* $t \mapsto q(\alpha, t) - \delta_0$ *and* $t_0 = 0$.

Proof. 1°, 2°. As in the preceding section, we mean here by $\mathbf{f}$ the normalized operator $\mathbf{f}'(x_0)^{-1}\mathbf{f}$. By the Newton–Leibnitz theorem,

$$\begin{aligned}
0 &= \mathbf{f}(x_*) \qquad\qquad (5.13)\\
&= \mathbf{f}(x_0) + \mathbf{f}'(x_0)(x_* - x_0) + \int_0^1 \Big[\mathbf{f}'(x_0 + t(x_* - x_0)) - \mathbf{f}'(x_0)\Big](x_* - x_0)dt
\end{aligned}$$

and so (with the abbreviation $\Delta := \|x_* - x_0\|$ and the agreement $\mathbf{f}'(x_0) = \mathbf{I}$ in mind)

$$|\delta_0 - \Delta| \le \|\mathbf{f}(x_0) + \mathbf{f}'(x_0)(x_* - x_0)\| \le \int_0^1 \|\mathbf{f}'(x_0 + t(x_* - x_0)) - \mathbf{f}'(x_0)\|\,\Delta\,dt\,.$$

Because of the ω-regular smoothness of $\mathbf{f}$, the norm in the integrand

$$\le \omega\Big(\omega^{-1}(\min\{\|\mathbf{f}'(x_0)\|\,,\|\mathbf{f}'(x_0+t(x_*-x_0))\|\})+t\Delta\Big)$$

$$-\min\{\|\mathbf{f}'(x_0)\|\,,\|\mathbf{f}'(x_0+t(x_*-x_0))\|\}$$

$$=\omega\Big(\min\{\alpha\,,\,\omega^{-1}(\|\mathbf{f}'(x_0+t(x_*-x_0))\|)\}+t\Delta\Big)$$

$$-\omega\Big(\min\{\alpha\,,\,\omega^{-1}(\|\mathbf{f}'(x_0+t(x_*-x_0))\|)\}\Big) \tag{5.14}$$

By Lemma 2.3, $\omega^{-1}(\|\mathbf{f}'(x_0+t(x_*-x_0))\|)\ge(\alpha-t\Delta)^+$ and so (due to the concavity of ω) the difference (5.14)

$$\le \omega\left(\min\{(\alpha-t\Delta)^+\,,\,\alpha\}+t\Delta\right)-\omega\left(\min\{(\alpha-t\Delta)^+\,,\,\alpha\}\right)$$
$$=\omega\big((\alpha-t\Delta)^++t\Delta\big)-\omega\big((\alpha-t\Delta)^+\big)=e(\alpha\,,t\Delta)\,.$$

Hence,

$$|\delta_0-\Delta|\le\int_0^1 e(\alpha\,,t\Delta)\Delta\,dt=\int_0^\Delta e(\alpha\,,t)\,dt=\Psi(\alpha\,,\Delta)\,,$$

or, equivalently,

$$p(\alpha\,,\Delta)\ge\delta_0\ge q(\alpha\,,\Delta)\,.$$

The first of these two inequalities can be rewritten as $\Delta\ge p^{-1}(\alpha\,,a)$, while the second means that

$$q(\alpha\,,\alpha)<\delta_0\bigvee\Big(q(\alpha\,,\alpha)\ge\delta_0\ \ \&\ \ \Big(\Delta\le q_-^{-1}(\alpha\,,\delta_0)\bigvee\Delta\ge q_+^{-1}(\alpha\,,\delta_0)\Big)\Big)\,.$$

If $q(\alpha\,,\alpha)<\delta_0$, then the equation $\mathbf{f}(x)=0$ may have no solutions in D, as the example of the function $t\mapsto q(\alpha\,,t)-\delta_0$ shows. In this case, the inequality $q(\alpha\,,\Delta)\le\delta_0$ contains no information about Δ, so that we are left only with the lower bound (5.11). If $q(\alpha\,,\alpha)\ge a$, then $p^{-1}(\alpha\,,\delta_0)\le\Delta\le q_-^{-1}(\alpha\,,\delta_0)$ and x_* is the only solution in the ball $B\big(x_0,q_+^{-1}(\alpha\,,\delta_0)\big)$.

3° As Proposition 5.2 states, if $w(\alpha)\ge\|\mathbf{f}(x)\|$ ($q(\alpha\,,\alpha)\ge a$ in our notations), then the sequence x_n generated by the method (5.5) from the starter x_0 converges to a solution x_∞ of the equation $\mathbf{f}(x)=0$, which, by 2°, has to be in the set (5.12).

4° For $t\in[0,\alpha]$,

$$q(\alpha,t)-\delta_0=0\Longleftrightarrow\delta_0\le w(\alpha)\ \ \&\ \ t=q_-^{-1}(\alpha,\delta_0)=\alpha-w^{-1}(w(\alpha)-\delta_0)\,.$$

□

The smoothness assumption makes this theorem inapplicable to equations with nondifferentiable operators. This shortcoming makes interesting also the following theorem, which, instead of regular smoothness of $\mathbf{f}$, requires only regular continuity of divided differences. Its proof depends on the convergence domain of the modified secant method.

5.4 Modified secant method

This term refers to the following method:

$$x_+ := x - [x_0, x_{-1} \,|\, \mathbf{f}]^{-1}\mathbf{f}(x)\,.$$

After normalization of **f**, it becomes

$$x_+ := x - \mathbf{f}(x)\,. \tag{5.15}$$

We analyze it here under the assumption that the selected divided difference of **f** is ω-regularly continuous on D relative to x_0 (Definition 3.3). As for the modified Newton method, we associate with the current iteration $x \in D$ the pair $(\bar{t}\,,\bar{\delta})$:

$$\bar{t} := \|x - x_0\|\ ,\ \bar{\delta} := \|x_+ - x\| = \big\|\mathbf{f}(x)\big\|\,.$$

The next lemma relates $(\bar{t}_+\,,\bar{\delta}_+)$ with $(\bar{t}\,,\bar{\delta})$.

Lemma 5.4. 1° $\bar{t}_+ \le \bar{t} + \bar{\delta}$.

2° *If the selected dd* $[x_1, x_2 \,|\, \mathbf{f}]$ *of the (normalized) operator* **f** *is* ω*-regularly continuous relative to* x_0*, then*

$$\bar{\delta}_+ \le \bar{\delta}\Big(\omega\big((\bar{a} - \bar{t}_+ - \bar{t}\,)^+ + \bar{\gamma} + \bar{t}_+ + \bar{t}\,\big) - \omega\big((\bar{a} - \bar{t}_+ - \bar{t}\,)^+\big)\Big)\,,$$

where $\bar{\gamma} := \|x_0 - x_{-1}\|$ *and* $\bar{a} := \omega^{-1}(1 - \underline{h}) - \bar{\gamma}$.

Proof. By the secant equation, $\mathbf{f}(x_+) - \mathbf{f}(x) = [x_+, x \,|\, \mathbf{f}](x_+ - x)$, so that

$$\begin{aligned}\bar{\delta}_+ &= \big\|\mathbf{f}(x_+)\big\| = \big\|\mathbf{f}(x_+) - \mathbf{f}(x) + \mathbf{f}(x)\big\| = \big\|[x_+, x \,|\, \mathbf{f}](x_+ - x) - (x_+ - x)\big\| \\ &\le \bar{\delta}\big\|[x_+, x \,|\, \mathbf{f}] - \mathbf{I}\big\| = \bar{\delta}\big\|[x_+, x \,|\, \mathbf{f}] - [x_0, x_{-1} \,|\, \mathbf{f}]\big\|,\end{aligned}$$

where, by Lemma 3.8, $\big\|[x_+, x \,|\, \mathbf{f}] - [x_0, x_{-1} \,|\, \mathbf{f}]\big\|$

$$\begin{aligned}&\le \omega\Big(\min\{\omega^{-1}(\big\|[x_+, x \,|\, \mathbf{f}]\big\| - \underline{h})\,, \omega^{-1}(\big\|[x_0, x_{-1} \,|\, \mathbf{f}]\big\| - \underline{h})\} + \\ &\qquad\qquad \|x_+ - x_0\| + \|x - x_{-1}\|\Big) - \\ &\quad \omega\Big(\min\{\omega^{-1}(\big\|[x_+, x \,|\, \mathbf{f}]\big\| - \underline{h})\,, \omega^{-1}(\big\|[x_0, x_{-1} \,|\, \mathbf{f}]\big\| - \underline{h})\}\Big)\,.\end{aligned}$$

In view of (3.14) and (4.5),

$$\begin{aligned}\omega^{-1}(\big\|[x_+, x \,|\, \mathbf{f}]\big\| - \underline{h}) &\ge \big(\omega^{-1}(1 - \underline{h}) - \|x_+ - x_0\| - \|x - x_{-1}\|\big)^+ \\ &\ge \big(\omega^{-1}(1 - \underline{h}) - \|x_+ - x_0\| - \|x - x_0\| - \|x_0 - x_{-1}\|\big)^+ \\ &= (\bar{a} - \bar{t}_+ - \bar{t})^+\,.\end{aligned}$$

So, because of the concavity and monotonicity of ω, $\big\|[x_+, x\,|\,\mathbf{f}] - [x_0, x_{-1}\,|\,\mathbf{f}]\big\|$

$$\le \omega\Big(\min\{(\bar{a} - \bar{t}_+ - \bar{t}\,)^+, \omega^{-1}(1 - \underline{h})\} + \bar{t}_+ + \bar{t} + \|x_0 - x_{-1}\|\Big) - \omega\Big(\min\{(\bar{a} - \bar{t}_+ - \bar{t}\,)^+, \omega^{-1}(1 - \underline{h})\}\Big)$$

$$= \omega\Big((\bar{a} - \bar{t}_+ - \bar{t}\,)^+ + \bar{t}_+ + \bar{t} + \bar{\gamma}\Big) - \omega\Big((\bar{a} - \bar{t}_+ - \bar{t}\,)^+\Big).$$

Therefore, $\bar{\delta}_+ \le \bar{\delta}\Big(\omega\big((\bar{a} - \bar{t}_+ - \bar{t}\,)^+ + \bar{t}_+ + \bar{t} + \bar{\gamma}\big) - \omega\big((\bar{a} - \bar{t}_+ - \bar{t}\,)^+\big)\Big)$. □

The lemma suggests the following majorant generator:

$$t_+ := t + \delta\,,\ \delta_+ := \delta\Big(\omega\big((a - 2t - \delta\,)^+ + 2t + \delta + \gamma\big) - \omega\big((a - 2t - \delta\,)^+\big)\Big), \tag{5.16}$$

where γ and a are an upper bounds for $\bar{\gamma}$ and a lower bound for $\bar{a}$, respectively.

Proposition 5.5.

1° *The sequence* (t_n, δ_n), *generated by the generator* (5.16) *from a starter* $(0\,, \delta_0)$, *converges (to* $(t_\infty, 0)$, $t_\infty < \infty$*) if and only if* $\delta_0 \le f_\infty(0)$, *where* $f_0(t) := t_\infty - t$ *and* $f_{n+1}(t)$ *is the (unique) solution for* δ *of the equation*

$$f_n(t + \delta) = \delta\Big(\omega\big((a - 2t - \delta)^+ + \gamma + 2t + \delta\big) - \omega\big((a - 2t - \delta)^+\big)\Big).$$

2° *The function* f_∞ *is the only nonzero solution of the system*

$$x\big(t + x(t)\big) = x(t)\Big(\omega\big((a - 2t - x(t))^+ + \gamma + 2t + x(t)\big) - \omega\big((a - 2t - x(t))^+\big)\Big),$$

$$x(t_\infty) = 0\,. \tag{5.17}$$

Proof. 1° Let $t_\infty < \infty$. Then $t_{n+1} < t_\infty \iff \delta_n < t_\infty - t_n =: f_0(t_n)$. Suppose that, for some $k \ge 0$, $t_{n+1} < t_\infty \iff \delta_{n-k} < f_k(t_{n-k})$, where f_k is decreasing and positive in $[0\,, t_\infty]$ and vanishes at t_∞. Let

$$F_k(t, \delta) := f_k(t + \delta) - \delta\Big(\omega\big((a - 2t - \delta)^+ + \gamma + 2t + \delta\big) - \omega\big((a - 2t - \delta)^+\big)\Big)$$

and (using (5.16)) rewrite the last inequality as

$$F_k(t_{n-k-1}\,, \delta_{n-k-1}) > 0\,. \tag{5.18}$$

As f_k is decreasing, F_k is decreasing in $\delta \in [0\,, t_\infty - t]$ from $F_k(t\,, 0) =$

$f_k(t) > 0$ to

$$\begin{aligned} &F_k(t\,, t_\infty - t) \\ &= f_k(t_\infty) - (t_\infty - t)\Big(\omega\big((a - 2t - t_\infty + t)^+ + \gamma + 2t + t_\infty - t\big) - \\ &\qquad\qquad \omega\big((a - 2t - t_\infty + t)^+\big)\Big) \\ &= -(t_\infty - t)\Big(\omega\big((a - t_\infty - t)^+ + \gamma + t_\infty + t\big) - \omega\big((a - t_\infty - t)^+\big)\Big) < 0\,. \end{aligned}$$

Therefore, the equation $F_k(t\,,\delta) = 0$ is uniquely solvable for $\delta \in (0\,, t_\infty - t)$. We denote the solution $f_{k+1}(t)$:

$$F_k\big(t\,, f_{k+1}(t)\big) = 0\,. \tag{5.19}$$

By definition, $f_{k+1}(t) \in (0\,, t_\infty - t)$, so that $f_{k+1}(t) > 0$ in $(0\,, t_\infty - t)$ and $f_{k+1}(t_\infty) = 0$. Besides, it is also decreasing there. Indeed, since F_k is decreasing in each of its two arguments,

$$\begin{aligned} 0 < t < t' < t_\infty &\Longrightarrow F_k\big(t\,, f_{k+1}(t)\big) = 0 = F_k\big(t'\,, f_{k+1}(t')\big) < F_k\big(t\,, f_{k+1}(t')\big) \\ &\Longrightarrow f_{k+1}(t) > f_{k+1}(t')\,. \end{aligned}$$

Thus, $t_{n+1} < t_\infty \iff \delta_{n-k} < f_k(t_{n-k})$ implies $t_{n+1} < t_\infty \iff \delta_{n-k-1} < f_{k+1}(t_{n-k-1})$. By induction, $t_{n+1} < t_\infty \iff \delta_0 < f_n(0)$ and $\underset{n}{\&}\, t_n < t_\infty \iff \delta_0 < \inf\limits_n f_n(0)$. The sequence f_n is pointwise decreasing:

$$\underset{n}{\&}\, f_{n+1}(t) < f_n(t)\ ,\ \forall\, t \in [0\,, t_\infty]\,. \tag{5.20}$$

This is verified inductively. First, we have to show that $f_1(t) < f_0(t)$ or, as F_0 is decreasing in δ, that $F_0\big(t, f_0(t)\big) < 0 = F_0\big(t\,, f_1(t)\big)$. By definition, $F_0\big(t\,, f_0(t)\big) = f_0\big(t + f_0(t)\big) -$

$$\begin{aligned} &f_0(t)\Big(\omega\big((a - 2t - f_0(t))^+ + \gamma + 2t + f_0(t)\big) - \omega\big((a - 2t - f_0(t))^+\big)\Big) \\ &= f_0(t_\infty) - (t_\infty - t)\Big(\omega\big((a - t_\infty - t)^+ + \gamma + t_\infty + t\big) - \omega\big((a - t_\infty - t)^+\big)\Big) \\ &= -(t_\infty - t)\Big(\omega\big((a - t_\infty - t)^+ + \gamma + t_\infty + t\big) - \omega\big((a - t_\infty - t)^+\big)\Big) < 0\,. \end{aligned}$$

Suppose that, for some $n \geq 1$, $f_n(t) < f_{n-1}(t)\,, \forall\, t \in [0\,, t_\infty]$. Then, by (5.19) and (5.20),

$$F_{n-1}\big(t, f_n(t)\big) = 0 = F_n\big(t, f_{n+1}(t)\big)$$

$$= f_n\big(t + f_{n+1}(t)\big) - f_{n+1}(t)\Big(\omega\big((a - 2t - f_{n+1}(t))^+ + \gamma + 2t + f_{n+1}(t)\big) - \omega\big((a - 2t - f_{n+1}(t))^+\big)\Big)$$

$$< f_{n-1}\big(t + f_{n+1}(t)\big) - f_{n+1}(t)\Big(\omega\big((a - 2t - f_{n+1}(t))^+ + \gamma + 2t + f_{n+1}(t)\big) - \omega\big((a - 2t - f_{n+1}(t))^+\big)\Big)$$

$$= F_{n-1}\big(t, f_{n+1}(t)\big)\,.$$

As F_{n-1} is decreasing in δ, $F_{n-1}\big(t, f_n(t)\big) < F_{n-1}\big(t, f_{n+1}(t)\big) \Longrightarrow f_n(t) > f_{n+1}(t)$. By induction, (5.20) is proved. It follows that $\inf_n f_n(u) = f_\infty(u)$. Taking limits in (5.19) and $f_n(t_\infty) = 0$ yields $F_\infty\big(t, f_\infty(t)\big) = 0$ & $f_\infty(t_\infty) = 0$, that is, f_∞ solves the system (5.17). Moreover, f_∞ is the only nonzero solution. To see it, let $x(t)$ be a nonzero solution and consider the generator $\mathbf{g} : (t, \delta) \mapsto (t_+, \delta_+)$ defined as follows:

$$t_+ := t + x(t)\ ,\ \delta_+ := x\big(t + x(t)\big)\,.$$

Then $\delta = x(t) \Longrightarrow t_+ = t + \delta$ &

$$\delta_+ = x(t)\Big(\omega\big((a - 2t - x(t))^+ + \gamma + 2t + x(t)\big) - \omega\big((a - 2t - x(t))^+\big)\Big)$$

$$= \delta\Big(\omega\big((a - 2t - \delta)^+ + \gamma + 2t + \delta\big) - \omega\big((a - 2t - \delta)^+\big)\Big)$$

and $\delta_+ = x(t + \delta) = x(t_+)$. Hence, $\mathbf{g}$ coincides with (5.16). It follows that $x = f_\infty$. □

The convergence theorem for the modified secant method (5.15) is obtained now as convergence theorems for other methods we have been busy at already.

Proposition 5.6. *Suppose that the selected dd* $[x_1, x_2 \,|\, \mathbf{f}]$ *of the (normalized) operator* $\mathbf{f}$ *in* (5.15) *is* ω*-regularly continuous on* D *relative to* x_0 *. If*

$$\|\mathbf{f}(x_0)\| \le \delta_0 \le f_\infty(0)\,,$$

where f_∞ *is the function of Proposition 5.5, then*

1° *the sequence* (t_n, δ_n)*, generated by the generator* (5.16) *from a starter* $(0\,, \delta_0)$*, converges (to* $(t_\infty, 0)$*,* $t_\infty < \infty$*);*

2° *the sequence* x_n, *generated by the method* (5.15) *from a starter* x_0, *converges to a solution* x_∞ *of the equation* $\mathbf{f}(x) = 0$;

3° $\underset{n}{\&} \|x_\infty - x_n\| \le t_\infty - t_n$.

When ω is linear: $\omega(t) = ct$, the generator (5.16) becomes

$$t_+ := t + \delta \ , \ \delta_+ := c\delta(2t + \delta + \gamma)\,. \tag{5.21}$$

For it, we have

Lemma 5.7. 1° *The function* $I(t, \delta) := t^2 - (c^{-1} - \gamma)t - c^{-1}\delta$ *is an invariant of the generator* (5.21).

2° *The sequence* $(t_n\,, \delta_n)$, *generated by the generator* (5.21) *from a starter* $(0\,, \delta_0)$, *converges* (*to* $(t_\infty, 0)\,, t_\infty < \infty$) *if and only if*

$$4c^{-1}\delta_0 \le \left(c^{-1} - \gamma\right)^2.$$

In this case,

$$\underset{n}{\&}\, t_n = \frac{1}{2}\left(c^{-1} - \gamma - \sqrt{\left(c^{-1} - \gamma\right)^2 - 4c^{-1}(\delta_0 - \delta_n)}\,\right) =: f(\delta_n)\,.$$

3° *The function* f *is the only continuous solution of the system*

$$x\big(c\delta(2x(\delta) + \delta + \gamma)\big) = x(\delta) + \delta \quad \& \quad x(0) = t_\infty\,. \tag{5.22}$$

Proof. 1°

$$\begin{aligned} I(t_+\,, \delta_+) &= t_+^2 - (c^{-1} - \gamma)t_+ - c^{-1}\delta_+ \\ &= (t + \delta)^2 - (c^{-1} - \gamma)(t + \delta) - c^{-1}c\delta(2t + \delta + \gamma) \\ &= t^2 + 2t\delta + \delta^2 - (c^{-1} - \gamma)t - (c^{-1} - \gamma)\delta - 2t\delta - \delta^2 - \gamma\delta \\ &= t^2 - (c^{-1} - \gamma)t - c^{-1}\delta = I(t\,, \delta)\,. \end{aligned}$$

2° By 1°, $\underset{n}{\&}\, I(t_n\,, \delta_n) = I(t_\infty\,, 0)$. In particular $I(0\,, \delta_0) = I(t_\infty\,, 0)$, that is, $t_\infty^2 - (c^{-1} - \gamma)t_\infty = -c^{-1}\delta_0$. This equation is solvable for t_∞ if and only if $4c^{-1}\delta_0 \le \left(c^{-1} - \gamma\right)^2$. Under this condition, the equation $I(t_n\,, \delta_n) = I(t_0\,, \delta_0)$ can be solved for t_n. The solution is

$$t_n = \frac{1}{2}\left(c^{-1} - \gamma - \sqrt{\left(c^{-1} - \gamma\right)^2 - 4c^{-1}\left(\delta_0 - \delta_n - t_0^2 + (c^{-1} - \gamma)t_0\right)}\,\right).$$

In particular, $t_0 = 0 \Longrightarrow t_n = 0.5\left(c^{-1} - \gamma - \sqrt{\left(c^{-1} - \gamma\right)^2 - 4c^{-1}\left(\delta_0 - \delta_n\right)}\,\right)$.

3° First, let us see that, no matter what the constant α is, the function

$$f(\delta) := \frac{1}{2}\left(c^{-1} - \gamma - \sqrt{\left(c^{-1} - \gamma\right)^2 + 4c^{-1}\delta - \alpha}\,\right)$$

satisfies the functional equation (5.22). To shorten the ensuing calculations, denote the last radical by the symbol sq: $sq := \sqrt{\left(c^{-1}-\gamma\right)^2 + 4c^{-1}\delta - \alpha}$. Then

$$
\begin{aligned}
& f(\delta)=0.5\left(c^{-1}-\gamma-sq\right) \\
\Longrightarrow\ & c\delta\left(2f(\delta)+\delta+\gamma\right) = c\delta\left(c^{-1}-sq+\delta\right) \\
\Longrightarrow\ & f\left(c\delta\left(2f(\delta)+\delta+\gamma\right)\right) = f\left(c\delta\left(c^{-1}-sq+\delta\right)\right) \\
& = \frac{1}{2}\left(c^{-1}-\gamma-\sqrt{\left(c^{-1}-\gamma\right)^2+4\delta\left(c^{-1}-sq+\delta\right)-\alpha}\right).
\end{aligned}
$$

On the other hand, $f(\delta)+\delta=0.5\left(c^{-1}-\gamma-sq\right)+\delta$. So,

$$
\begin{aligned}
& f\left(c\delta\left(2f(\delta)+\delta+\gamma\right)\right) = f(\delta)+\delta \\
\Longleftrightarrow\ & c^{-1}-\gamma-\sqrt{\left(c^{-1}-\gamma\right)^2+4\delta\left(c^{-1}-sq+\delta\right)-\alpha} = c^{-1}-\gamma-sq+2\delta \\
\Longleftrightarrow\ & \sqrt{\left(c^{-1}-\gamma\right)^2+4\delta\left(c^{-1}-sq+\delta\right)-\alpha} = sq-2\delta \\
\Longleftrightarrow\ & \left(c^{-1}-\gamma\right)^2+4\delta\left(c^{-1}-sq+\delta\right)-\alpha \\
& \qquad = \left(c^{-1}-\gamma\right)^2+4c^{-1}\delta-\alpha-4\delta\, sq+4\delta^2,
\end{aligned}
$$

which is true. The end condition in (5.22) determines the value of α:

$$
\begin{aligned}
f(0)=t_\infty \Longleftrightarrow\ & \frac{1}{2}\left(c^{-1}-\gamma-\sqrt{\left(c^{-1}-\gamma\right)^2-\alpha}\right)=t_\infty \\
\Longleftrightarrow\ & \sqrt{\left(c^{-1}-\gamma\right)^2-\alpha}=c^{-1}-\gamma-2t_\infty \\
\Longleftrightarrow\ & \left(c^{-1}-\gamma\right)^2-\alpha=\left(c^{-1}-\gamma\right)^2-4t_\infty\left(c^{-1}-\gamma\right)+4t_\infty^2 \\
\Longleftrightarrow\ & \alpha=4t_\infty\left(c^{-1}-\gamma\right)+4t_\infty^2=4c^{-1}\delta_0\,.
\end{aligned}
$$

Thus, f is a solution of (5.22). To see that there is no other solution, let x be a solution and consider the generator $\mathbf{g}:(u,v)\to(u_+,v_+)$ defined as follows:

$$
u_+ := x(v)+v\ ,\quad v_+ := cv\left(2x(v)+v+\gamma\right).
$$

Then $(u,v)=(t,\delta)\ \&\ u=x(v)\Longrightarrow u=x(\delta)=t$,

$$
v_+ := cv\left(2x(v)+v+\gamma\right) = c\delta\left(2x(\delta)+\delta+\gamma\right) = c\delta\left(2t+\delta+\gamma\right)=\delta_+\,,
$$

and

$$
u_+ := x(v)+v = x(\delta)+\delta = x\left(c\delta\left(2x(\delta)+\delta+\gamma\right)\right) = x(\delta_+) = x(v_+)\,.
$$

At the same time, $u_+ := x(v) + v = x(\delta) + \delta = t + \delta = t_+$. Thus,

$$(u,v) = (t,\delta) \ \ \& \ \ u = x(v) \Longrightarrow (u_+,v_+) = (t_+,\delta_+) \ \ \& \ \ u_+ = x(v_+)\,.$$

This means that the generator $\mathbf{g}$ coincides with (5.21). It follows that $x = f$. □

Application of this lemma leads to the following corollary of Proposition 5.6:

Corollary 5.8. *Suppose that the selected dd $[x_1,x_2\,|\,\mathbf{f}]$ of the operator $\mathbf{f}$ in (5.15) is Lipschitz continuous on D relative to x_0:*

$$\big\|[x_1,x_2\,|\,\mathbf{f}]-[x_0,x_0\,|\,\mathbf{f}]\big\| \le c\big(\|x_1-x_0\|+\|x_2-x_0\|\big)\ ,\ \forall\, x_1,x_2\in D.$$

If

$$\|\mathbf{f}(x_0)\| \le \delta_0 \le c\left(\frac{c^{-1}-\gamma}{2}\right)^2 ,$$

then

1° *the sequence (t_n,δ_n), generated by the generator (5.21) from a starter $(0\,,\delta_0)$, converges to $t_\infty = 0.5\Big(c^{-1}-\gamma-\sqrt{(c^{-1}-\gamma)^2-4c^{-1}\delta_0}\Big)$ and $\delta_\infty = 0$;*

2° *the sequence x_n, generated by the method (5.15) from a starter x_0, converges to a solution x_∞ of the equation $\mathbf{f}(x) = 0$;*

3° $\underset{n}{\&}\ \|x_\infty - x_n\| \le \dfrac{2\delta_n}{\sqrt{(c^{-1}-\gamma)^2-4c^{-1}(\delta_0-\delta_n)}+\sqrt{(c^{-1}-\gamma)^2-4c^{-1}\delta_0}}\,.$

5.5 Existence and uniqueness of solutions (regular continuity).

Suppose that $\mathbf{f}$ is continuous on $D \subset \mathbb{X}$ and its selected dd $[x_1,x_2\,|\,\mathbf{f}]$ is boundedly invertible at x_0, x_{-1}: $\exists\,[x_0,x_{-1}\,|\,\mathbf{f}]^{-1} \in \mathcal{L}(\mathbb{Y},\mathbb{X})$ (the space of bounded linear operators acting from $\mathbb{Y}$ into $\mathbb{X}$). Then again we can conveniently assume (without loss of generality) that $\mathbf{f}$ is already normalized:

$$[x_0,x_{-1}\,|\,\mathbf{f}] = \mathbf{I}\,. \tag{5.23}$$

In what follows we use the abbreviations

$$\alpha_0 := \|\mathbf{f}(x_0)\|\ ,\quad \delta_0 := \|x_0 - x_{-1}\|\,.$$

Let the selected dd $[x_1,x_2\,|\,\mathbf{f}]$ of $\mathbf{f}$ be ω-regularly continuous on D in the sense of (3.8) with $\underline{h}=0$ (for simplicity) and denote $\omega^{-1}(1)-\delta_0$ by a (for brevity):

$$a := \omega^{-1}(1) - \delta_0 \,.$$

The existence and uniqueness set is described below through the functions

$$p(t) := t\Big(1+\omega\big((a-t)^+ + \delta_0 + t\big) - \omega\big((a-t)^+\big)\Big) = \begin{cases} t(2-\omega(a-t)) \,, \\ \qquad\qquad \text{if } \ 0 \le t \le a \,, \\ t(1+\omega(\delta_0+t)) \,, \\ \qquad\qquad \text{if } \ t \ge a \,, \end{cases}$$

and

$$q(t) := t\Big(1-\omega\big((a-t)^+ + \delta_0 + t\big) + \omega\big((a-t)^+\big)\Big) = \begin{cases} t\omega(a-t) \,, \text{ if } \ 0 \le t \le a \,, \\ t(1-\omega(\delta_0+t)) \,, \text{ if } \ t \ge a \,. \end{cases}$$

As p is increasing in $[0,\infty)$ from 0 to ∞, the equation $p(t)=\alpha_0$ is uniquely solvable for t whatever $\alpha_0 > 0$: $t = p^{-1}(\alpha_0)$. As to q, it is increasing in $[0,\tau)$, where τ is some point in $(0\,,a)$, and decreasing in $(\tau\,,\infty)$. To see it, consider the derivative

$$q'(t) = \begin{cases} \omega(a-t) - t\,\omega'(a-t) \,, & \text{if } \ 0 \le t < a \,, \\ 1-\omega(\delta_0+t) - t\,\omega'(\delta_0+t) \,, & \text{if } \ t > a \,. \end{cases}$$

In $[0\,,a)$, it is decreasing from $\omega(a) > 0$ to $-a\omega'(0) < 0$. So, there is a $\tau \in (0\,,a)$ such that $q'(t) > 0$, $\forall\, t \in (0,\tau)$, and $q'(t) < 0$, $\forall\, t > \tau$. Hence, q is increasing in $[0,\tau)$ from zero to $\tau\omega(a-\tau)$ and decreasing in $(\tau,a]$. Besides, $t > a \Longrightarrow \omega(\delta_0+t) \ge \omega(\delta_0+a) = 1 \Longrightarrow q'(t) \le -t\,\omega'(\delta_0+t) \le 0$. Thus, q is increasing in $(0,\tau)$ and decreasing beyond. It follows that the equation $q(t) = \alpha_0$ has no solutions if $\alpha_0 > q(\tau)$. Otherwise, it has two: $t = q_-^{-1}(\alpha_0) \le \tau$ and $t = q_+^{-1}(\alpha_0) \ge \tau$, where q_- and q_+ are the restrictions of q to $[0,\tau)$ and (τ,∞), respectively. These solutions coalesce if $\tau\omega(a-\tau) = \alpha_0$.

If, for example, $\omega(t) = ct^\nu$, $0 < \nu \le 1$, then $a = c^{-1/\nu} - \delta_0$,

$$p(t) = \begin{cases} t\left(2 - c\left(c^{-1/\nu} - \delta_0 - t\right)^\nu\right) \,, \text{ if } \ t \in \left[0, c^{-1/\nu} - \delta_0\right] \,, \\ t(1+c(\delta_0+t)^\nu) \,, \text{ if } \ t \ge c^{-1/\nu} - \delta_0 \,, \end{cases}$$

$$q(t) = \begin{cases} ct\left(c^{-1/\nu} - \delta_0 - t\right)^\nu \,, \text{ if } \ t \in \left[0, c^{-1/\nu} - \delta_0\right] \,, \\ t(1-c(\delta_0+t)^\nu) \,, \text{ if } \ t \ge c^{-1/\nu} - \delta_0 \,, \end{cases}$$

$$\tau = \frac{c^{-1/\nu} - \delta_0}{\nu+1} \,, \quad q(\tau) = c\nu^\nu \left(\frac{c^{-1/\nu} - \delta_0}{\nu+1}\right)^{\nu+1} \,. \tag{5.24}$$

Theorem 5.9. *Let the selected dd* $[x_1, x_2\,|\,\mathbf{f}]$ *of* $\mathbf{f}$ *be* ω*-regularly continuous on* D*.*

1° *If the equation* $\mathbf{f}(x) = 0$ *has a solution* $x_* \in D$*, then*

$$\|x_0 - x_*\| \ge p^{-1}(\alpha_0) \,.$$

2° *If it has a solution* $x_* \in D$ *and* $\alpha_0 \le q(\tau)$, *then*

$$p^{-1}(\alpha_0) \le \|x_0 - x_*\| \le q_-^{-1}(\alpha_0)$$

and x_* *is the only solution in the ball* $B\big(x_0\,,\, q_+^{-1}(\alpha_0)\big)$.

3° *If* $\alpha_0 \le \min\{f_\infty(0)\,, q(\tau)\}$, *where* f_∞ *is the function of Proposition 5.5, then the equation* $\mathbf{f}(x) = 0$ *has a solution in the set*

$$D(x_0, x_{-1}) := \left\{ x \in D \;\middle|\; p^{-1}(\alpha_0) \le \|x - x_0\| \le q_-^{-1}(\alpha_0) \right\}. \tag{5.25}$$

Proof. 1°,2° By the secant equation

$$\mathbf{f}(x_0) = \mathbf{f}(x_0) - \mathbf{f}(x_*) = [x_0, x_* \,|\, \mathbf{f}](x_0 - x_*),$$

so that

$$\begin{aligned} \big|\|\mathbf{f}(x_0)\| - \|x_0 - x_*\|\big| \le \|\mathbf{f}(x_0) - (x_0 - x_*)\| &= \|([x_0, x_* \,|\, \mathbf{f}] - \mathbf{I})(x_0 - x_*)\| \\ &\le \|x_0 - x_*\| \cdot \|\mathbf{I} - [x_0, x_* \,|\, \mathbf{f}]\| \\ &= \|x_0 - x_*\| \cdot \|[x_0, x_{-1} \,|\, \mathbf{f}] - [x_0, x_* \,|\, \mathbf{f}]\| \end{aligned}$$

by (5.23). Because of the ω-regular continuity of the dd, $\|[x_0, x_{-1} \,|\, \mathbf{f}] - [x_0, x_* \,|\, \mathbf{f}]\|$

$$\begin{aligned} &\le \omega\big(\omega^{-1}(\min\{1\,,\, \|[x_0, x_* \,|\, \mathbf{f}]\|\}) + \|x_{-1} - x_*\|\big) - \min\{1\,,\, \|[x_0, x_* \,|\, \mathbf{f}]\|\} \\ &= \omega\big(\min\big\{\omega^{-1}(1)\,,\, \omega^{-1}(\|[x_0, x_* \,|\, \mathbf{f}]\|)\big\} + \|x_{-1} - x_*\|\big) - \\ &\qquad\qquad \omega\big(\min\big\{\omega^{-1}(1)\,,\, \omega^{-1}(\|[x_0, x_* \,|\, \mathbf{f}]\|)\big\}\big)\,. \end{aligned} \tag{5.26}$$

By (3.14), $\omega^{-1}(\|[x_0, x_* \,|\, \mathbf{f}]\|) \ge \big(\omega^{-1}(1) - \|x_{-1} - x_*\|\big)^+$ and so (thanks to the concavity of ω) the difference (5.26)

$$\begin{aligned} &\le \omega\Big(\min\Big\{\omega^{-1}(1)\,,\, \big(\omega^{-1}(1) - \|x_{-1} - x_*\|\big)^+\Big\} + \|x_{-1} - x_*\|\Big) - \\ &\qquad\qquad \omega\Big(\min\Big\{\omega^{-1}(1)\,,\, \big(\omega^{-1}(1) - \|x_{-1} - x_*\|\big)^+\Big\}\Big) \\ &= \omega\Big(\big(\omega^{-1}(1) - \|x_{-1} - x_*\|\big)^+ + \|x_{-1} - x_*\|\Big) - \omega\Big(\big(\omega^{-1}(1) - \|x_{-1} - x_*\|\big)^+\Big) \\ &\le \omega\Big(\big(\omega^{-1}(1) - \|x_{-1} - x_0\| - \|x_0 - x_*\|\big)^+ + \|x_{-1} - x_0\| + \|x_0 - x_*\|\Big) \\ &\qquad\qquad - \omega\Big(\big(\omega^{-1}(1) - \|x_{-1} - x_0\| - \|x_0 - x_*\|\big)^+\Big)\,. \end{aligned}$$

Therefore, for $d := \|x_0 - x_*\|$, we have

$$|\alpha_0 - d| \le d\Big(\omega\big((a - d)^+ + \delta_0 + d\big) - \omega\big((a - d)^+\big)\Big)$$

or, equivalently,

$$q(d) \le \alpha_0 \le p(d)\,.$$

The second inequality can be rewritten as $d \ge p^{-1}(\alpha_0)$, while the first means that

$$\alpha_0 > q(\tau) \bigvee \Big(\alpha_0 \le q(\tau) \ \ \& \ \ \big(d \le q_-^{-1}(\alpha_0) \bigvee d \ge q_+^{-1}(\alpha_0)\big)\Big)\,.$$

If $\alpha_0 > q(\tau)$, then the inequality $\alpha_0 \ge q(d)$ says nothing about the whereabouts of x_*, so that we can assert only that $d \ge p^{-1}(\alpha_0)$ (no solutions in the ball $B\big(x_0, p^{-1}(\alpha_0)\big)$). If $\alpha_0 \le q(\tau)$, then $q(d) \le \alpha_0 \iff d \le q_-^{-1}(\alpha_0) \bigvee d \ge q_+^{-1}(\alpha_0)$. Thus, if x_* is a solution and $\alpha_0 \le q(\tau)$, then it must satisfy

$$d \ge p^{-1}(\alpha_0) \ \ \& \ \ \big(d \le q_-^{-1}(\alpha_0) \bigvee d \ge q_+^{-1}(\alpha_0)\big)\ ,$$

that is, either $p^{-1}(\alpha_0) \le d \le q_-^{-1}(\alpha_0)$ (the set (5.25) contains a solution) or $d \ge \max\{p^{-1}(\alpha_0)\,,\, q_+^{-1}(\alpha_0)\} = q_+^{-1}(\alpha_0)$ (no other solutions in the ball $B\big(x_0\,,\, q_+^{-1}(\alpha_0)\big)$).

3° If $\alpha_0 \le \min\{q(\tau)\,, f_\infty(0,\delta_0)\}$, then Proposition 5.6 guarantees convergence of the sequence x_n generated by the modified secant method to a solution x_∞ of the equation $\mathbf{f}(x) = 0$. By 2°, this solution has to be in the set (5.25). □

For linear ω ($\omega(t) = ct$), $p(t) = t\big(1+c(\delta_0+t)\big)$ and $q(t) = t\big(1-c(\delta_0+t)\big)$), so that

$$p^{-1}(\alpha_0) = \frac{\sqrt{(c^{-1}+\delta_0)^2 + 4c^{-1}\alpha_0} - c^{-1} - \delta_0}{2}\,,$$

$$q_-^{-1}(\alpha_0) = \frac{c^{-1} - \delta_0 - \sqrt{(c^{-1}-\delta_0)^2 - 4c^{-1}\alpha_0}}{2}\,,$$

$$q_+^{-1}(\alpha_0) = \frac{c^{-1} - \delta_0 + \sqrt{(c^{-1}-\delta_0)^2 - 4c^{-1}\alpha_0}}{2}\,.$$

Besides, when ω is linear, both $f_\infty(0)$ of Proposition 5.6 and $q(\tau)$ become $0.25c\big(c^{-1}-\delta_0\big)^2$ (see Corollary 5.8 and (5.24)). Therefore, for Lipschitz continuous dd we have

Corollary 5.10. *Let the selected dd* $[x_1, x_2 \,|\, \mathbf{f}]$ *of* $\mathbf{f}$ *be Lipschitz continuous on D:* $\forall\, x_1, x_2, u_1, u_2 \in D$

$$\big\|[x_1, x_2 \,|\, \mathbf{f}] - [u_1, u_2 \,|\, \mathbf{f}]\big\| \le c\big(\|x_1 - u_1\| + \|x_2 - u_2\|\big)\,. \quad (5.27)$$

1° *If the equation* $\mathbf{f}(x) = 0$ *has a solution* x_* *in D, then*

$$\|x_* - x_0\| \ge \frac{\sqrt{(c^{-1}+\delta_0)^2 + 4c^{-1}\alpha_0} - c^{-1} - \delta_0}{2}\,.$$

2° *If it has a solution* x_* *in* D *and* $4c^{-1}\alpha_0 \le \left(c^{-1} - \delta_0\right)^2$, *then*

$$\frac{\sqrt{(c^{-1}+\delta_0)^2+4c^{-1}\alpha_0}-c^{-1}-\delta_0}{2} \le \|x_*-x_0\| \le \frac{c^{-1}-\delta_0-\sqrt{(c^{-1}-\delta_0)^2-4c^{-1}\alpha_0}}{2}$$

and x_* *is the only solution in the ball*

$$B\left(x_0\,,\frac{c^{-1}-\delta_0+\sqrt{(c^{-1}-\delta_0)^2-4c^{-1}\alpha_0}}{2}\right).$$

3° *If* $4c^{-1}\alpha_0 \le \left(c^{-1} - \delta_0\right)^2$, *then the equation* $\mathbf{f}(x) = 0$ *has a solution in the set* $D(x_0\,, x_{-1})$ *of all* $x \in D$ *satisfying*

$$\frac{\sqrt{(c^{-1}+\delta_0)^2+4c^{-1}\alpha_0}-c^{-1}-\delta_0}{2} \le \|x-x_0\| \le \frac{c^{-1}-\delta_0-\sqrt{(c^{-1}-\delta_0)^2-4c^{-1}\alpha_0}}{2}. \tag{5.28}$$

4° *The bounds in* (5.28) *are sharp.*

5.6 Secant updates of low rank

Let for brevity $y := \mathbf{f}(x_+) - \mathbf{f}(x)$. If $\mathbf{B}$ in (5.1) is of rank 1, $\mathbf{B}x := u\langle v\,, x\rangle$, $u, v \in \mathbb{H}$, then the inverse secant equation $(\mathbf{A}+u\langle v\,, \cdot\rangle)y = x_+ - x$ implies

$$\langle v\,, y\rangle \ne 0 \;\;\&\;\; u = -\frac{\mathbf{Af}(x_+)}{\langle v\,, y\rangle}\,,$$

so that

$$\mathbf{A}_+ = \mathbf{A} - \frac{\mathbf{Af}(x_+)}{\langle v\,, y\rangle}\langle v\,, \cdot\rangle\,. \tag{5.29}$$

By the Sherman–Morrison lemma (Lemma 1.3), if $\mathbf{A}$ is invertible, then $\mathbf{A}_+$ is invertible $\Longleftrightarrow \langle v\,, \mathbf{f}(x)\rangle \ne 0$. This condition suggests the choice $v = \lambda\mathbf{f}(x)$, whereas taking $v = \lambda\mathbf{A}^*\mathbf{Af}(x)$ yields Broyden's update (4.1).

Consider next rank 2 updates: $\mathbf{B} := p\langle q\,, \cdot\rangle + u\langle v\,, \cdot\rangle$. The inverse secant equation $(\mathbf{A}+p\langle q\,, \cdot\rangle + u\langle v\,, \cdot\rangle)y = x_+ - x$ implies either

$$\langle v\,, y\rangle = 0 \ne \langle q\,, y\rangle \;\&\; p = -\frac{\mathbf{Af}(x_+)}{\langle q\,, y\rangle} \quad\text{or}\quad \langle v\,, y\rangle \ne 0 \;\&\; u = -\frac{\mathbf{Af}(x_+)+p\langle q\,, y\rangle}{\langle v\,, y\rangle}\,.$$

The first possibility leads again to rank 1 updates, whereas the second results in

$$\begin{aligned}\mathbf{A}_+ &= \mathbf{A} + p\langle q\,,\cdot\rangle - \frac{\mathbf{Af}(x_+) + p\langle q\,,y\rangle}{\langle v\,,y\rangle}\langle v\,,\cdot\rangle \\ &= \mathbf{A} + \mathbf{Af}(x_+)\left\langle -\frac{v}{\langle v\,,y\rangle}\,,\cdot\right\rangle + p\left\langle q - \frac{\langle q\,,y\rangle}{\langle v\,,y\rangle}v\,,\cdot\right\rangle. \end{aligned} \tag{5.30}$$

If $\mathbf{A}$ is invertible, then $\mathbf{A}p/\|\mathbf{A}p\|$ is as arbitrary as p. Hence, the update

$$\mathbf{Af}(x_+)\left\langle -\frac{v}{\langle v\,,y\rangle}\,,\cdot\right\rangle + \mathbf{A}p\left\langle q - \frac{\langle q\,,y\rangle}{\langle v\,,y\rangle}v\,,\cdot\right\rangle\ ,\ \|\mathbf{A}p\| = 1\,,$$

is as general as that in (5.30). Moreover, we can safely assume that $\mathbf{A}p$ is orthogonal to $u := \mathbf{Af}(x_+)$, for

$$\mathbf{A}p = \frac{\langle \mathbf{A}p\,,u\rangle}{\|u\|^2}\,u + w\ ,\ w := \mathbf{A}p - \frac{\langle \mathbf{A}p\,,u\rangle}{\|u\|^2}\,u\,, \tag{5.31}$$

implies $w \perp u$ and $u\langle -\langle v\,,y\rangle^{-1}v\,,\cdot\rangle + \mathbf{A}p\langle q - \langle v\,,y\rangle^{-1}\langle q\,,y\rangle v\,,\cdot\rangle$

$$\begin{aligned} &= u\left\langle -\frac{v}{\langle v\,,y\rangle}\,,\cdot\right\rangle + \left(\frac{\langle \mathbf{A}p\,,u\rangle}{\|u\|^2}\,u + w\right)\left\langle q - \frac{\langle q\,,y\rangle}{\langle v\,,y\rangle}v\,,\cdot\right\rangle \\ &= u\left\langle -\frac{v}{\langle v\,,y\rangle}\,,\cdot\right\rangle + \frac{\langle \mathbf{A}p\,,u\rangle}{\|u\|^2}\,u\left\langle q - \frac{\langle q\,,y\rangle}{\langle v\,,y\rangle}v\,,\cdot\right\rangle + w\left\langle q - \frac{\langle q\,,y\rangle}{\langle v\,,y\rangle}v\,,\cdot\right\rangle \\ &= u\left\langle -\frac{v}{\langle v\,,y\rangle} + \frac{\langle \mathbf{A}p\,,u\rangle}{\|u\|^2}\left(q - \frac{\langle q\,,y\rangle}{\langle v\,,y\rangle}v\right)\,,\cdot\right\rangle + w\left\langle q - \frac{\langle q\,,y\rangle}{\langle v\,,y\rangle}v\,,\cdot\right\rangle. \end{aligned}$$

The vector $q - \big(\langle q\,,y\rangle/\langle v\,,y\rangle\big)v$ is as arbitrary as q and so can be replaced by q, which means that, without loss of generality, $q \perp y$. Likewise, the vector

$$\frac{\langle \mathbf{A}p\,,u\rangle}{\|u\|^2}q - \frac{v}{\langle v\,,y\rangle}$$

is as arbitrary as v and so we can safely assume that $\langle v\,,y\rangle = -1$. These remarks justify adoption of the formula

$$\mathbf{B} = \mathbf{Af}(x_+)\langle v\,,\cdot\rangle + \mathbf{A}p\langle q\,,\cdot\rangle\,, \tag{5.32}$$

where

$$p, q, v \in \mathbb{H}\ ,\ \|\mathbf{A}p\| = 1\ \ \&\ \ \mathbf{A}p \perp \mathbf{Af}(x_+)\ \ \&\ \ q \perp y\ \ \&\ \ \langle v\,,y\rangle = -1\,, \tag{5.33}$$

as the general form of rank 2 secant updates satisfying the inverse secant equation. Broyden's update in (4.1) (which is of rank 1) is obtained by choosing any p with $\|\mathbf{A}p\| = 1$, $q = 0$, and $v = -\mathbf{A}^*\mathbf{Af}(x)/\big\langle \mathbf{A}^*\mathbf{Af}(x)\,,y\big\rangle$. To identify the parameters $p\,,q\,,v$ corresponding to the BFGS update (see (0.6))

$$\mathbf{A}_+^{-1} = \mathbf{A}^{-1}\left(\mathbf{I} + \mathbf{A}y\left\langle \frac{y}{\langle y\,,s\rangle}\,,\cdot\right\rangle + s\left\langle -\frac{\mathbf{A}^{-1}s}{\langle \mathbf{A}^{-1}s\,,s\rangle}\,,\cdot\right\rangle\right)$$

the following lemma, which uses the Sherman–Morrison formula (Lemma 1.3) to derive a criterion for invertibility of an operator modified by the general rank 2 update, will be helpful.

Lemma 5.11. *Let* a, b, c, d *be arbitrary vectors of a Hilbert space* $\mathbb{H}$. *The operator* $\mathbf{I} + a\langle b\,,\,\cdot\rangle + c\langle d\,,\,\cdot\rangle$ *is invertible if and only if*

$$det := \big(1 + \langle a\,,\, b\rangle\big)\big(1 + \langle c\,,\, d\rangle\big) - \langle a\,,\, d\rangle\,\langle c\,,\, b\rangle \neq 0\,.$$

In this case,

$$(\mathbf{I} + a\langle b\,,\,\cdot\rangle + c\langle d\,,\,\cdot\rangle)^{-1} = \mathbf{I} + a'\langle b\,,\cdot\rangle + c'\langle d\,,\,\cdot\rangle\,, \qquad (5.34)$$

where

$$a' := \frac{c\langle a\,,d\rangle - a(1 + \langle c\,,d\rangle)}{det}\,, c' := \frac{a\langle c\,,b\rangle - c(1 + \langle a\,,b\rangle)}{det}. \qquad (5.35)$$

Proof.

$$\begin{aligned}
&(\mathbf{I} + a\langle b\,,\,\cdot\rangle + c\langle d\,,\,\cdot\rangle)(\mathbf{I} + a'\langle b\,,\,\cdot\rangle + c'\langle d\,,\,\cdot\rangle) \\
&= \mathbf{I} + a\langle b\,,\,\cdot\rangle + c\langle d\,,\,\cdot\rangle + (a' + a\langle b\,,a'\rangle + c\langle d\,,a'\rangle)\langle b\,,\,\cdot\rangle + \\
&\qquad\qquad (c' + a\langle b\,,c'\rangle + c\langle d\,,c'\rangle)\langle d\,,\,\cdot\rangle \\
&= \mathbf{I} + (a' + a(1 + \langle b\,,a'\rangle) + c\langle d\,,a'\rangle)\langle b\,,\,\cdot\rangle + (c' + a\langle b\,,c'\rangle + \\
&\qquad\qquad c(1 + \langle d\,,c'\rangle))\langle d\,,\,\cdot\rangle\,.
\end{aligned}$$

By (5.35), $a' + a(1 + \langle b\,,a'\rangle) + c\langle d\,,a'\rangle$

$$\begin{aligned}
&= \frac{c\langle a\,,d\rangle - a(1 + \langle c\,,d\rangle)}{det} + a\left(1 + \left\langle b\,,\, \frac{c\langle a\,,d\rangle - a(1 + \langle c\,,d\rangle)}{det}\right\rangle\right) \\
&\qquad\qquad + c\left\langle d\,,\, \frac{c\langle a\,,d\rangle - a(1 + \langle c\,,d\rangle)}{det}\right\rangle \\
&= \frac{c\langle a\,,d\rangle - a(1 + \langle c\,,d\rangle)}{det} \\
&\quad + a\frac{(1 + \langle a\,,b\rangle)(1 + \langle c\,,d\rangle) - \langle a\,,d\rangle\langle b\,,c\rangle + \langle a\,,d\rangle\langle b\,,c\rangle - \langle a\,,b\rangle(1 + \langle c\,,d\rangle)}{det} \\
&\qquad\qquad + c\frac{\langle c\,,d\rangle\langle a\,,d\rangle - \langle a\,,d\rangle(1 + \langle c\,,d\rangle)}{det} \\
&= \frac{c\langle a\,,d\rangle - a(1 + \langle c\,,d\rangle)}{det} + a\frac{1 + \langle c\,,d\rangle}{det} - c\frac{\langle a\,,d\rangle}{det} = 0
\end{aligned}$$

and $c' + a\langle b, c'\rangle) + c(1 + \langle d, c'\rangle)$

$$= \frac{a\langle b, c\rangle - c(1 + \langle a, b\rangle)}{det} + a\left\langle b, \frac{a\langle b, c\rangle - c(1 + \langle a, b\rangle)}{det}\right\rangle$$

$$c\left(1 + \left\langle d, \frac{a\langle b, c\rangle) - c(1 + \langle a, b\rangle)}{det}\right\rangle\right)$$

$$= \frac{a\langle b, c\rangle - c(1 + \langle a, b\rangle)}{det} + a\frac{\langle a, b\rangle\langle b, c\rangle - \langle b, c\rangle(1 + \langle a, b\rangle)}{det}$$

$$+ c\frac{(1 + \langle a, b\rangle)(1 + \langle c, d\rangle) - \langle a, d\rangle\langle b, c\rangle + \langle a, d\rangle\langle b, c\rangle - \langle c, d\rangle(1 + \langle a, b\rangle)}{det}$$

$$= \frac{a\langle b, c\rangle - c(1 + \langle a, b\rangle)}{det} - a\frac{\langle b, c\rangle}{det} + c\frac{1 + \langle a, b\rangle}{det} = 0 .$$

Thus, $(\mathbf{I} + a\langle b, \cdot\rangle + c\langle d, \cdot\rangle)(\mathbf{I} + a'\langle b, \cdot\rangle + c'\langle d, \cdot\rangle) = \mathbf{I}$. Similarly, one can verify that $(\mathbf{I} + a'\langle b, \cdot\rangle + c'\langle d, \cdot\rangle)(\mathbf{I} + a\langle b, \cdot\rangle + c\langle d, \cdot\rangle) = \mathbf{I}$. So, (5.34). □

For the BFGS update, we have

$$a = \mathbf{A}y \;, \; b = \frac{y}{\langle y, s\rangle} \;, \; c = s \;, \; d = -\frac{\mathbf{A}^{-1}s}{\langle \mathbf{A}^{-1}s, s\rangle} \,,$$

so that $\langle c, b\rangle = 1$, $\langle c, d\rangle = -1$, $\langle a, d\rangle = -\langle \mathbf{A}y, \mathbf{A}^{-1}s\rangle / \langle \mathbf{A}^{-1}s, s\rangle$, and $det = \langle \mathbf{A}y, \mathbf{A}^{-1}s\rangle / \langle \mathbf{A}^{-1}s, s\rangle$. So, application of the lemma results in

$$\mathbf{A}_+ = \left(\mathbf{I} + a'\left\langle \frac{y}{\langle y, s\rangle}, \cdot\right\rangle - c'\left\langle \frac{\mathbf{A}^{-1}s}{\langle \mathbf{A}^{-1}s, s\rangle}, \cdot\right\rangle\right)\mathbf{A} \,,$$

where

$$a' = -s \;, \; c' = \frac{\langle \mathbf{A}^{-1}s, s\rangle}{\langle y, s\rangle\langle \mathbf{A}^{-1}s, \mathbf{A}y\rangle}\left(\langle y, s\rangle\mathbf{A}y - \langle s + \mathbf{A}y, y\rangle s\right) .$$

Correspondingly, the inversion-free form of the BFGS update is

$$\mathbf{B} = -\frac{s}{\langle y, s\rangle}\langle \mathbf{A}^*y, \cdot\rangle - \frac{\langle y, s\rangle\mathbf{A}y - \langle s + \mathbf{A}y, y\rangle s}{\langle y, s\rangle\,\langle \mathbf{A}^{-1}s, \mathbf{A}y\rangle}\langle \mathbf{A}^*\mathbf{A}^{-1}s \;, \cdot\rangle \,.$$

Let for brevity $u := \mathbf{Af}(x_+)$, $\lambda := 1/\left(\langle y, s\rangle\,\langle \mathbf{A}^{-1}s, \mathbf{A}y\rangle\right)$ and note that $\mathbf{A}y = u + s$. So, $\langle y, s\rangle\mathbf{A}y - \langle s + \mathbf{A}y, y\rangle s = \langle y, s\rangle u - \langle \mathbf{A}y, y\rangle s$ and

$$\begin{aligned}
\mathbf{B} &= \lambda\left[(\mathbf{A}y, y\rangle s - \langle y, s\rangle u)\langle \mathbf{A}^*\mathbf{A}^{-1}s, \cdot\rangle - s\langle \mathbf{A}y, \mathbf{A}^{-1}s\rangle\langle \mathbf{A}^*y, \cdot\rangle\right] \\
&= u\langle -\lambda\langle y, s\rangle\mathbf{A}^*\mathbf{A}^{-1}s, \cdot\rangle + s\langle \lambda\langle \mathbf{A}y, y\rangle\mathbf{A}^*\mathbf{A}^{-1}s - \lambda\langle \mathbf{A}y, \mathbf{A}^{-1}s\rangle\mathbf{A}^*y, \cdot\rangle \\
&= u\langle w, \cdot\rangle + s\langle r, \cdot\rangle , \qquad (5.36)
\end{aligned}$$

where (for brevity)

$$w := -\lambda\langle y\,,s\rangle \mathbf{A}^*\mathbf{A}^{-1}s\ ,\ r := \lambda\mathbf{A}^*\big(\langle \mathbf{A}y\,,y\rangle\mathbf{A}^{-1}s - \langle \mathbf{A}y\,,\mathbf{A}^{-1}s\rangle y\big)\,.$$

Although $r \perp y$, this representation of the BFGS update is still not in the canonical form (5.32), (5.33), for the requirements (5.33) are not fulfilled. To squeeze it into the frame (5.33), we use the identity (cf. (5.31))

$$s = \frac{\langle s\,,u\rangle}{\|u\|^2}u + s'\,,\ s' := s - \frac{\langle s\,,u\rangle}{\|u\|^2}u$$

to rewrite (5.36) as $u\langle w',\cdot\rangle + \mathbf{A}p\langle q\,,\cdot\rangle$, where

$$w' := w + \frac{\langle s\,,u\rangle}{\|u\|^2}r\ ,\ p := \frac{-\mathbf{f}(x) - \|u\|^{-2}\langle s\,,u\rangle\mathbf{f}(x_+)}{\|s'\|}\ ,\ q := \|s'\|r\,,$$

and

$$\|\mathbf{A}p\| = 1\ \ \&\ \ \mathbf{A}p \perp u\ \ \&\ \ q \perp y\,.$$

Besides, $v := -w/\langle w\,,y\rangle \Longrightarrow \langle v\,,y\rangle = -1$. Thus, the BFGS update is obtained from (5.32), (5.33) by choosing

$$p = \frac{-\mathbf{f}(x) - \|u\|^{-2}\langle s\,,u\rangle\mathbf{f}(x_+)}{\|s'\|}\ ,\ q = \|s'\|r\ ,\ v = -\frac{w}{\langle w\,,y\rangle}\,. \tag{5.37}$$

If $\mathbf{A}$ is invertible, then the operator

$$\mathbf{A}_+ = \mathbf{A}\Big(\mathbf{I} + p\langle q\,,\cdot\rangle + \mathbf{f}(x_+)\langle v\,,\cdot\rangle\Big)$$

updated by the update (5.32) is invertible simultaneously with the operator in parentheses. The latter is invertible by the lemma if and only if

$$\big(1 + \langle p\,,q\rangle\big)\big(1 + \langle \mathbf{f}(x_+)\,,v\rangle\big) \neq \langle p\,,v\rangle\big\langle \mathbf{f}(x_+)\,,q\big\rangle\,.$$

As $q \perp y = 0 \Longleftrightarrow \big\langle \mathbf{f}(x_+)\,,q\big\rangle = \big\langle \mathbf{f}(x)\,,q\big\rangle$. The result is the following.

Corollary 5.12. *Let the operator* $\mathbf{A} \in \mathcal{L}(\mathbb{H})$ *be invertible. If the update* $\mathbf{B}$ *is as in* (5.32)*, then* $\mathbf{A}+\mathbf{B}$ *is invertible if and only if*

$$\Big\langle \langle \mathbf{f}(x)\,,q\rangle\, p - \big(1 + \langle p\,,q\rangle\big)\,\mathbf{f}(x_+)\ ,\ v\Big\rangle \neq 1 + \langle p\,,q\rangle\,. \tag{5.38}$$

5.7 Optimal secant updates of rank 2

The search for optimal secant updates can be based on Theorem 5.3 or Theorem 5.9. In this section, the analysis based on Corollary 5.10 is presented.

The presentation is along the lines of [23]. Theorem 5.3 has not been used for this purpose so far.

Let the selected dd $[x_1, x_2 \,|\, \mathbf{f}]$ of the operator $\mathbf{f}$ be Lipschitz continuous on D:

$$\big\|[x_1, x_2 \,|\, \mathbf{f}] - [u_1, u_2 \,|\, \mathbf{f}]\big\| \le c\big(\|x_1 - u_1\| + \|x_2 - u_2\|\big)\ ,\ \forall\, x_1, x_2, u_1, u_2 \in D\,. \tag{5.39}$$

Since the operator $\mathbf{f}$ can always be replaced by $c^{-1}\mathbf{f}$, which has the same zeros as $\mathbf{f}$ and whose dd $[x_1, x_2 \,|\, c^{-1}\mathbf{f}] = c^{-1}[x_1, x_2 \,|\, \mathbf{f}]$ (Proposition 3.2), we can assume (without loss of generality) that the constant c in (5.39) is 1. This observation shortens subsequent calculations considerably.

Suppose that the current iteration $(x, \mathbf{A}) \in D \times \mathcal{L}(\mathbb{H})$ of the method

$$x_+ := x - \mathbf{A}\mathbf{f}(x)\ ,\ y := \mathbf{f}(x_+) - \mathbf{f}(x)\ ,\ \mathbf{A}_+ := \mathbf{A}\Big(\mathbf{I} + p\langle q\,,\cdot\rangle + \mathbf{f}(x_+)\langle v\,,\cdot\rangle\Big)\,, \tag{5.40}$$

where

$$\mathbf{A}p \perp \mathbf{A}\mathbf{f}(x_+)\ \ \&\ \ \|\mathbf{A}p\| = 1\ \ \&\ \ q \perp y\ \ \&\ \ \langle v\,, y\rangle = -1\,, \tag{5.41}$$

has been obtained with boundedly invertible $\mathbf{A}$. If

$$\bar\alpha := \|\mathbf{f}(x_+)\|\ ,\quad \bar\delta := \|x_+ - x\|\ ,$$

are such that

$$\bar\delta + 2\sqrt{\bar\alpha} \le 1\ , \tag{5.42}$$

then Corollary 5.10 guarantees the existence (and uniqueness) of a solution x_* in the set

$$D(x_+, x) := \Big\{x' \in D \;\Big|\; \xi \le \|x' - x_+\| \le \eta\Big\}$$

where

$$\xi := \xi(4\bar\alpha\,, \bar\delta) := \frac{\sqrt{\big(1+\bar\delta\big)^2 + 4\bar\alpha} - 1 - \bar\delta}{2}\,,$$

$$\eta := \eta(4\bar\alpha\,, \bar\delta) := \frac{1 - \bar\delta - \sqrt{\big(1-\bar\delta\big)^2 - 4\bar\alpha}}{2}\,. \tag{5.43}$$

We have now to choose the parameters p, q, v of the next rank 2 update in (5.40). Each choice results in

$$\begin{aligned} x_{++} &= x_+ - \mathbf{A}_+\mathbf{f}(x_+) = x_+ - \Big(\mathbf{A} + \mathbf{A}\mathbf{f}(x_+)\langle v\,,\cdot\rangle + \mathbf{A}p\langle q\,,\cdot\rangle\Big)\mathbf{f}(x_+) \\ &= x_+ - \mathbf{A}\mathbf{f}(x_+)\Big(1 + \big\langle v\,, \mathbf{f}(x_+)\big\rangle\Big) - \mathbf{A}p\big\langle q\,, \mathbf{f}(x_+)\big\rangle \\ &= x_+ - \mathbf{A}\mathbf{f}(x_+)\big\langle v\,, \mathbf{f}(x)\big\rangle - \mathbf{A}p\big\langle q\,, \mathbf{f}(x)\big\rangle\,, \end{aligned}$$

(since $\langle v, y\rangle = -1 \iff 1 + \langle v, \mathbf{f}(x_+)\rangle = \langle v, \mathbf{f}(x)\rangle$ and $\langle q, y\rangle = 0 \iff \langle q, \mathbf{f}(x_+)\rangle = \langle q, \mathbf{f}(x)\rangle$) and

$$\bar{\alpha}_+ := \|\mathbf{f}(x_{++})\| \ , \ \bar{\delta}_+ := \|x_{++} - x_+\| = \big\|\mathbf{A}\mathbf{f}(x_+)\langle v, \mathbf{f}(x)\rangle + \mathbf{A}p\langle q, \mathbf{f}(x)\rangle\big\| \ . \tag{5.44}$$

Clearly, x_{++} must remain in $D(x_+, x)$:

$$\xi \le \bar{\delta}_+ \le \eta \ . \tag{5.45}$$

The value of $\bar{\alpha}_+$ is conditioned by the Lipschitz continuity of the dd $[x_1, x_2 \,|\, \mathbf{f}]$. Namely, as $\mathbf{f}(x_{++}) = \mathbf{f}(x_+) + [x_{++}, x_+ \,|\, \mathbf{f}\,](x_{++} - x_+)$ by the secant equation and

$$\mathbf{f}(x_+) = -\mathbf{A}_+^{-1}(x_{++} - x_+) = -[x_+, x \,|\, \mathbf{f}\,](x_{++} - x_+)$$

by (5.40), we get $\mathbf{f}(x_{++}) = \big([x_{++}, x_+ \,|\, \mathbf{f}\,] - [x_+, x \,|\, \mathbf{f}\,]\big)(x_{++} - x_+)$ and so

$$\begin{aligned} \big\|\mathbf{f}(x_{++})\big\| &\le \big\|[x_{++}, x_+ \,|\, \mathbf{f}\,] - [x_+, x \,|\, \mathbf{f}\,]\big\| \cdot \|x_{++} - x_+\| \\ &\le \|x_{++} - x_+\|\big(\|x_{++} - x_+\| + \|x_+ - x\|\big) \end{aligned}$$

by (5.39) (with $c = 1$). Hence,

$$\bar{\alpha}_+ \le \bar{\delta}_+ \left(\bar{\delta}_+ + \bar{\delta}\right) \ . \tag{5.46}$$

Besides, we require that $\bar{\alpha}_+$ and $\bar{\delta}_+$ satisfy

$$\bar{\delta}_+ + 2\sqrt{\bar{\alpha}_+} \le 1 \ , \tag{5.47}$$

since then x_{++} induces its own existence set

$$D(x_{++}, x_+) = \left\{x \in D \;\middle|\; \xi\big(4\bar{\alpha}_+, \bar{\delta}_+\big) \le \|x - x_{++}\| \le \eta\big(4\bar{\alpha}_+, \bar{\delta}_+\big)\right\} . \tag{5.48}$$

Consequently, the solution is $x_* \in D(x_+, x) \cap D(x_{++}, x_+)$. If we accept that x_* is distributed in this set uniformly (no other distribution seems more reasonable), then the entropy of its position within this set can be measured by the set's size, which is not decreasing when the size of $D(x_{++}, x_+)$ or, equivalently, its thickness

$$th(4\bar{\alpha}_+, \bar{\delta}_+) := 1 - \frac{\sqrt{\left(1 + \bar{\delta}_+\right)^2 + 4\bar{\alpha}_+} + \sqrt{\left(1 - \bar{\delta}_+\right)^2 - 4\bar{\alpha}_+}}{2} \tag{5.49}$$

is increasing. Then, which p, q, v do we choose in order to minimize $th(4\bar{\alpha}_+, \bar{\delta}_+)$? Inasmuch as it depends on $\bar{\alpha}_+$, whose value is unknown at the time the decision is made, this decision has to be based on some hypothesis about possible values of $\bar{\alpha}_+$. One such hypothesis (worst case scenario) is that the operator's response will be the least desirable regarding the value of $th(4\bar{\alpha}_+, \bar{\delta}_+)$. Another, aimed at optimality "on average", assumes that the response will be most probable for the operator at hand. The analysis in this section is based on the first hypothesis. The second is explored in the next section.

According to the worst case approach, we expect that the thickness (5.49) will get its maximum value $Th(\bar{\delta}_+)$ feasible under the constraints (5.45), (5.46), (5.47). Therefore, the best choice one can make in this situation is that, which minimizes $Th(\bar{\delta}_+)$. Thus, our task is first to maximize the thickness (5.49) over all $\bar{\alpha}_+$ satisfying constraints (5.45), (5.46), (5.47) and then to minimize the function $Th(\bar{\delta}_+)$ that results from maximization over all $\bar{\delta}_+ \in [\xi\,,\eta]$.

While evaluating Th, we use the abbreviations

$$s := 4\bar{\alpha}_+\ ,\quad t := \bar{\delta}_+\ . \tag{5.50}$$

With these abbreviations, (5.45) is rewritten as $\xi \le t \le \eta$, (5.46) $\iff s \le 4t(t+\bar{\delta})$, (5.47) $\iff s \le (1-t)^2$, and $th(s\,,t) = 1 - 0.5\Big(\sqrt{(1+t)^2+s}+\sqrt{(1-t)^2-s}\Big)$. It follows that

$$\begin{aligned} Th(t) &:= \max_s\Big\{th(s\,,t)\ \Big|\ \xi \le t \le \eta\ \ \&\ \ s \le 4t(t+\bar{\delta})\ \ \&\ \ s \le (1-t)^2\Big\} \\ &= 1 - 0.5F(t)\,, \end{aligned}$$

where

$$F(t) := \begin{cases} \min\limits_s\Big\{\sqrt{(1+t)^2+s}+\sqrt{(1-t)^2-s}\ \Big|\ 0 \le s \le \min\big\{4t(t+\bar{\delta})\,,\,(1-t)^2\big\}\Big\} \\ \qquad\qquad\qquad\qquad\qquad\qquad\qquad\qquad \text{if } \xi \le t \le \eta\,, \\ \infty,\ \text{otherwise}\,. \end{cases}$$

As the objective is decreasing in s, $F(t)$

$$= \begin{cases} \sqrt{(1+t)^2+\min\big\{4t(t+\bar{\delta})\,,(1-t)^2\big\}}+\sqrt{(1-t)^2-\min\big\{4t(t+\bar{\delta})\,,(1-t)^2\big\}} \\ \qquad\qquad\qquad\qquad\qquad\qquad\qquad\qquad \text{if } \xi \le t \le \eta\,, \\ \infty,\ \text{otherwise}\,. \end{cases} \tag{5.51}$$

For $t \in [\xi\,,\eta]$, $4t(t+\bar{\delta}) \ge (1-t)^2 \iff \max\{\xi\,,\beta\} \le t \le \eta$, where

$$\beta := \beta(\bar{\delta}) := \frac{2\sqrt{1+\bar{\delta}+\bar{\delta}^2}-1-2\bar{\delta}}{3} = \frac{1}{2\sqrt{1+\bar{\delta}+\bar{\delta}^2}+1+2\bar{\delta}}\ . \tag{5.52}$$

Lemma 5.13. *Let* $0 \le \bar{\delta} \le 1\ \ \&\ \ 4\bar{\alpha} \le \big(1-\bar{\delta}\big)^2$ *(as in* (5.45)*). Then*

$1^\circ\ \beta > \xi$.

$2^\circ\ \beta \ge \eta \iff 0 \le \bar{\delta} \le 0.6\ \ \&\ \ 4\bar{\alpha} \le 4\beta(1-\bar{\delta}-\beta)\ \bigvee\ 0.6 \le \bar{\delta} \le 1$.

Proof. 1° Otherwise, we would have

$$\frac{\sqrt{(1+\bar\delta)^2+4\bar\alpha}-1-\bar\delta}{2} \geq \frac{2\sqrt{1+\bar\delta+\bar\delta^{\,2}}-1-2\bar\delta}{3}$$

$$\iff 3\sqrt{(1+\bar\delta)^2+4\bar\alpha}-3-3\bar\delta \geq 4\sqrt{1+\bar\delta+\bar\delta^{\,2}}-2-4\bar\delta$$

$$\iff 3\sqrt{(1+\bar\delta)^2+4\bar\alpha} \geq 4\sqrt{1+\bar\delta+\bar\delta^{\,2}}+1-\bar\delta$$

$$\iff 9(1+\bar\delta)^2+36\bar\alpha \geq 16\left(1+\bar\delta+\bar\delta^{\,2}\right)+8(1-\bar\delta)\sqrt{1+\bar\delta+\bar\delta^{\,2}}+(1-\bar\delta)^2$$

$$\iff 36\bar\alpha \geq 8-4\bar\delta+8\bar\delta^{\,2}+8(1-\bar\delta)\sqrt{1+\bar\delta+\bar\delta^{\,2}}\,.$$

On the other hand, by assumption, $36\bar\alpha \leq 9(1-\bar\delta)^2$. So, the hypothesis $\xi \geq \beta$ implies

$$8-4\bar\delta+8\bar\delta^{\,2}+8(1-\bar\delta)\sqrt{1+\bar\delta+\bar\delta^{\,2}} \leq 9(1-\bar\delta)^2$$

or, equivalently,

$$14\bar\delta-\bar\delta^{\,2}+8(1-\bar\delta)\sqrt{1+\bar\delta+\bar\delta^{\,2}} \leq 1\,. \tag{5.53}$$

Differentiate twice the function $g(t) := 14t-t^2+8(1-t)\sqrt{1+t+t^2}$ on the left:

$$g'(t)=2\left(7-t-2\frac{1+t+4t^2}{\sqrt{1+t+t^2}}\right)\ ,\ g''(t)=-2\left(1+\frac{1+15t+12t^2+8t^3}{(1+t+t^2)^{3/2}}\right).$$

As g'' is negative in $[0\,,1]$, g' is decreasing there, which means that g is concave. Then $\min\limits_{0\leq t\leq 1} g(t)=\min\{g(0)\,,g(1)\}=\min\{8\,,13\}=8$, contrary to (5.53).

2° $\beta \geq \eta$

$$\iff 2\beta \geq 1-\bar\delta-\sqrt{(1-\bar\delta)^2-4\bar\alpha} = \sqrt{(1-\bar\delta)^2-4\bar\alpha} \geq 1-\bar\delta-2\beta$$

$$\iff 2\beta \geq 1-\bar\delta \ \bigvee\ 2\beta \leq 1-\bar\delta \ \ \&\ \ (1-\bar\delta)^2-4\bar\alpha \geq (1-\bar\delta)^2-4\beta\,(1-\bar\delta)+4\beta^2$$

$$\iff 2\beta \geq 1-\bar\delta \ \bigvee\ 2\beta \leq 1-\bar\delta \ \ \&\ \ 4\bar\alpha \leq 4\beta(1-\bar\delta-\beta)$$

and $2\beta \geq 1-\bar\delta$

$$\iff 4\sqrt{1+\bar\delta+\bar\delta^2}-2-4\bar\delta \geq 3-3\bar\delta \iff 4\sqrt{1+\bar\delta+\bar\delta^2} \geq 5+\bar\delta$$

$$\iff 16\left(1+\bar\delta+\bar\delta^2\right) \geq 25+10\bar\delta+\bar\delta^2 \iff 15\bar\delta^2+6\bar\delta-9 \geq 0$$

$$\iff \bar\delta \geq 0.6\,.$$

Thus, $\beta \geq \eta \iff 0.6 \leq \bar\delta \leq 1 \ \bigvee\ 0 \leq \bar\delta \leq 0.6 \ \ \&\ \ 4\bar\alpha \leq 4\beta(1-\bar\delta-\beta)\,.$ □

By the lemma, for $t \in [\xi\,,\eta]$,

$$\min\left\{4t(t+\bar\delta), (1-t)^2\right\} = \begin{cases} (1-t)^2\,, \text{ if } \beta < \eta \ \& \ \beta \le t \le \eta\,, \\ 4t(t+\bar\delta)\,, \text{ if } \beta < \eta \ \& \ \xi \le t \le \beta \ \bigvee \ \beta \ge \eta\,. \end{cases}$$

So,

$$F(t) = \begin{cases} \sqrt{(1+t)^2 + (1-t)^2} = \sqrt{2+2t^2}\,, \ \text{ if } \beta < \eta \ \& \ \beta \le t \le \eta\,, \\ \sqrt{(1+t)^2 + 4t(t+\bar\delta)} + \sqrt{(1-t)^2 - 4t(t+\bar\delta)}\,, \\ \qquad\qquad\qquad \text{if } \beta < \eta \ \& \ \xi \le t \le \beta \ \bigvee \ \beta \ge \eta\,. \end{cases}$$

and $\beta < \eta$

$$\Longrightarrow F(t) = \begin{cases} \sqrt{(1+t)^2 + 4t(t+\bar\delta)} + \sqrt{(1-t)^2 - 4t(t+\bar\delta)}\,, \ \text{ if } \xi \le t \le \beta\,, \\ \sqrt{2+2t^2}\,, \ \text{ if } \beta \le t \le \eta\,. \end{cases}$$

$$\Longrightarrow \max_{\xi \le t \le \eta} F(t) = \max\left\{\max_{\xi \le t \le \beta} F(t)\,,\ \max_{\beta \le t \le \eta} F(t)\right\} = \max\{m_1\,, m_2\}\,,$$

where

$$m_1 := \max_{\xi \le t \le \beta}\left(\sqrt{(1+t)^2 + 4t(t+\bar\delta)} + \sqrt{(1-t)^2 - 4t(t+\bar\delta)}\right),$$

$$m_2 := \max_{\beta \le t \le \eta} \sqrt{2+2t^2} = \sqrt{2+2\eta^2}\,.$$

To evaluate m_1, we have to analyze the behavior of the objective

$$g(t,\bar\delta) = \sqrt{1+t^2+2t\left(2t+1+2\bar\delta\right)} + \sqrt{1+t^2-2t\left(2t+1+2\bar\delta\right)} \qquad (5.54)$$

with respect to $t \in [\xi\,, \beta]$.

Lemma 5.14. *For all* $\bar\delta \in [0\,,1]$, *the function* $t \mapsto g(t,\bar\delta)$ *is decreasing in* $[\xi\,,\infty)$.

Proof. Let for brevity $p(t) := 1+t^2$, $q(t) := 2t\left(2t+1+2\bar\delta\right)$, and differentiate $g(t,\bar\delta) = \sqrt{p(t)+q(t)} + \sqrt{p(t)-q(t)}$ with respect to t:

$$\frac{\partial g}{\partial t} = \frac{p' + q'}{2\sqrt{p+q}} - \frac{q' - p'}{2\sqrt{p-q}}\,.$$

It follows that

$$\frac{\partial g}{\partial t} < 0 \Longleftrightarrow \frac{p' + q'}{2\sqrt{p+q}} < \frac{q' - p'}{2\sqrt{p-q}}\,.$$

As $q' > p'$ in $(0\,,\infty)$, we infer that

$$\frac{\partial g}{\partial t} < 0 \Longleftrightarrow (p'^2 + 2p'q' + q'^2)(p-q) < (q'^2 - 2p'q' + p'^2)(p+q)$$
$$\Longleftrightarrow 2pp'q' < q\left(p'^2 + q'^2\right)\,.$$

Recalling definitions of p and q, we see that the inequality $\partial g/\partial t < 0$ is equivalent to

$$2(1+t^2)\,2t\big(8t+2\,(1+2\bar\delta)\big) < 2t(2t+1+2\bar\delta)\Big(4t^2+\big(8t+2\,(1+2\bar\delta)\big)^2\Big)$$

$$\Longleftrightarrow (1+t^2)\,(4t+1+2\bar\delta) < (2t+1+2\bar\delta)\,\Big(t^2+\big(4t+1+2\bar\delta\big)^2\Big)$$

$$\Longleftrightarrow 4t^3+(1+2\bar\delta)t^2+4t+1+2\bar\delta <$$

$$2t^3+(1+2\bar\delta)t^2+(2t+1+2\bar\delta)\Big(17t^2+8\,(1+2\bar\delta)\,t+\big(1+2\bar\delta\big)^2\Big)$$

$$\Longleftrightarrow 2t^3+4t+1+2\bar\delta < 34t^3+33(1+2\bar\delta)t^2+10\big(1+2\bar\delta\big)^2 t+\big(1+2\bar\delta\big)^3,$$

which is obviously true. □

By the lemma, $m_1 = g(\xi,\bar\delta)$ and so

$$\beta<\eta \Longrightarrow \max_{\xi\le t\le\eta} F(t) = \max\Big\{g(\xi,\bar\delta)\,,\sqrt{2+2\eta^2}\Big\}.$$

Lemma 5.15. $\beta<\eta \Longrightarrow \max\Big\{g(\xi,\bar\delta)\,,\sqrt{2+2\eta^2}\Big\} = g(\xi,\bar\delta)$.

Proof. As $\xi(s,\bar\delta)$ and $\eta(s,\bar\delta)$ are increasing in s and $t\mapsto g(t,\bar\delta)$ is decreasing in $(0,\infty)$, the function

$$f(s,\bar\delta) := \sqrt{2+2\eta(s,\bar\delta)^2} - g\big(\xi(s,\bar\delta),\bar\delta\big)$$

is increasing in s, so that $\sup_s\Big\{f(s,\bar\delta)\ \Big|\ 0\le s\le \big(1-\bar\delta\big)^2\Big\}$

$$= f\Big(\big(1-\bar\delta\big)^2,\bar\delta\Big) = \sqrt{2+2\left(\frac{1-\bar\delta}{2}\right)^2} - g\left(\frac{\sqrt{2+2\bar\delta^{\,2}}-1-\bar\delta}{2},\bar\delta\right).$$

The function $t\mapsto\sqrt{2+2t^2}-t$ is decreasing in $(0\,,1)$. Therefore,

$$\bar\delta\mapsto g\Big(0.5\,\Big(\sqrt{2+2\bar\delta^{\,2}}-1-\bar\delta\Big)\,,\bar\delta\Big)$$

is incresing and $\bar\delta\mapsto f\Big(\big(1-\bar\delta\big)^2,\bar\delta\Big)$ is decreasing. By Lemma 5.13, $\beta<\eta\Longrightarrow$ $0\le\bar\delta\le 0.6$ and so

$$\beta<\eta\Longrightarrow f\Big(\big(1-\bar\delta\big)^2,\bar\delta\Big) \le \sup_{0\le t\le 0.6} f\big((1-t)^2,t\big) = f(1\,,0)$$

$$= \sqrt{2.5} - g\Big(0.5\big(\sqrt{2}-1\big),0\Big)$$

$$= 1.58... - 1.95... < 0\,.$$

It follows that $\forall\, s \in \left[0\,, (1-\bar{\delta})^2\right]\;\; f(s, \bar{\delta}) < 0\,$, i.e.,

$$\sqrt{2 + 2\eta(s, \bar{\delta})^2} < g(\xi(s, \bar{\delta}), \bar{\delta})\,.$$

□

By the lemma, $\beta < \eta \Longrightarrow \min\limits_{\xi \le t \le \eta} Th(t) = 1 - 0.5g(\xi)$. If $\beta \ge \eta$, then $Th(t) = 1 - 0.5g(t)$ and $\min\limits_{\xi \le t \le \eta} Th(t) = 1 - 0.5g(\xi)$, by Lemma 5.14. Thus, $\min\limits_{\xi \le t \le \eta} Th(t) = 1 - 0.5g(\xi)$ and $\arg\min\limits_{\xi \le t \le \eta} Th(t) = \xi$, no matter how β relates to η. By (5.44),

$$t = \bar{\delta}_+ = \left\|\mathbf{Af}(x_+)\langle v\,, \mathbf{f}(x)\rangle + \mathbf{A}p\langle q\,, \mathbf{f}(x)\rangle\right\|$$

and so the optimal values of $p\,, q\,, v$ are those that satisfy (in addition to (5.33))

$$\left\|\mathbf{Af}(x_+)\langle v\,, \mathbf{f}(x)\rangle + \mathbf{A}p\langle q\,, \mathbf{f}(x)\rangle\right\| = \xi\,.$$

In view of (5.33), this condition can be rewritten as

$$\left\|\mathbf{Af}(x_+)\right\|^2 \langle v\,, \mathbf{f}(x)\rangle^2 + \langle q\,, \mathbf{f}(x)\rangle^2 = \xi^2. \tag{5.55}$$

We state the result as

Proposition 5.16. *The parameter triple* (p, q, v) *in* (5.40), (5.41) *is optimal with respect to the entropy (worst case) optimality criterion if it satisfies the condition* (5.55).

Note that the condition (5.55) does not depend on p and so any value of p that agrees with (5.33) is optimal.

Taking $q = 0$ in (5.55) produces an optimality condition for rank 1 secant updates:

$$\langle v\,, y\rangle = -1 \;\;\&\;\; \left|\langle v\,, \mathbf{f}(x)\rangle\right| = \pm\frac{\xi}{\left\|\mathbf{Af}(x_+)\right\|} =: \pm\lambda\,.$$

The system

$$\langle v\,, y\rangle = -1 \;\;\&\;\; \langle v\,, \mathbf{f}(x)\rangle = \lambda$$

is equivalent to

$$\langle v\,, a\rangle = \alpha \;\;\&\;\; \langle v\,, b\rangle = \beta\,, \tag{5.56}$$

where

$$a := \frac{y}{\|y\|}\,,\; \alpha := -\frac{1}{\|y\|}\,,\; b := \frac{\mathbf{f}(x) - \langle \mathbf{f}(x)\,, a\rangle a}{\left\|\mathbf{f}(x) - \langle \mathbf{f}(x)\,, a\rangle a\right\|}\,,\; \beta := \frac{\lambda - \langle \mathbf{f}(x)\,, a\rangle \alpha}{\left\|\mathbf{f}(x) - \langle \mathbf{f}(x)\,, a\rangle a\right\|}\,. \tag{5.57}$$

By Lemma 1.5, 4°, the general solution of the system (5.56) is

$$v = \alpha a + \beta b + z - \langle a\,, z\rangle a - \langle b\,, z\rangle b\;,\;\; z \in \mathbb{H},$$

and $\alpha a + \beta b$ is the solution of minimum norm. Hence

Corollary 5.17. *A rank 1 secant update* $-\mathbf{Af}(x_+)\langle v\,,\cdot\rangle$ *is optimal if*

$$v = \alpha a + \beta b + z - \langle a\,,z\rangle a - \langle b\,,z\rangle b\,,$$

where z *is any vector of* $\mathbb{H}$ *and* α, a, β, b *are as in* (5.57). *The vector* $\alpha a + \beta b$ *is the optimal value of* v *of minimum norm.*

The corollary provokes the question: is Broyden's choice $\mathbf{A}^*\mathbf{Af}(x)$ of v (entropy) optimal? By the corollary, the value $\mathbf{A}^*\mathbf{Af}(x)$ of v is optimal if

$$\big\langle \mathbf{A}^*\mathbf{Af}(x)\,,y\big\rangle = -1 \;\;\&\;\; \big|\big\langle \mathbf{A}^*\mathbf{Af}(x)\,,\mathbf{f}(x)\big\rangle\big| = \frac{\xi}{\big\|\mathbf{Af}(x_+)\big\|}\,.$$

The first condition is equivalent to $\big\|\mathbf{Af}(x)\big\|^2 = \big\langle \mathbf{Af}(x_+)\,,\mathbf{Af}(x)\big\rangle + 1$, while the second $\iff \big\|\mathbf{Af}(x)\big\|^2 = \xi\,/\big\|\mathbf{Af}(x_+)\big\|$. Hence, Broyden's choice is optimal if

$$\begin{aligned}
&\big\|\mathbf{Af}(x_+)\big\|\Big(\big\langle \mathbf{Af}(x_+)\,,\mathbf{Af}(x)\big\rangle + 1\Big) = \xi \\
&= 0.5\left(\sqrt{\big(1+\big\|\mathbf{Af}(x)\big\|\big)^2 + 4\big\|\mathbf{f}(x_+)\big\|} - 1 - \big\|\mathbf{Af}(x)\big\|\right) \\
&= \frac{2\big\|\mathbf{f}(x_+)\big\|}{\sqrt{\big(1+\big\|\mathbf{Af}(x)\big\|\big)^2 + 4\big\|\mathbf{f}(x_+)\big\|} + 1 + \big\|\mathbf{Af}(x)\big\|}\,,
\end{aligned}$$

or, equivalently, if

$$\frac{\big\|\mathbf{Af}(x_+)\big\|}{\big\|\mathbf{f}(x_+)\big\|}\Big(\big\langle \mathbf{Af}(x_+)\,,\mathbf{Af}(x)\big\rangle + 1\Big)\cdot \left(\sqrt{\big(1+\big\|\mathbf{Af}(x)\big\|\big)^2 + 4\big\|\mathbf{f}(x_+)\big\|} + 1 + \big\|\mathbf{Af}(x)\big\|\right) = 2\,.$$

Let us rewrite this equality in terms of the vector $s := -\mathbf{Af}(x)$:

$$\frac{\big\|\mathbf{Af}(x+s)\big\|}{\big\|\mathbf{f}(x+s)\big\|}\Big(1 - \big\langle \mathbf{Af}(x+s)\,,s\big\rangle\Big)\left(\sqrt{\big(1+\|s\|\big)^2 + 4\big\|\mathbf{f}(x+s)\big\|} + 1 + \|s\|\right) = 2\,. \tag{5.58}$$

Given s, the left side can still vary depending on $\mathbf{A}$ and $\mathbf{f}$ and so this condition cannot be held in general. Therefore, the answer to our question is no, Broyden's choice is not optimal.

As we already know the (5.33)-triple $(p\,,q\,,v)$ for the BFGS update (see (5.37), (5.36)), Proposition 5.16 allows us to pose a similar question about this particular rank 2 update. By the proposition, it is optimal if

$$\|u\|^2\langle v\,,\mathbf{f}(x)\rangle^2 + \big\langle q\,,\mathbf{f}(x)\big\rangle^2 = \alpha^2\,, \tag{5.59}$$

where $u := \mathbf{A}\mathbf{f}(x_+)$ and, by (5.37), (5.36),

$$q = \frac{\|s - \|u\|^{-2}\langle s\,,u\rangle\, u\|}{\langle s\,,y\rangle\langle \mathbf{A}y\,,\mathbf{f}(x)\rangle}\mathbf{A}^*\Big(\langle \mathbf{A}y\,,\mathbf{f}(x)\rangle y - \langle \mathbf{A}y\,,y\rangle\mathbf{f}(x)\Big)\ ,\ v = -\frac{\mathbf{A}^*\mathbf{f}(x)}{\langle \mathbf{A}y\,,\mathbf{f}(x)\rangle}\,.$$

Substitution of these values into (5.59) results in

$$\|u\|^2\frac{\langle s\,,\mathbf{f}(x)\rangle^2}{\langle u+s\,,\mathbf{f}(x)\rangle^2}+\frac{\|s\|^2-\langle s\,,u\rangle^2}{\langle s\,,y\rangle^2\langle u+s\,,\mathbf{f}(x)\rangle^2}\Big\langle\langle s,\mathbf{f}(x)\rangle y-\langle y\,,s\rangle\mathbf{f}(x)\,,u+s\Big\rangle^2=\alpha^2.$$

Now, arguing as for Broyden's update, we arrive at the same answer: the BFGS update is not optimal.

5.8 Optimal on average rank 2 secant updates

According to the optimality on average philosophy, we should expect that the operator's response $\bar\alpha_+$ to our choice of the parameters p, q, v in (5.40) will be most probable, i.e., equal to its mathematical expectation. Under the assumption that the solution x_* is distributed in the set (5.48) uniformly, the mathematical expectation of $\bar\alpha_+$ is the center of the range of its feasible values. By (5.46) and (5.47), this range is $\left[0\,,\min\left\{\bar\delta_+(\bar\delta_+ + \bar\delta)\,,0.25\,(1-\bar\delta_+)^2\right\}\right]$, so that the expected value of $4\bar\alpha_+$ is

$$\min\left\{2\bar\delta_+(\bar\delta_+ + \bar\delta)\,,0.5\,(1-\bar\delta_+)^2\right\}.$$

Therefore, this time, the best choice of p, q, v is that, which minimizes the thickness (5.49) under the constraints

$$4\bar\alpha_+ = \min\left\{2\bar\delta_+(\bar\delta_+ + \bar\delta)\,,0.5\,(1-\bar\delta_+)^2\right\}\quad \&\quad \xi\le\bar\delta_+\le\eta$$

(see (5.45)). Equivalently, we have to maximize

$$F(t) := \sqrt{(1+t)^2+\min\{2t(t+b)\,,0.5(1-t)^2\}}+\sqrt{(1-t)^2-\min\{2t(t+b)\,,0.5(1-t)^2\}}$$

(here $t := \bar\delta_+$ as in (5.50)) over the segment $[\xi\,,\eta]$. As we have seen in the preceding section, the minimum under the radicals $= 0.5(1-t)^2$, if $\beta < \eta\ \&\ \beta\le t\le\eta$, and $= 2t(t+\bar\delta)$, if $\beta<\eta\ \&\ \xi\le t\le\beta\ \bigvee\ \beta\ge\eta$. So,

$$\begin{aligned}\beta<\eta\ \&\ \beta\le t\le\eta \Longrightarrow F(t) &= \sqrt{(1+t)^2+0.5(1-t)^2}+\sqrt{(1-t)^2-0.5(1-t)^2}\\ &=\frac{1}{\sqrt2}\left(\sqrt{3+2t+3t^2}+1-t\right)\end{aligned}$$

and $\beta < \eta$ & $\xi \le t \le \beta \bigvee \beta \ge \eta \Longrightarrow$

$$F(t) = \sqrt{(1+t)^2 + 2t\big(t+\bar\delta\big)} + \sqrt{(1-t)^2 - 2t\big(t+\bar\delta\big)}$$
$$= \sqrt{1+t^2+2t\big(t+1+\bar\delta\big)} + \sqrt{1+t^2-2t\big(t+1+\bar\delta\big)}.$$

Therefore, if $\beta < \eta$ (and so $0 \le \bar\delta < 0.6$ by Lemma 5.13), then $\max_{\xi \le t \le \eta} F(t) = \max\{m_1, m_2\}$, where

$$m_1 := \frac{1}{\sqrt2}\left(1 + \max_{\beta \le t \le \eta}\left(\sqrt{3+2t+3t^2} - t\right)\right)$$

$$m_2 := \max_{\xi \le t \le \beta}\left(\sqrt{1+t^2+2t\big(t+1+\bar\delta\big)} + \sqrt{1+t^2-2t\big(t+1+\bar\delta\big)}\right).$$

Differentiate the first objective $f_1(t) := \sqrt{3+2t+3t^2} - t$:

$$f_1'(t) = \frac{1+3t}{\sqrt{3+2t+3t^2}} - 1\,, \quad f_1''(t) = \frac{8}{\left(3+2t+3t^2\right)^{3/2}}\,.$$

We see that f_1' is increasing in $(0, \infty)$, so that f_1 is convex, $\max_{\beta \le t \le \eta} f_1(t) = \max\{f_1(\beta), f_1(\eta)\}$, and

$$m_1 = \frac{1}{\sqrt2}\Big(1 + \max\{f_1(\beta), f_1(\eta)\}\Big).$$

Consider next the second objective

$$f_2(t, \bar\delta) := \sqrt{1+t^2+2t\big(1+\bar\delta+t\big)} + \sqrt{1+t^2-2t\big(1+\bar\delta+t\big)}\,.$$

Lemma 5.18. *For all $\bar\delta \in [0, 1]$, the function $t \mapsto f_2(t, \bar\delta)$ is decreasing in $(0, \infty)$.*

Proof. Let for brevity $p(t) := 1+t^2$ and $q(t) := 2t\big(1+b+t\big)$. The derivative

$$f_2'(t) = \frac{q'+p'}{2\sqrt{p+q}} - \frac{q'-p'}{2\sqrt{p-q}}$$

is negative $\Longleftrightarrow 2pp'q' < q\big(p'^2 + q'^2\big)$ (as in the proof of Lemma 5.14), where $p'(t) = 2t$ and $q'(t) = 2\big(1+b+2t\big)$. Hence, $f_2'(t) < 0$

$$\Longleftrightarrow 2(1+t^2)\cdot 2t \cdot 2(1+b+2t) < 2t(1+b+t)\cdot\left(4t^2 + 4(1+b+2t)^2\right)$$

$$\Longleftrightarrow \left(1+t^2\right)(1+b+2t) < (1+b+t)\left(t^2 + (1+b)^2 + 4t(1+b) + 4t^2\right)$$

$$\Longleftrightarrow 2t^3 + (1+b)t^2 + 2t + 1 + b < 5t^3 + 9(1+b)t^2 + 5(1+b)^2 t + (1+b)^3,$$

which is clearly true. ☐

It follows that $m_2 = f_2(\xi\,,\bar\delta)$ and so

$$\beta<\eta \Longrightarrow \max_{\xi\le t\le\eta} F(t)=\max\left\{\frac{1}{\sqrt2}\Big(1+\max\{f_1(\beta)\,,\ f_1(\eta)\}\Big)\,,f_2(\xi\,,\bar\delta)\right\}.$$

As we are going to demonstrate now, if $\bar\delta+2\sqrt{\bar\alpha}<1$ (as assumed), then this maximum is equal to $f_2(\xi\,,\bar\delta)$. For this purpose, we have to consider $\xi\,,\eta\,,\beta$ as functions of the variables $4\bar\alpha$ and $\bar\delta$ (see (5.50), (5.52)). When $\sigma := 4\bar\alpha$ scans the range $\big[0\,,(1-\bar\delta)^2\big]$, the function $\sigma\mapsto\eta(\sigma,\bar\delta)$ is increasing from $\eta(0\,,\bar\delta)=0$ to $\eta\big((1-\bar\delta)^2,\bar\delta\big)=(1-\bar\delta)/\,2$. So,

$$\begin{aligned}\max_\sigma\Big\{f_1\big(\eta(\sigma,\bar\delta)\big)\,\Big|\,0\le\sigma\le(1-\bar\delta)^2\Big\}&=\max_t\Big\{f_1(t)\;\Big|\;0\le t\le(1-\bar\delta)/\,2\Big\}\\&=\max\Big\{f_1(0)\,,f_1\big((1-\bar\delta)/\,2\big)\Big\}\end{aligned}$$

(because f_1 is convex) and

$$\max\Big\{f_1\big(\beta(\bar\delta)\big)\,,\ f_1\big(\eta(\sigma\,,\bar\delta)\big)\Big\}\le\max\Big\{f_1\big(\beta(\bar\delta)\big)\,,\ f_1\big((1-\bar\delta)/\,2\big)\,,\ f_1(0)\Big\}.$$

Therefore,

$$\frac{1}{\sqrt2}\Big(1+\max\{f_1\big(\beta(\bar\delta)\big)\,,\ f_1\big(\eta(\sigma\,,\bar\delta)\big)\}\Big)\le$$

$$\frac{1}{\sqrt2}\Big(1+\max\{f_1\big(\beta(\bar\delta)\big)\,,\ f_1\big((1-\bar\delta)/\,2\big)\,,\ f_1(0)\}\Big).$$

Examination of plots of the functions $f_1\big(\beta(\bar\delta)\big)$ and $f_1\big((1-\bar\delta)/\,2\big)$ over the segment $[0\,,0.6]$ shows that the last maximum $=f_1(0)=\sqrt3$ and so

$$\frac{1}{\sqrt2}\Big(1+\max\big\{f_1\big(\beta(\bar\delta)\big)\,,f_1\big((1-\bar\delta)/2\big)\,,f_1(0),\big\}\Big)=\frac{1}{\sqrt2}\left(1+\sqrt3\right)=1.93\ldots\,.$$

At the same time, $\forall\,\sigma\in\big[0\,,\ (1-\bar\delta)^2\big]$, $f_2\big(\xi(\sigma,\bar\delta)\,,\bar\delta\big)$

$$\begin{aligned}&\ge\min_\sigma\Big\{f_2\big(\xi(\sigma,\bar\delta)\,,\bar\delta\big)\ \Big|\ 0\le\sigma\le(1-\bar\delta)^2\Big\}=f_2\Big(\xi\big((1-\bar\delta)^2,\bar\delta\big),\bar\delta\Big)\\&=f_2\left(\frac{\sqrt{2+2\bar\delta^2}-1-\bar\delta}{2}\,,\bar\delta\right),\end{aligned}$$

since f_2 is decreasing in t by Lemma 5.18. The function $\bar\delta\mapsto\sqrt{2+2\bar\delta^2}-1-\bar\delta$ is decreasing in $[0\,,0.6]$. Then, $\bar\delta\mapsto f_2\left(0.5\left(\sqrt{2+2\bar\delta^2}-1-\bar\delta\right),\bar\delta\right)$ is increasing and so $\displaystyle\min_{0\le\bar\delta\le0.6} f_2\left(0.5\left(\sqrt{2+2\bar\delta^2}-1-\bar\delta\right),\bar\delta\right)=f_2\left(0.5\left(\sqrt2-1\right),0\right)=$ $1.97\ldots$. We see that

$$0\le\bar\delta\le0.6\ \ \&\ \ 0\le\sigma\le(1-\bar\delta)^2\Longrightarrow\frac{1}{\sqrt2}\left(1+\max\Big\{f_1\big(\beta(\bar\delta)\big)\,,f_1\big(\eta(\sigma,\bar\delta)\big)\Big\}\right)$$

$$\leq \frac{1}{\sqrt{2}}\left(1+\max\left\{f_1(\beta(\bar{\delta}))\,,\max_t\left\{f_1(\eta(t,\bar{\delta}))\;\middle|\;0\leq t\leq(1-\bar{\delta})^2\right\}\right\}\right)=1.93...$$
$$<1.97...=\min_t\left\{f_2(\xi(t,\bar{\delta})\,,\bar{\delta})\;\middle|\;0\leq t\leq(1-\bar{\delta})^2\right\}\leq f_2(\xi(\sigma,\bar{\delta})\,,\bar{\delta}).$$

Thus, $\beta<\eta\Longrightarrow\max\limits_{\xi\leq t\leq\eta}F(t)=f_2(\xi\,,\bar{\delta})$.

Now we address the second possibility: $\beta\geq\eta$. Then $F(t)=f_2\big(t\,,\bar{\delta}\big)$, $\forall t\in[\xi\,,\eta]$, so that $\max\limits_{\xi\leq t\leq\eta}F(t)=\max\limits_{\xi\leq t\leq\eta}f_2\big(t\,,\bar{\delta}\big)=f_2\big(\xi\,,\bar{\delta}\big)$. So, independently of β, $\max\limits_{\xi\leq t\leq\eta}F(t)=f_2\big(\xi\,,\bar{\delta}\big)$ and $\arg\max\limits_{\xi\leq t\leq\eta}F(t)=\xi$. Hence, Proposition 5.16 remains in force for optimal on average (5.33)-triples.

As seen from Proposition 5.16, optimal triples $p\,,q\,,v$ constitute a rich subset of those admissible by (5.33). So, we can try to single out from it triples that promise further improvement of the method (5.40). One of desirable traits of a method of the family (5.40) is reduced condition number

$$\kappa:=\|\mathbf{A}_+\|\cdot\|\mathbf{A}_+^{-1}\|$$

of the updated operator $\mathbf{A}_+$. When $\mathbf{A}$ is given, evaluation of κ requires evaluation of the norm of the operator $\mathbf{A}+\mathbf{Af}(x_+)\langle v\,,\cdot\rangle+\mathbf{A}p\langle q\,,\cdot\rangle$ and of its inverse. The norm

$$\big\|\mathbf{A}(\mathbf{I}+\mathbf{f}(x_+)\langle v\,,\cdot\rangle+p\langle q\,,\cdot\rangle)\big\|\leq\|\mathbf{A}\|\cdot\big\|\mathbf{I}+\mathbf{f}(x_+)\langle v\,,\cdot\rangle+p\langle q\,,\cdot\rangle\big\|\,,$$

of whose two multipliers we control only the second. So, trying to minimize

$$\big\|\mathbf{A}(\mathbf{I}+\mathbf{f}(x_+)\langle v\,,\cdot\rangle+p\langle q\,,\cdot\rangle)\big\|,$$

we should direct our efforts to minimization of $\big\|\mathbf{I}+\mathbf{f}(x_+)\langle v\,,\cdot\rangle+p\langle q\,,\cdot\rangle\big\|$. Unfortunately, for $q\neq0$ this problem defies analytical solution. But for zero q, the condition number κ can be successfully minimized. Its minimization is carried out in the next section.

5.9 Minimum condition number of $\mathbf{I}+\mathbf{f}(x_+)\langle v\,,\cdot\rangle$

Applying the Sherman–Morrison lemma (Lemma 1.3) and bearing in mind that $\langle v\,,y\rangle=-1\Longrightarrow\big\langle v\,,\mathbf{f}(x_+)\big\rangle+1=\big\langle v\,,\mathbf{f}(x)\big\rangle$, we obtain that

$$\big\langle v\,,\mathbf{f}(x)\big\rangle\neq0\Longrightarrow\big(\mathbf{I}+\mathbf{f}(x_+)\langle v\,,\cdot\rangle\big)^{-1}=\mathbf{I}+\mathbf{f}(x_+)\left\langle-\frac{v}{\langle v\,,\mathbf{f}(x)\rangle}\,,\cdot\right\rangle.$$

So, for v satisfying $\big\langle v\,,\mathbf{f}(x)\big\rangle\neq0$, the condition number

$$\kappa(v)=\big\|\mathbf{I}+\mathbf{f}(x_+)\langle v\,,\cdot\rangle\big\|\cdot\left\|\mathbf{I}+\mathbf{f}(x_+)\left\langle-\frac{v}{\langle v\,,\mathbf{f}(x)\rangle}\,,\cdot\right\rangle\right\|.\tag{5.60}$$

Our task is to determine which vector v minimizes $\kappa(v)$ subject to

$$\langle v\,,y\rangle = -1 \quad \& \quad \left|\langle v\,,\mathbf{f}(x)\rangle\right| = \frac{\xi}{\left\|\mathbf{A}\mathbf{f}(x_+)\right\|} \tag{5.61}$$

(Corollary 5.17). We begin with minimizing the first norm. Then, we will see that its minimizer minimizes the second also. For the sake of brevity, it is advisable to adopt some abbreviations:

$$a := \mathbf{f}(x_+)\ ,\ b := \mathbf{f}(x)\ ,\ y := \mathbf{f}(x_+) - \mathbf{f}(x)\ ,\ \alpha := \|a\|\ ,\ \beta := \|b\|\ ,\ \nu := \|y\|\ ,$$

$$\zeta := \langle b\,,y\rangle\ ,\ det := \beta^2\nu^2 - \zeta^2\ ,\ \rho := \frac{\xi}{\left\|\mathbf{A}\mathbf{f}(x_+)\right\|}. \tag{5.62}$$

The first (squared) minimum

$$\begin{aligned} m_1(y,b,\rho) &:= \min_v\left\{\left\|\mathbf{I} + a\langle v\,,\cdot\rangle\right\|^2 \;\middle|\; \langle v\,,y\rangle = -1 \;\;\&\;\; \left|\langle b\,,v\rangle\right| = \rho\right\} \\ &= \min\left\{m_{11}\,,m_{12}\right\}, \end{aligned} \tag{5.63}$$

where

$$\begin{aligned} m_{11} &:= \min_v\left\{\left\|\mathbf{I} + a\langle v\,,\cdot\rangle\right\|^2 \;\middle|\; \langle v\,,y\rangle = -1 \;\;\&\;\; \langle b\,,v\rangle = \rho\right\}, \\ m_{12} &:= \min_v\left\{\left\|\mathbf{I} + a\langle v\,,\cdot\rangle\right\|^2 \;\middle|\; \langle v\,,y\rangle = -1 \;\;\&\;\; \langle b\,,v\rangle = -\rho\right\}. \end{aligned}$$

By Lemma 4.1,

$$\left\|\mathbf{I} + a\langle v\,,\cdot\rangle\right\|^2 = 1 + \langle a\,,v\rangle + 0.5\alpha\|v\|\left(\alpha\|v\| + \sqrt{\alpha^2\|v\|^2 + 4(1 + \langle a\,,v\rangle)}\right), \tag{5.64}$$

where $1 + \langle a\,,v\rangle = \langle b\,,v\rangle$. So, m_{11}

$$= \min_v\left\{\rho + 0.5\alpha\|v\|\left(\alpha\|v\| + \sqrt{\alpha^2\|v\|^2 + 4\rho}\right) \;\middle|\; \langle v\,,y\rangle = -1 \;\;\&\;\; \langle v\,,b\rangle = \rho\right\}$$

$$= \rho + 0.5\min_v\left\{\alpha\|v\|\left(\alpha\|v\| + \sqrt{\alpha^2\|v\|^2 + 4\rho}\right) \;\middle|\; \langle v\,,y\rangle = -1 \;\;\&\;\; \langle v\,,b\rangle = \rho\right\}$$

and $2(m_{11} - \rho)$

$$= \min_v\left\{\alpha\|v\|\left(\alpha\|v\| + \sqrt{\alpha^2\|v\|^2 + 4\rho}\right) \;\middle|\; \langle v\,,y\rangle = -1 \;\;\&\;\; \langle v\,,b\rangle = \rho\right\}$$

$$= \min_{(\tau,v)\in TV} \tau\left(\tau + \sqrt{\tau^2 + 4\rho}\right),$$

where $TV := \left\{(\tau, v) \;\middle|\; \langle v\,,y\rangle = -1 \;\;\&\;\; \langle v\,,b\rangle = \rho \;\;\&\;\; \alpha\|v\| = \tau\right\}$. By the lemma on sections (Lemma 1.4),

$$2(m_{11} - \rho) = \min_{\tau\in T}\ \min_{v\in V(\tau)} \tau\left(\tau + \sqrt{\tau^2 + 4\rho}\right) = \min_{\tau\in T}\tau\left(\tau + \sqrt{\tau^2 + 4\rho}\right),$$

where

$$V(\tau) := \{v \mid (\tau, v) \in TV\} = \{v \mid \langle v\,, y\rangle = -1 \ \ \& \ \ \langle v\,, b\rangle = \rho \ \ \& \ \ \alpha\|v\| = \tau\}$$

and $T := \{\tau \mid V(\tau) \neq \emptyset\}$. As follows from Corollary 1.7,

$$V(\tau) \neq \emptyset \iff \|\rho y + b\|^2 \leq \frac{\tau^2}{\alpha^2}\, det$$

and

$$det > 0 \Longrightarrow V(\tau) = \left\{ \frac{\rho\nu^2 + \zeta}{det}\, b - \frac{\beta^2 + \rho\zeta}{det}\, y + z \;\middle|\; \begin{array}{c} \langle y\,, z\rangle = \langle b\,, z\rangle = 0 \\ \|z\|^2 \leq \dfrac{\tau^2}{\alpha^2} - \dfrac{\|\rho y + b\|^2}{det} \end{array} \right\}, \tag{5.65}$$

$$det = 0 \Longrightarrow V(\tau) = \left\{ z - \frac{\langle y\,, z\rangle + 1}{\nu^2}\, y \;\middle|\; \left\| z - \frac{\langle y\,, z\rangle}{\nu^2}\, y \right\|^2 = \frac{\tau^2}{\alpha^2} - \frac{1}{\nu^2} \right\}. \tag{5.66}$$

In the first case, $T = \left\{\tau \;\middle|\; \tau \geq \alpha\|\rho y + b\| \big/ \sqrt{det}\right\}$,

$$\begin{aligned} 2(m_{11} - \rho) &= \min_{\tau} \left\{ \tau\left(\tau + \sqrt{\tau^2 + 4\rho}\right) \;\middle|\; \tau \geq \alpha \frac{\|\rho y + b\|}{\sqrt{det}} \right\} \\ &= \alpha \frac{\|\rho y + b\|}{\sqrt{det}} \left(\alpha \frac{\|\rho y + b\|}{\sqrt{det}} + \sqrt{\alpha^2 \frac{\|\rho y + b\|^2}{det} + 4\rho} \right), \end{aligned}$$

and

$$m_{11} = \rho + \alpha \frac{\|\rho y + b\|}{2\sqrt{det}} \left(\alpha \frac{\|\rho y + b\|}{\sqrt{det}} + \sqrt{\alpha^2 \frac{\|\rho y + b\|^2}{det} + 4\rho} \right) =: f(y, b, \rho). \tag{5.67}$$

Then

$$m_{12} = f(y, b, -\rho) = -\rho + \alpha \frac{\|b - \rho y\|}{2\sqrt{det}} \left(\alpha \frac{\|b - \rho y\|}{\sqrt{det}} + \sqrt{\alpha^2 \frac{\|b - \rho y\|^2}{det} - 4\rho} \right)$$

and

$$m_1(y, b, \rho) = \min\{f(y, b, \rho)\,,\, f(y, b, -\rho)\}\,.$$

The minimizer constitutes the singleton

$$V\left(\alpha \frac{\|\pm\rho y + b\|}{\sqrt{det}}\right) = \left\{ v \;\middle|\; v = \frac{\pm\rho\nu^2 + \zeta}{det}\, b - \frac{\beta^2 \pm \rho\zeta}{det}\, y \right\},$$

where $\pm$ is $+$ if $f(y, b, \rho) < f(y, b, -\rho)$ and $-$ otherwise. Let us see that the vector

$$v_* := v_*(y, b, \rho) := \frac{\pm\rho\nu^2 + \zeta}{det}\, b - \frac{\beta^2 \pm \rho\zeta}{det}\, y \tag{5.68}$$

is really the minimizer, that is, it satisfies the constraints (5.61) and

$$\left\|\mathbf{I} + a\langle v_*\,,\cdot\rangle\right\|^2 = \begin{cases} f(y,b,\rho)\,, \text{ if } f(y,b,\rho) \le f(y,b,-\rho)\,, \\ f(y,b,-\rho)\,, \text{ if } f(y,b,\rho) \ge f(y,b,-\rho). \end{cases}$$

Indeed,

$$\begin{aligned} \langle v_*\,, y\rangle &= \left\langle \frac{\pm\rho\nu^2 + \zeta}{det}\, b - \frac{\beta^2 \pm \rho\zeta}{det}\, y\,, y \right\rangle = \frac{1}{det}\Big((\pm\rho\nu^2 + \zeta)\zeta - (\beta^2 \pm \rho\zeta)\nu^2 \Big) \\ &= \frac{1}{det}\left(\pm\rho\nu^2\zeta + \zeta^2 - \beta^2\nu^2 \mp \rho\zeta\nu^2 \right) = -1\,, \\ \langle v_*\,, b\rangle &= \left\langle \frac{\pm\rho\nu^2 + \zeta}{det}\, b - \frac{\beta^2 \pm \rho\zeta}{det}\, y\,, b \right\rangle = \frac{1}{det}\Big((\pm\rho\nu^2 + \zeta)\beta^2 - (\beta^2 \pm \rho\zeta)\zeta \Big) \\ &= \frac{1}{det}\Big(\pm\rho\nu^2\beta^2 + \zeta\beta^2 - \beta^2\zeta \mp \rho\zeta^2 \Big) = \pm\rho\,, \end{aligned}$$

and, as in (5.64),

$$\begin{aligned} \left\|\mathbf{I} + a\langle v_*\,,\cdot\rangle\right\|^2 &= \langle b\,, v_*\rangle + 0.5\alpha\|v_*\| \left(\alpha\|v_*\| + \sqrt{\alpha^2\|v_*\|^2 + 4\langle b\,, v_*\rangle} \right) \\ &= \pm\rho + 0.5\alpha\|v_*\| \left(\alpha\|v_*\| + \sqrt{\alpha^2\|v_*\|^2 \pm 4\rho} \right), \end{aligned}$$

where

$$\begin{aligned} \|v_*\|^2 &= \left\| \frac{\pm\rho\nu^2 + \zeta}{det}\, b - \frac{\beta^2 \pm \rho\zeta}{det}\, y \right\|^2 \\ &= \frac{1}{det^2}\Big((\pm\rho\nu^2 + \zeta)^2\beta^2 - 2(\pm\rho\nu^2 + \zeta)(\beta^2 \pm \rho\zeta)\zeta + (\beta^2 \pm \rho\zeta)^2\nu^2 \Big) \\ &= \frac{1}{det^2}\Big(\rho^2\nu^2(\beta^2\nu^2 - \zeta^2) \pm 2\rho\zeta(\beta^2\nu^2 - \zeta^2) + \beta^2(\beta^2\nu^2 - \zeta^2) \Big) \\ &= \frac{\rho^2\nu^2 \pm 2\rho\zeta + \beta^2}{det} = \frac{\|\pm\rho y + b\|^2}{det}\,. \end{aligned}$$

Hence, $\left\|\mathbf{I} + a\langle v_*\,,\cdot\rangle\right\|^2$

$$\begin{aligned} &= \pm\rho + \alpha\frac{\|\pm\rho y + b\|}{2\sqrt{det}} \left(\alpha\frac{\|\pm\rho y + b\|}{\sqrt{det}} + \sqrt{\alpha^2\frac{\|\pm\rho y + b\|^2}{det} \pm 4\rho} \right) \\ &= f(y,b,\pm\rho)\,. \end{aligned}$$

We now address the second possibility $det = 0$, that is, $\beta^2\nu^2 = \zeta^2$. This means that y and b are linearly dependent, $\rho y + b = 0$, and, by (5.66), $V(\tau) \ne \emptyset \iff \tau \ge \alpha/\nu$. So,

$$T = \begin{cases} \{\tau \mid \tau \ge \alpha/\nu\}\,, \text{ if } \rho^2\nu^2 = \zeta^2 = \beta^2\nu^2\,, \\ \emptyset\,, \text{ otherwise (that is, if the constraints (5.61) are inconsistent),} \end{cases}$$

$$m_{11} = \begin{cases} \rho + \dfrac{\alpha}{2\nu}\left(\dfrac{\alpha}{\nu} + \sqrt{\dfrac{\alpha^2}{\nu^2} + 4\rho}\right), & \text{if } \zeta = \rho\nu^2 = \beta\nu\,, \\ \infty\,, \text{ otherwise,} \end{cases}$$

$$m_{12} = \begin{cases} -\rho + \dfrac{\alpha}{2\nu}\left(\dfrac{\alpha}{\nu} + \sqrt{\dfrac{\alpha^2}{\nu^2} - 4\rho}\right), & \text{if } \zeta = -\rho\nu^2 = -\beta\nu\,, \\ \infty\,, \text{ otherwise,} \end{cases}$$

and

$$m_1(y, b, \rho) = \begin{cases} \rho + \dfrac{\alpha}{2\nu}\left(\dfrac{\alpha}{\nu} + \sqrt{\dfrac{\alpha^2}{\nu^2} + 4\rho}\right), & \text{if } \zeta = \rho\nu^2 = \beta\nu\,, \\ -\rho + \dfrac{\alpha}{2\nu}\left(\dfrac{\alpha}{\nu} + \sqrt{\dfrac{\alpha^2}{\nu^2} - 4\rho}\right), & \text{if } \zeta = -\rho\nu^2 = -\beta\nu\,, \\ \infty\,, \text{ otherwise.} \end{cases} \tag{5.69}$$

The minimizer v_* is the only vector of the set (see (5.66)) $V(\alpha/\nu) = \{v \mid v = y/\nu^2\}$, i.e.,

$$det = 0 \Longrightarrow v_* := v_*(y, b, \rho) = \nu^{-2} y\,.$$

Now consider now the second minimum

$$m_2(y, b, \rho) := \min_v \left\{ \left\| \mathbf{I} + a\left\langle -\frac{v}{\langle v\,, b\rangle}\,, \cdot \right\rangle \right\|^2 \;\middle|\; \langle v\,, y\rangle = -1 \;\;\&\;\; \big|\langle v\,, b\rangle\big| = \rho \right\}. \tag{5.70}$$

If v obeys the constraints, then, for $w := -v/\langle v\,, b\rangle$, we have $\big|\langle w\,, y\rangle\big| = 1/\rho$ and $\langle w\,, b\rangle = -1$, so that

$$\begin{aligned} m_2(y, b, \rho) &= \min_w \left\{ \|\mathbf{I} + a\langle w\,, \cdot\rangle\|^2 \;\middle|\; \big|\langle w\,, y\rangle\big| = \frac{1}{\rho} \;\;\&\;\; \langle w\,, b\rangle = -1 \right\} \\ &= m_1(b, y, 1/\rho)\,. \end{aligned}$$

If $det > 0$, then, according to the definition (5.67) of the function f,

$$f(b,y,1/\rho) = \frac{1}{\rho} + \alpha\frac{\|\rho^{-1}b+y\|}{2\sqrt{det}}\left(\alpha\frac{\|\rho^{-1}b+y\|}{\sqrt{det}} + \sqrt{\alpha^2\frac{\|\rho^{-1}b+y\|^2}{det} + \frac{4}{\rho}}\right)$$

$$= \frac{1}{\rho} + \alpha\frac{\|\rho y+b\|}{2\rho\sqrt{det}}\left(\alpha\frac{\|\rho y+b\|}{\rho\sqrt{det}} + \sqrt{\alpha^2\frac{\|\rho y+b\|^2}{\rho^2 det} + \frac{4}{\rho}}\right)$$

$$= \frac{1}{\rho^2}\left(\rho + \alpha\frac{\|\rho y+b\|}{2\sqrt{det}}\left(\alpha\frac{\|\rho y+b\|}{\sqrt{det}} + \sqrt{\alpha^2\frac{\|\rho y+b\|^2}{det} + 4\rho}\right)\right)$$

$$= \frac{1}{\rho^2}f(y,b,\rho)$$

and, analogously, $f(b,y,-1/\rho) = \rho^{-2}f(y,b,-\rho)$. Hence,

$$m_2(y,b,\rho) = \frac{1}{\rho^2}\min\{f(b,y,\rho)\,,\,f(b,y,-\rho)\} = \frac{1}{\rho^2}m_1(y,b,\rho)\,.$$

It follows that the minimum (5.70) is attained on the same vector v_* (see (5.68)) as the minimum (5.63):

$$m_2(y,b,\rho) = \left\|\mathbf{I} + a\left\langle -\frac{v_*}{\langle v_*\,,b\rangle}\,,\cdot\right\rangle\right\|\,,$$

and that

$$\min_v\left\{\kappa(v)\;\middle|\;\langle v\,,y\rangle = -1\;\;\&\;\;|\langle v\,,b\rangle| = \rho\right\} \geq \sqrt{m_1(y,b,\,\rho)\,m_2(y,b,\,\rho)}$$

$$= \frac{1}{\rho}m_1(y,b,\,\rho) = \frac{1}{\rho}\left\|\mathbf{I} + a\langle v_*\,,\cdot\rangle\right\|^2.$$

On the other hand, as has been shown above, v_* satisfies the constraints and so this minimum

$$\leq \kappa(v_*) = \left\|\mathbf{I}+a\langle v_*\,,\cdot\rangle\right\|\cdot\left\|\mathbf{I} + a\left\langle -\frac{v_*}{\langle v_*\,,\mathbf{f}(x)\rangle}\,,\cdot\right\rangle\right\| = \sqrt{m_1(y,b,\,\rho)\,m_2(y,b,\,\rho)}.$$

Thus,

$$det > 0 \Longrightarrow \min_v\left\{\kappa(v)\;\middle|\;\langle v\,,y\rangle = -1\;\;\&\;\;|\langle v\,,b\rangle| = \rho\right\} = \kappa(v_*)\,.$$

If $det = 0$, then, by (5.69),

$$m_1(b,y,1/\rho) = \begin{cases} \dfrac{1}{\rho} + \dfrac{\alpha}{2\beta}\left(\dfrac{\alpha}{\beta} + \sqrt{\dfrac{\alpha^2}{\beta^2} + \dfrac{4}{\rho}}\right), & \text{if } \zeta = \beta^2/\rho = \beta\nu\,, \\ -\dfrac{1}{\rho} + \dfrac{\alpha}{2\beta}\left(\dfrac{\alpha}{\beta} + \sqrt{\dfrac{\alpha^2}{\beta^2} - \dfrac{4}{\rho}}\right), & \text{if } \zeta = -\beta^2/\rho = -\beta\nu\,, \\ \infty\,, \ \text{otherwise.} \end{cases}$$

In particular, $\zeta = \beta^2/\rho = \beta\nu \Longrightarrow \beta/\rho = \nu \Longrightarrow$

$$\begin{aligned} m_1(b,y,1/\rho) &= \frac{1}{\rho^2}\left(\rho + \frac{\alpha}{2\beta/\rho}\left(\frac{\alpha}{\beta/\rho} + \sqrt{\frac{\alpha^2}{\beta^2/\rho^2} + 4\rho}\right)\right) \\ &= \frac{1}{\rho^2}\left(\rho + \frac{\alpha}{2\nu}\left(\frac{\alpha}{\nu} + \sqrt{\frac{\alpha^2}{\nu^2} + 4\rho}\right)\right) = \frac{1}{\rho^2}m_1(y,b,\rho) \end{aligned}$$

and likewise if $\zeta = -\beta^2/\rho = -\beta\nu$. Hence, as in the case $det > 0$, $m_2(y,b,\rho) = \rho^{-2}m_1(y,b,\rho)$ and

$$\min_v\Big\{\kappa(v)\ \Big|\ \langle v\,,y\rangle = -1\ \ \&\ \ \big|\langle v\,,b\rangle\big| = \rho\Big\} = \kappa(v_*)\,.$$

Summing up the analysis of this section, we can state its result (with the abbreviations (5.62) and the definition (5.67) of the function f still in force) as

Proposition 5.19. *The minimum condition number of the operator* $\mathbf{I} + a\langle v\,,\cdot\rangle$ *over all* v *satisfying*

$$\langle v\,,y\rangle = -1\ \ \&\ \ \big|\langle v\,,b\rangle\big| = \rho$$

is attained for

$$v = v_* := \begin{cases} \dfrac{\rho\nu^2 + \zeta}{det}\,b - \dfrac{\beta^2 + \rho\zeta}{det}\,y\,, & \textit{if } det > 0\ \ \&\ \ f(y,b,\rho) \le f(y,b,-\rho)\,, \\ \dfrac{-\rho\nu^2 + \zeta}{det}\,b - \dfrac{\beta^2 - \rho\zeta}{det}\,y\,, & \textit{if } det > 0\ \ \&\ \ f(y,b,\rho) \ge f(y,b,-\rho)\,, \\ \nu^{-2}y\,, & \textit{if } det = 0\,. \end{cases}$$

The numerical performance of the secant update method

$$x_+ := x - \mathbf{Af}(x)\ ,\ \ \mathbf{A}_+ := \mathbf{A} + \mathbf{Af}(x_+)\big\langle v_*\,,\cdot\big\rangle\,, \tag{5.71}$$

based on this proposition can be expected to be more stable if we rephrase the definition of v_* in terms of a instead of $y = a - b$. Substituting $a - b$ for y gives

$$v_* = \frac{1}{det}\Big(\big(\pm\rho\nu^2 + \zeta + \beta^2 \pm \rho\zeta\big)b - \big(\beta^2 \pm \rho\zeta\big)a\Big)\,,$$

where $\nu^2 = \|a-b\|^2 = \alpha^2 - 2\langle a\,,b\rangle + \beta^2$, $\zeta = \langle b\,,a-b\rangle = \langle a\,,b\rangle - \beta^2$, and so $\pm\rho\nu^2 + \zeta + \beta^2 \pm \rho\zeta = \pm\rho\alpha^2 + (1\mp\rho)\langle a\,,b\rangle$ and $\beta^2 \pm \rho\zeta = \pm\rho\langle a\,,b\rangle + (1\mp\rho)\beta^2$. The expression of det changes correspondingly:

$$\beta^2\nu^2 - \zeta^2 = \beta^2\big(\alpha^2 - 2\langle a\,,b\rangle + \beta^2\big) - \big(\langle a\,,b\rangle - \beta^2\big)^2 = \alpha^2\beta^2 - \langle a\,,b\rangle^2 .$$

As $\rho y + b = \rho a + (1-\rho)b$, the definition (5.67) of f is rewritten as $f(a-b, b, \rho)$

$$= \rho + \alpha\frac{\|\rho a + (1-\rho)b\|}{2\sqrt{det}}\left(\alpha\frac{\|\rho a + (1-\rho)b\|}{\sqrt{det}} + \sqrt{\alpha^2\frac{\|\rho a + (1-\rho)b\|^2}{det} + 4\rho}\,\right). \tag{5.72}$$

Thus, in terms of $a := \mathbf{f}(x_+)$ and $b := \mathbf{f}(x)$,

$$v_* = \begin{cases} \dfrac{\rho\alpha^2 + (1-\rho)\langle a\,,b\rangle}{det}\,b - \dfrac{\rho\langle a\,,b\rangle + (1-\rho)\beta^2}{det}\,a\,, \\ \qquad\qquad \text{if } det>0 \ \ \& \ \ f(a-b,b,\rho) \le f(a-b,b,-\rho)\,, \\ \dfrac{-\rho\alpha^2 + (1+\rho)\langle a\,,b\rangle}{det}\,b + \dfrac{\rho\langle a\,,b\rangle - (1+\rho)\beta^2}{det}\,a\,, \\ \qquad\qquad \text{if } det>0 \ \ \& \ \ f(a-b,b,\rho) \ge f(a-b,b,-\rho)\,, \\ \dfrac{a-b}{\|a-b\|^2}\,, \ \text{ if } det = 0\,. \end{cases} \tag{5.73}$$

If the current iteration $\big(x\,,\mathbf{A}\,,\mathbf{f}(x)\big)$ of the method (5.71) satisfies the condition

$$4\alpha \le \big(1 - \|\mathbf{A}b\|\big)^2,$$

then the proposition suggests constructing the next iteration $(x_+, \mathbf{A}_+)$ according to the following.

Algorithm

1: $x_+ := x - \mathbf{A}\mathbf{f}(x)$;
2: Evaluate the vectors $a := \mathbf{f}(x_+)$, $\mathbf{A}\mathbf{f}(x_+)$;
3: Evaluate the norms $\alpha := \|\mathbf{f}(x_+)\|$, $\beta := \|\mathbf{f}(x)\|$, $\delta := \|\mathbf{A}\mathbf{f}(x)\|$,
$\lambda := \|\mathbf{A}\mathbf{f}(x_+)\|$, and the product $\mu := \langle a\,,b\rangle$;
4: $\xi := 0.5\Big(\sqrt{(1+\delta)^2 + 4\alpha} - 1 - \delta\Big)$, $\rho := \xi/\lambda$;
5: Evaluate $f(a-b, b, \rho)$ as prescribed by (5.72);
6: Evaluate v_* as prescribed by (5.73);
7: $\mathbf{A}_+ := \mathbf{A} + \mathbf{A}\mathbf{f}(x_+)\langle v_*\,,\cdot\rangle$.

The method (5.71) seems to be new and as such it deserves a look into its convergence properties.

5.10 Research projects

Theorems 5.3 and 5.9 analyze different situations. The first deals with differentiable operators, while the second admits nondifferentiable ones. On the other hand, Theorem 5.3 requires only regular continuity of the derivative vs. Lipschitz continuity of the selected dd in Theorem 5.9. The optimality condition established by Proposition 5.16 is based on Theorem 5.9. Getting its analog for regularly smooth operators seems an interesting problem. Another is to extend Theorem 5.3 to operators with regularly continuous dd's. Such an extension would subsume both theorems. It would become a significant contribution to research of solvability of operator equations thinly represented in literature.

In Section 5.7, we found that Broyden's method is not (entropy) optimal. This observation provokes the question: what is? The rank 1 optimal secant update method corresponding to the secant update developed in the last section looks promising. However, its merits (or faults) must be substantiated either theoretically or experimentally (or both). A similar question should be asked about the BFGS update. If it is not optimal, which optimal (according to Proposition 5.16) rank 2 secant update is? To answer the question, one has to specify his choice of optimal triple (p, q, v) and to study convergence properties of the resulting iterative method.

As seen from Proposition 5.16, the optimal triples (p, q, v) constitute a rich set. This fact prompts the question: which triple among the optimal ones determines a secant update method with maximum convergence domain, that is, the set of all starters x_0, from which the method generates a convergent sequence x_n of successive approximations? I do not know the answer. A theorem that would provide it will result in an interesting iterative method.

Methods of the type (5.1) can be looked upon as controllable discrete dynamical systems with the operator $\mathbf{A}$ being the control parameter. Under such a point of view, the problem of optimization of an iterative method becomes a problem of optimal control of a discrete dynamical system subject to the constraints

$$\mathop{\&}_{k=0}^{n} x_k \in D := dom(\mathbf{f})\,. \tag{5.74}$$

The natural objective of such a problem is the norm of $\mathbf{f}(x_n)$ that, given a starter $x_0 \in D$, is achievable by a choice of controls $\mathbf{A}_0, \mathbf{A}_1, \dots, \mathbf{A}_{n-1}$. We should wish to minimize this norm and to determine the optimal control $\left(\mathbf{A}_0^{(n)}, \dots, \mathbf{A}_{n-1}^{(n)}\right)$ that results in minimum $\left\|\mathbf{f}(x_n)\right\|$:

$$\left(\mathbf{A}_0^{(n)}, \dots, \mathbf{A}_{n-1}^{(n)}\right) = \arg \min_{\mathbf{A}_0, \dots, \mathbf{A}_{n-1}} \left\|\mathbf{f}(x_n)\right\|$$

subject to

$$\mathop{\&}_{k=0}^{n-1}\Big(x_{k+1}=x_k-\mathbf{A}_k\mathbf{f}(x_k)\ \ \&\ \ \mathbf{A}_k\mathbf{f}(x_k)\in D-x_k\Big)\,.$$

Using the lemma on sections (Lemma 1.4) and induction, one can prove that

$$\mathop{\&}_{k=0}^{n-1}\mathbf{A}_k^{(n)}=\arg\min_{\mathbf{A}}\Big\{\big\|\mathbf{f}(x_{k-1}-\mathbf{A}\mathbf{f}(x_{k-1}))\big\|\ \Big|\ \mathbf{A}\mathbf{f}(x_{k-1})\in D-x_{k-1}\Big\}\,. \quad (5.75)$$

Unlike the approach we have pursued in this chapter, here there is no need to assume the existence of a solution of the equation $\mathbf{f}(x)=0$. Moreover, solution of the problem (5.75) guarantees the inequality $\|\mathbf{f}(x_+)\|\le\|\mathbf{f}(x)\|$, which in turn makes probable convergence of the iterations x_n to a minimizer of the norm $\|\mathbf{f}(x)\|$ over D. The rate of convergence is a separate problem. It is not inconceivable that it will prove to be prohibitively slow.

Another worthy goal is the maximum convergence domain of a method (5.1). The difficulty here is that we do not know how to measure the size of this set. The volume? The diameter? Somebody has to suggest a computable measure of the size.

To make a method (5.1) to obey the constraint (5.74) at each iteration (*globalize* it) is an additional challenge. One approach to globalization of a method is to replace the operator $\mathbf{f}$ by the composition $\mathbf{f}\circ\mathbf{P}_D$, where $\mathbf{P}_D$ is the metric projector onto D. Another is to employ a line search in the direction prescribed by the operator $\mathbf{A}$. Each one has its disadvantage. Composition with metric projection most often strips $\mathbf{f}$ of its differentiability, which is a desirable property of an operator. Besides, metric projection may stick the iterations in an undesirable fix point x of the composed operator, where $\mathbf{f}(x)\ne 0$.

Chapter 6

Optimal secant-type methods

By the term secant-type methods we mean all methods of the form

$$z_+ := F\big(z\,, \mathbf{f}(z)\big)\,, \tag{6.1}$$

which, like the generic secant method (0.3), require only one evaluation of the operator $\mathbf{f}$ per iteration (no derivatives). Such are, for example, the methods (3.4) and (4.1):

$$F\left(\begin{bmatrix} x \\ \mathbf{A} \end{bmatrix}, \mathbf{f}\left(\begin{bmatrix} x \\ \mathbf{A} \end{bmatrix}\right)\right) = \begin{bmatrix} x - \mathbf{A}\mathbf{f}(x) \\ 2\mathbf{A} - \mathbf{A}\big[x - \mathbf{A}\mathbf{f}(x), x \,|\, \mathbf{f}\big]\mathbf{A} \end{bmatrix}$$

for Ulm's method, and

$$F\left(\begin{bmatrix} x \\ \mathbf{A} \end{bmatrix}, \mathbf{f}\left(\begin{bmatrix} x \\ \mathbf{A} \end{bmatrix}\right)\right) = \begin{bmatrix} x - \mathbf{A}\mathbf{f}(x) \\ \mathbf{A} - \dfrac{\mathbf{A}\mathbf{f}(x_+)}{\big\langle \mathbf{A}^*\mathbf{A}\mathbf{f}(x)\,, \mathbf{f}(x_+) - \mathbf{f}(x)\big\rangle}\big\langle \mathbf{A}^*\mathbf{A}\mathbf{f}(x)\,, \cdot\big\rangle \end{bmatrix}$$

for Broyden's. Various members of this class differ from one another by the mapping F used to generate the next approximation z_+ from the current iteration $\big(z\,, \mathbf{f}(z)\big)$.

6.1 Motivation

In this chapter, as in the preceding one, we are motivated by the natural desire to choose among all methods of a certain class the most preferable one using the same (entropy) optimality criterion.

6.2 Existence and uniqueness of solutions (scalar equations)

To my knowledge, the class of methods (6.1) has not been discussed in the literature so far. So, we first should ask ourselves what can be gained by these more general kinds of methods in the simpliest case of scalar equations. The answer, which is not trivial in itself, can give valuable insight into what is feasible in designing general iterative methods. Besides, considering the case of scalar equations may help to determine the best strategy in line search, since the restriction of an operator to the ray in a selected direction is a scalar function. In order to shed some light on the question, we will need a variant of Theorem 5.3 for scalar equations. As expected, for scalar equations the existence and uniqueness conditions admit a more precise form than in the general case. Therefore, the corresponding proposition below is not a simple corollary of Theorem 5.3 and needs a proof. We state it for functions defined on $\mathbb{R}$. This is not a serious limitation, since one always can extend an ω-regularly smooth function f defined on a segment $[\underline{x}\,,\,\overline{x}]$ to $\mathbb{R}$ by letting

$$f(x) := \begin{cases} f(\underline{x}) + f'(\underline{x})(x - \underline{x}), & \text{if } x < \underline{x}\,, \\ f(\overline{x}) + f`(\overline{x})(x - \overline{x}), & \text{if } x > \overline{x}\,. \end{cases}$$

The functions w, p, q are as in Theorem 5.3:

$$w(t) := \int_0^t \omega(\tau)\,d\tau\,,\ \Psi(\alpha,t) = \begin{cases} t\omega(\alpha) - w(\alpha) + w(\alpha - t)\,, & \text{if } 0 \le t \le \alpha\,, \\ \alpha\omega(\alpha) - 2w(\alpha) + w(t)\,, & \text{if } t \ge \alpha \ge 0\,. \end{cases}$$

$$p(\alpha\,,t) := t\,\omega(\alpha) + \Psi(\alpha\,,t)\,,\ q(\alpha\,,t) := t\,\omega(\alpha) - \Psi(\alpha\,,t)\,.$$

Proposition 6.1. *Let a function $f : \mathbb{R} \to \mathbb{R}$ be differentiable everywhere and ω-regularly smooth on $\mathbb{R}$ in the sense of (2.2). Denote:*

$$a := f(x_0)\,,\ b := f'(x_0)\,,\ \kappa := \omega^{-1}(|b|)\,.$$

1° *If $|a| > w(\kappa)$ & $ab > 0$, then f has no zeros in the interval*

$$\left(x_0 - p^{-1}(\kappa\,, |a|)\,,\, x_0 + q_+^{-1}(\kappa\,, -|a|)\right)\,.$$

2° *If $|a| > w(\kappa)$ & $ab < 0$, then f has no zeros in the interval*

$$\left(x_0 - q_+^{-1}(\kappa\,, -|a|)\,,\, x_0 + p^{-1}(\kappa\,, |a|)\right)\,.$$

3° *If $|a| \le w(\kappa)$ & $ab > 0$, then f has a zero in the segment*

$$\left[x_0 - q_-^{-1}(\kappa\,, |a|)\,,\, x_0 - p^{-1}(\kappa\,, |a|)\right]\,,$$

which is the only zero of f *in the interval*

$$\left(x_0 - q_+^{-1}(\kappa\,, |a|)\,,\ x_0 + q_+^{-1}(\kappa\,, -|a|)\right).$$

4° *If* $|a| \le w(\kappa)$ & $ab < 0$, *then* f *has a zero in*

$$\left[x_0 + p^{-1}(\kappa\,, |a|)\,,\ x_0 + q_-^{-1}(\kappa\,, |a|)\right]$$

and no other zeros in $\left(x_0 - q_+^{-1}(\kappa\,, -|a|)\,,\ x_0 + q_+^{-1}(\kappa\,, |a|)\right)$.

Proof. Let x_* be a zero of f. Then by the Newton–Leibnitz theorem,

$$\begin{aligned} 0 = f(x_*) &= f(x_0) + f'(x_0)(x_* - x_0) + \int_{x_0}^{x_*} \left(f'(t) - f'(x_0)\right) dt \\ &= a + b(x_* - x_0) + \int_{x_0}^{x_*} \left(f'(t) - f'(x_0)\right) dt \end{aligned}$$

and so

$$\left|a + b(x_* - x_0)\right| \le \int_{\min\{x_0, x_*\}}^{\max\{x_0, x_*\}} \left|f'(t) - f'(x_0)\right| dt\,.$$

Arguing as in the proof of the theorem, we get the inequality

$$\int_{\min\{x_0, x_*\}}^{\max\{x_0, x_*\}} \left|f'(t) - f'(x_0)\right| dt \le \Psi\left(\kappa\,, |x_* - x_0|\right).$$

So (with the abbreviation $\Delta := x_* - x_0$), $|a + b\Delta| \le \Psi\left(\kappa\,, |\Delta|\right)$. As is easy to verify, $|a + b\Delta| = \left||a| sign(ab) + |b|\Delta\right|$. Then $\left||a| sign(ab) + |b|\Delta\right| \le \Psi\left(\kappa\,, |\Delta|\right)$

$$\Longleftrightarrow -|b|\Delta - \Psi\left(\kappa\,, \Delta|\right) \le |a| sign(ab) \le -|b|\Delta + \Psi\left(\kappa\,, \Delta|\right).$$

In particular (because $|b| = \omega(\kappa)$ by the definition of κ),

$$\begin{aligned} ab > 0 &\Longrightarrow -\Delta\omega(\kappa) - \Psi\left(\kappa\,, |\Delta|\right) \le |a| \le -\Delta\omega(\kappa) + \Psi\left(\kappa\,, |\Delta|\right) \\ &\Longleftrightarrow \begin{cases} |a| \le -q(\kappa\,, \Delta)\,, & \text{if } \Delta > 0\,, \\ q\left(\kappa\,, |\Delta|\right) \le |a| \le p\left(\kappa\,, |\Delta|\right), & \text{if } \Delta < 0\,. \end{cases} \end{aligned}$$

According to Section 5.3, the function $t \mapsto p(\alpha, t)$ is increasing in $(0\,, \infty)$, so that the inequality $|a| \le p\left(\kappa\,, |\Delta|\right)$ is equvivalent to $p^{-1}\left(\kappa\,, |a|\right) \le |\Delta|$. Likewise, the function $t \mapsto -q(\kappa\,, \Delta)$ is increasing in $(\kappa\,, \infty)$ and so, for $\Delta > \kappa$,

$$|a| \le -q(\kappa\,, \Delta) \Longleftrightarrow q(\kappa\,, \Delta) \le -|a| \Longleftrightarrow \Delta \le q_+^{-1}\left(\kappa\,, -|a|\right).$$

The inequality $q\left(\kappa\,, |\Delta|\right) \le |a|$ is trivial if $|a| > q(\kappa\,, \kappa) = w(\kappa)$. Otherwise, it is equivalent to

$$|\Delta| \le q_-^{-1}\left(\kappa\,, |a|\right) \bigvee |\Delta| \ge q_+^{-1}\left(\kappa\,, |a|\right).$$

Thus,

$$ab > 0 \;\&\; |a| > w(\kappa) \Longrightarrow x_* - x_0 \le -p^{-1}(\kappa\,,|a|) \bigvee x_* - x_0 \ge q_+^{-1}(\kappa\,,-|a|)\,,$$

$\Big($meaning that the equation $f(x) = 0$ has no solutions in the interval $\big(x_0 - p^{-1}(\kappa\,,-|a|)\,,\, x_0 + q_+^{-1}(\kappa\,,-|a|)\big)\Big)$, while $ab > 0 \;\&\; |a| \le w(\kappa)$ implies

$$\Delta \le -q_+^{-1}(\kappa\,,|a|) \bigvee -q_-^{-1}(\kappa\,,|a|) \le \Delta \le -p^{-1}(\kappa\,,|a|) \bigvee \Delta \ge q_+^{-1}(\kappa\,,-|a|)$$

$\Big($i.e., there is a solution in the segment $\big[x_0 - q_-^{-1}(\kappa\,,|a|)\,,\, x_0 - p^{-1}(\kappa\,,|a|)\big]$ and no other solutions in the interval $\big(x_0 - q_+^{-1}(\kappa\,,|a|)\,,\, x_0 + q_+^{-1}(\kappa\,,-|a|)\big)\Big)$. Similarly,

$$\begin{aligned} ab < 0 \Longrightarrow\ & \Delta\omega(\kappa) - \Psi(\kappa\,,|\Delta|) \le |a| \le \Delta\omega(\kappa) + \Psi(\kappa\,,|\Delta|) \\ \Longleftrightarrow\ & \begin{cases} q(\kappa\,,\Delta) \le |a| \le p(\kappa\,,\Delta)\,, & \text{if } \Delta > 0\,, \\ |a| \le -q(\kappa\,,|\Delta|)\,, & \text{if } \Delta < 0\,, \end{cases} \end{aligned}$$

so that

$$ab < 0 \;\&\; |a| > w(\kappa) \Longrightarrow \Delta \le -q_+^{-1}(\kappa\,,-|a|) \bigvee \Delta \ge p^{-1}(\kappa\,,|a|)\,,$$

$\Big($no solutions in $\big(x_0 - q_+^{-1}(\kappa\,,-|a|)\,,\, x_0 + p^{-1}(\kappa\,,|a|)\big)\Big)$ and $ab < 0 \;\&\; |a| \le w(\kappa) \Longrightarrow$

$$\Delta \le -q_+^{-1}(\kappa\,,-|a|) \bigvee p^{-1}(\kappa\,,|a|) \le \Delta \le q_-^{-1}(\kappa\,,|a|) \bigvee \Delta \ge q_+^{-1}(\kappa\,,|a|)$$

$\Big($a solution in $\big(x_0 + p^{-1}(\kappa\,,|a|)\,,\, x_0 + q_-^{-1}(\kappa\,,|a|)\big)$ and no other solution in $\big(x_0 - q_+^{-1}(\kappa\,,-|a|)\,,\, x_0 + q_+^{-1}(\kappa\,,|a|)\big)\Big)$. □

For Lipschitz smooth functions $\big(\omega(t) = ct\big)$, the functions w, p, q symplify:

$$w(t) = 0.5ct^2\,,\quad p(\alpha\,,t) = ct(\alpha + 0.5t)\,,\quad q(\alpha\,,t) = ct(\alpha - 0.5t)\,,$$

so that $p^{-1}(\kappa\,,|a|) = \sqrt{\kappa^2 + 2c^{-1}|a|} - \kappa\,,$

$$q_-^{-1}(\kappa\,,|a|) = \kappa - \sqrt{\kappa^2 - 2c^{-1}|a|}\,,\quad q_+^{-1}(\kappa\,,|a|) = \kappa + \sqrt{\kappa^2 - 2c^{-1}|a|}\,.$$

Correspondingly, for Lipschitz smooth scalar functions, we have

Corollary 6.2. *Let a function* $f : \mathbb{R} \to \mathbb{R}$ *be Lipschitz smooth on* $\mathbb{R}$:

$$\big|f'(x_1) - f'(x_2)\big| \le c|x_1 - x_2|\,,\quad \forall\, x_1, x_2\,. \tag{6.2}$$

Denote:

$$a := f(x_0) \ , \ b := f'(x_0) \ , \ \kappa := c^{-1}(|b|) \ .$$

1° *If* $|a| > 0.5c\kappa^2$ & $ab > 0$, *then* f *has no zeroes in the interval*

$$\left(x_0 - \sqrt{\kappa^2 + 2c^{-1}|a|} - \kappa \, , \, x_0 + \kappa + \sqrt{\kappa^2 - 2c^{-1}|a|}\right) .$$

2° *If* $|a| > 0.5c\kappa^2$ & $ab < 0$, *then* f *has no zeroes in the interval*

$$\left(x_0 - \kappa - \sqrt{\kappa^2 + 2c^{-1}|a|} \, , \, x_0 + \sqrt{\kappa^2 + 2c^{-1}|a|} - \kappa\right) .$$

3° *If* $|a| \le 0.5c\kappa^2$ & $ab > 0$, *then* f *has a zero in the segment*

$$\left[x_0 - \kappa + \sqrt{\kappa^2 - 2c^{-1}|a|} \, , \, x_0 - \sqrt{\kappa^2 + 2c^{-1}|a|} + \kappa\right] ,$$

which is the only zero of f *in the interval*

$$\left(x_0 - \kappa - \sqrt{\kappa^2 - 2c^{-1}|a|} \, , \, x_0 + \kappa + \sqrt{\kappa^2 + 2c^{-1}|a|}\right) .$$

4° *If* $|a| \le 0.5c\kappa^2$ & $ab < 0$, *then* f *has a zero in*

$$\left[x_0 + \sqrt{\kappa^2 + 2c^{-1}|a|} - \kappa \, , \, x_0 + \kappa - \sqrt{\kappa^2 - 2c^{-1}|a|}\right]$$

and no other zeroes in $\left(x_0 - \kappa - \sqrt{\kappa^2 + 2c^{-1}|a|}, x_0 + \kappa + \sqrt{\kappa^2 - 2c^{-1}|a|}\right)$.

6.3 Optimal methods for scalar equations (Lipschitz smoothness)

Keeping the notations of Lemma 6.2 in force, let the function f be Lipschitz smooth on $\mathbb{R}$. We can, without loss of generality, take the Lipschitz constant c to be 1, since the normalized function $c^{-1}f$ has the same zeroes as f. Elimination of c, that is, its replacement by 1, shortens ensuing calculations considerably. Moreover, the normalization of f does not change the existence condition $|a| \le 0.5c\kappa^2$.

Suppose that an x_0 is known such that $|a| \le 0.5\kappa^2$. Then the existence of a zero of f either in the segment $\left[x_0 - \kappa + \sqrt{\kappa^2 - 2|a|} \, , \, x_0 - \sqrt{\kappa^2 + 2|a|} + \kappa\right]$, if $ab > 0$, or in $\left[x_0 + \sqrt{\kappa^2 + 2|a|} - \kappa \, , \, x_0 + \kappa - \sqrt{\kappa^2 - 2|a|}\right]$, if $ab < 0$, is guaranteed by Lemma 6.2, and we need to choose from the corresponding segment the next approximation x_1 to that zero. Two different situations are possible:

$$(i)\ ab > 0, \quad (ii)\ ab < 0.$$

It suffices to consider the first case. The second is analyzed quite similarly and its analysis produces the same result.

Thus, in the case $ab > 0$, we should look for a zero between $x_0 - \kappa + \sqrt{\kappa^2 - 2|a|}$ and $x_0 - \sqrt{\kappa^2 + 2|a|} + \kappa$:

$$x_0 - \kappa + \sqrt{\kappa^2 - 2|a|} < x_1 < x_0 - \sqrt{\kappa^2 + 2|a|} + \kappa\,. \tag{6.3}$$

Naturally, we want to subject our choice of x_1 to the condition that the new existence segment

$$\left[x_1 - \kappa_1 + \sqrt{\kappa_1^2 - 2|a_1|}\ ,\ x_1 - \sqrt{\kappa_1^2 + 2|a_1|} + \kappa_1\right] \tag{6.4}$$

(here $a_1 := f(x_1)$, $b_1 := f\,'(x_1)$, and $\kappa_1 := |b_1|$), if $a_1 b_1 > 0$, or

$$\left[x_1 + \sqrt{\kappa_1^2 + 2|a_1|} - \kappa_1\ ,\ x_1 + \kappa_1 - \sqrt{\kappa_1^2 - 2|a_1|}\right], \tag{6.5}$$

if $a_1 b_1 < 0$, be defined and confined within the appropriate part of the old:

$$|a_1| \le 0.5\kappa_1^2, \tag{6.6}$$

and

$$x_1 - \kappa_1 + \sqrt{\kappa_1^2 - 2|a_1|} \ge x_0 - \kappa + \sqrt{\kappa^2 - 2|a|}\ \ \&\ \ x_1 - \sqrt{\kappa^2 + 2|a|} + \kappa < x_1\,, \tag{6.7}$$

if $a_1 b_1 > 0$, or

$$x_1 + \sqrt{\kappa^2 + 2|a|} - \kappa > x_1\ \ \&\ \ x_1 + \kappa_1 - \sqrt{\kappa_1^2 - 2|a_1|} \le x_0 - \sqrt{\kappa^2 + 2|a|} + \kappa\,, \tag{6.8}$$

if $a_1 b_1 < 0$. Besides, a_1 and b_1 are conditioned by the Lipschitz smoothness of f. By (6.2),

$$|\kappa - \kappa_1| = \big||b| - |b_1|\big| \le |b - b_1| = \left|f\,'(x_0) - f\,'(x_1)\right| \le |x_0 - x_1|\,. \tag{6.9}$$

Moreover, by the Newton–Leibnitz theorem,

$$f(x_1) = f(x_0) + f\,'(x_0)(x_1 - x_0) + \int_{x_0}^{x_1} [f\,'(t) - f\,'(x_0)]dt$$

and

$$|f(x_1) - f(x_0) - f\,'(x_0)(x_1 - x_0)| \le \int_{\min\{x_0,x_1\}}^{\max\{x_0,x_1\}} |f\,'(t) - f\,'(x_0)|dt\ ,$$

where $|f'(t) - f'(x_0)| \le |t - x_0|$ because of the Lipschitz smoothness of f. So,

$$\begin{aligned}\big|a_1 - a - b(x_1 - x_0)\big| &\le \int_{\min\{x_0,x_1\}}^{\max\{x_0,x_1\}} |t - x_0|\, dt \\ &= \int_{\min\{x_0,x_1\}}^{x_0} (x_0 - t)dt + \int_{x_0}^{\max\{x_0,x_1\}} (t - x_0)dt \\ &= \frac{1}{2}(x_1 - x_0)^2\end{aligned}$$

and

$$a + b\,(x_1 - x_0) - \frac{1}{2}\,(x_1 - x_0)^2 \le a_1 \le a + b\,(x_1 - x_0) + \frac{1}{2}\,(x_1 - x_0)^2. \quad (6.10)$$

Our hypothesis is that the position of the zero is distributed in each of the segments (6.22) or (6.23) uniformly. Then the entropy of the zero's position is proportional to its length

$$l(\kappa_1\,, |a_1|) := 2\kappa_1 - \sqrt{\kappa_1^2 + 2|a_1|} - \sqrt{\kappa_1^2 - 2|a_1|}\,.$$

According to the worst case philosophy, we have to find x_1 in either the range

$$\Big[x_0 - \kappa + \sqrt{\kappa^2 - 2|a|}\ ,\ x_0 - \sqrt{\kappa^2 + 2|a|} + \kappa\Big]\,, \quad (6.11)$$

if $ab > 0$, or in $\Big[x_0 + \sqrt{\kappa^2 + 2|a|} - \kappa\,,\, x_0 + \kappa - \sqrt{\kappa^2 - 2|a|}\Big]$, if $ab < 0$, that minimizes the maximum value L of $l(\kappa_1\,, a_1)$ over all pairs $(\kappa_1\,, a_1)$ satisfying the constraints (6.6)–(6.10) or, symbolically, belonging to one of two sets

$$KA(\rho) := \left\{(\kappa_1\,, a_1)\ \middle|\ \begin{array}{r} |a_1| \le 0.5\kappa_1^2\ \ \&\ \ |\kappa_1 - \kappa| \le |\delta|\ \ \&\ \ \alpha \le a_1 \le \beta \\ \kappa_1 - \sqrt{\kappa_1^2 - 2|a_1|} \le \rho \end{array}\right\}, \quad (6.12)$$

$$KA(\sigma) := \left\{(\kappa_1\,, a_1)\ \middle|\ \begin{array}{r} |a_1| \le 0.5\kappa_1^2\ \ \&\ \ |\kappa_1 - \kappa| \le |\delta|\ \ \&\ \ \alpha \le a_1 \le \beta \\ \kappa_1 - \sqrt{\kappa_1^2 - 2|a_1|} \le \sigma \end{array}\right\}, \quad (6.13)$$

where (for brevity)

$$\delta := x_1 - x_0\ ,\ \rho := \delta + \kappa - \sqrt{\kappa^2 - 2|a|}\ ,\ \sigma := -\delta - \sqrt{\kappa^2 + 2|a|} + \kappa\,,$$

$$\alpha := a + b\,\delta - \frac{1}{2}\,\delta^2\ ,\ \beta := a + b\,\delta + \frac{1}{2}\,\delta^2\,.$$

Note that, by (6.3),

$$-\Big(\kappa - \sqrt{\kappa^2 - 2|a|}\Big) < \delta < -\Big(\sqrt{\kappa^2 + 2|a|} - \kappa\Big)$$

and so

$$\sqrt{\kappa^2 + 2|a|} - \kappa \le |\delta| \le \kappa - \sqrt{\kappa^2 - 2|a|}\,, \tag{6.14}$$

$0 < \rho < l(\kappa\,, a)$, and $0 < \sigma < l(\kappa\,, a)$. Clearly, $L = \max\{L(\rho)\,, L(\sigma)\}$, where $L(\rho) := \max\limits_{(\kappa_1, a_1) \in KA(\rho)} l(\kappa_1\,, a_1)$. While evaluating $L(\rho)$, it is expedient to use the abbreviated symbols for the variables $s := \kappa_1$ and $t := a_1$. In terms of s and t,

$$L(\rho) = \max_{(s,t) \in ST(\rho)} l(s\,, t)\,,$$

$$ST(\rho) := \left\{ (s\,, t) \;\middle|\; \begin{array}{r} |t| \le 0.5s^2 \;\;\&\;\; |s - \kappa| \le |\delta| \;\;\&\;\; \alpha \le t \le \beta \\ s - \sqrt{s^2 - 2|t|} \le \rho \end{array} \right\}.$$

By the lemma on sections (Lemma 1.4),

$$L(\rho) = \max_{s \in S(\rho)} \max_{t \in T(s\,,\rho)} l(s\,, t)\,, \tag{6.15}$$

where

$$T(s\,,\rho) := \left\{ t \;\middle|\; (s, t) \in ST(\rho) \right\}$$
$$= \begin{cases} \left\{ t \;\middle|\; |t| \le 0.5s^2 \;\;\&\;\; \alpha \le t \le \beta \;\;\&\;\; s - \sqrt{s^2 - 2|t|} \le \rho \right\} \\ \hfill \text{if } |s - \kappa| \le |\delta|\,, \\ \varnothing\,, \text{ otherwise} \end{cases}$$

and $S(\rho) := \{ s \mid T(s\,,\rho) \ne \varnothing \}$. If $s \le \rho$, the constraint $\sqrt{s^2 - 2|t|} \ge s - \rho$ is trivial and can be dropped. So,

$$|s - \kappa| \le |\delta| \;\;\&\;\; s \le \rho \Longrightarrow T(s\,,\rho) = \left\{ t \;\middle|\; |t| \le 0.5s^2 \;\;\&\;\; \alpha \le t \le \beta \right\}$$
$$= \left\{ t \;\middle|\; \max\{-0.5s^2, \alpha\} \le t \le \min\{0.5s^2, \beta\} \right\}$$

and $T(s\,,\rho) \ne \varnothing \Longleftrightarrow \alpha \le 0.5s^2 \;\;\&\;\; -0.5s^2 \le \beta$. If $s \ge \rho$, then $0.5\rho(2s - \rho) \ge 0$,

$$\sqrt{s^2 - 2|t|} \ge s - \rho \Longleftrightarrow |t| \le 0.5\rho(2s - \rho)\,,$$

and $|s - \kappa| \le |\delta| \;\;\&\;\; s \ge \rho \Longrightarrow$

$$T(s\,,\rho) = \left\{ t \;\middle|\; |t| \le 0.5s^2 \;\;\&\;\; \alpha \le t \le \beta \;\&\; |t| \le 0.5\rho(2s - \rho) \right\}$$
$$= \left\{ t \;\middle|\; |t| \le 0.5\rho(2s - \rho) \;\;\&\;\; \alpha \le t \le \beta \right\},$$
$$= \left\{ t \;\middle|\; \max\{-0.5\rho(2s - \rho)\,, \alpha\} \le t \le \min\{0.5\rho(2s - \rho)\,, \beta\} \right\},$$

so that $T(s\,,\rho)\neq\varnothing \iff -0.5\rho(2s-\rho)\leq\beta$ & $\alpha\leq 0.5\rho(2s-\rho)$. It follows that

$$S(\rho)=\Big\{s \;\Big|\; |s-\kappa|\leq|\delta| \;\&\; \Big(s\leq\rho \;\&\; 0.5s^2\geq\max\{\alpha\,,-\beta\} \bigvee$$

$$s\geq\rho \;\&\; 0.5\rho(2s-\rho)\geq\max\{\alpha\,,-\beta\}\Big)\Big\}$$

$$=\Big\{s \;\Big|\; |s-\kappa|\leq|\delta| \;\&\; \Big(s\leq\rho \;\&\; s^2\geq 2\gamma \bigvee s\geq\rho \;\&\; \rho s\geq 0.5\rho^2+\gamma\Big)\Big\},$$

where (for brevity) $\gamma:=\max\{\alpha\,,-\beta\}$. If $\gamma\leq 0$, then the constraint $s^2\geq 2\gamma$ is redundant and

$$s\leq\rho \;\&\; s^2\geq 2\gamma \bigvee s\geq\rho \;\&\; \rho s\geq\frac{\rho^2}{2}+\gamma \iff s\leq\rho \bigvee s\geq\rho \;\&\; \rho s\geq\frac{\rho^2}{2}+\gamma$$

$$\iff s\leq\rho \bigvee s\geq\frac{\rho}{2}+\frac{\gamma}{\rho}.$$

This condition is met by any real s. So,

$$\gamma\leq 0 \Longrightarrow S(\rho)=\Big\{s \;\Big|\; |s-\kappa|\leq|\delta|\Big\}=\big[\kappa-|\delta|\,,\kappa+|\delta|\big]\neq\varnothing.$$

Lemma 6.3. $\gamma\leq 0$.

Proof. As $\gamma:=\max\{\alpha\,,-\beta\}$, the claim is equivalent to $\alpha\leq 0\leq\beta$. Since δ is negative, $0\geq\alpha:=a+b\delta-0.5\delta^2=a-b|\delta|-0.5|\delta|^2$

$$\iff |\delta|^2+2b|\delta|-2a\geq 0$$

$$\iff |\delta|\leq -b-\sqrt{b^2+2a} \bigvee |\delta|\geq -b+\sqrt{b^2+2a}\,.$$

If $b>0$ (and so $a>0$), then $\sqrt{\kappa^2+2|a|}-\kappa=-b+\sqrt{b^2+2a}$ and (6.14) $\Longrightarrow \alpha\leq 0$. Else, $-b-\sqrt{b^2+2a}=|b|-\sqrt{|b|^2-2|a|}=\kappa-\sqrt{\kappa^2-2|a|}$ and the claim $\alpha\leq 0$ follows from the second inequality in (6.14). Similarly, $0\leq\beta=a-b|\delta|+0.5|\delta|^2$

$$\iff |\delta|^2-2b|\delta|+2a\geq 0$$

$$\iff |\delta|\leq b-\sqrt{b^2-2a} \bigvee |\delta|\geq b+\sqrt{b^2-2a}\,.$$

If $b>0$ and $a>0$, then $b-\sqrt{b^2-2a}=\kappa-\sqrt{\kappa^2-2|a|}$ and (6.14) $\Longrightarrow \beta\geq 0$. Otherwise, $b-\sqrt{b^2-2a}=-\kappa+\sqrt{\kappa^2+2|a|}$ and again (6.14) $\Longrightarrow \beta\geq 0$. $\square$

Inasmush as the function $t\mapsto l(s\,,t)=2s-\sqrt{s^2+2t}-\sqrt{s^2-2t}$ is increasing in $\big[0\,,0.5s^2\big]$ (differentiate it with respect to t), the interior maximum in (6.15) $\big($let us denote it $F_\rho(s)\big)$

$$=\begin{cases} l\big(s\,,\min\{0.5s^2\,,\beta\}\big)\,, & \text{if } s\leq\rho\,,\\ l\big(s\,,\min\{0.5\rho(2s-\rho)\,,\beta\}\big)\,, & \text{if } s\geq\rho\,,\end{cases}$$

or, in more detail,

$$F_\rho(s) = \begin{cases} l(s\,,0.5s^2)\,, \text{ if } s \le \rho \;\;\&\;\; 0.5s^2 \le \beta\,, \\ l(s\,,\beta)\,, \text{ if } s{\le}\rho \;\;\&\;\; 0.5s^2 \ge \beta \bigvee s \ge \rho \;\;\&\;\; 0.5\rho(2s-\rho) \ge \beta\,, \\ l(s\,,0.5\rho(2s-\rho))\,, \text{ if } s \ge \rho \;\;\&\;\; 0.5\rho(2s-\rho) \le \beta\,. \end{cases}$$

$$= \begin{cases} (2-\sqrt{2})\, s\,, \text{ if } s{\le}\min\left\{\rho\,,\sqrt{2\beta}\right\}\,, \\ 2s-\sqrt{s^2+2\beta}-\sqrt{s^2-2\beta}\,, \text{ if } \sqrt{2\beta}{\le}s{\le}\rho \bigvee s{\ge}\max\left\{\rho\,,\dfrac{\rho}{2}+\dfrac{\beta}{\rho}\right\}\,, \\ \dfrac{2\rho^2}{s+\rho+\sqrt{s^2+2\rho s-\rho^2}}\,, \text{ if } \rho{\le}s{\le}\dfrac{\rho}{2}+\dfrac{\beta}{\rho}\,. \end{cases}$$

The function $s \mapsto 2\rho^2\Big/\left(s+\rho+\sqrt{s^2+\rho s-\rho^2}\right) = s+\rho-\sqrt{s^2+2\rho s-\rho^2}$ is clearly decreasing. The function $s \mapsto 2s-\sqrt{s^2+2\beta}-\sqrt{s^2-2\beta}$ is decreasing too. To verify this, differentiate it with respect to s twice:

$$\frac{\partial l}{\partial s} = 2 - \frac{s}{\sqrt{s^2+2\beta}} - \frac{s}{\sqrt{s^2-2\beta}}\,,$$

$$\frac{\partial^2 l}{\partial s^2} = 2\beta\left(\frac{1}{\left(s^2-2\beta\right)^{3/2}} - \frac{1}{\left(s^2+2\beta\right)^{3/2}}\right).$$

We see that $\partial^2 l/\partial s^2 > 0$ and so $\partial l/\partial s$ is increasing in $(\sqrt{2\beta}\,,\infty)$ from $\partial l/\partial s(\sqrt{2\beta}\,,\beta) = -\infty$ to $\partial l/\partial s(\infty\,,\beta) = 0$. Therefore, $\partial l/\partial s < 0$ and $s \mapsto l(s\,,\beta)$ is decreasing. Thus, F_ρ is increasing in the segment

$$\left\{s \mid 0 \le s \le \rho \;\;\&\;\; 0.5s^2 \le \beta\right\} = \left[0\,,\min\left\{\rho\,,\sqrt{2\beta}\right\}\right]$$

and decreasing beyond, so that its maximum in $[0\,,\infty)$ is attained at

$$s_\rho := \min\left\{\rho\,,\sqrt{2\beta}\right\}\,.$$

It follows that

$$L(\rho) = \max_{s\in S(\rho)} F_\rho(s) = \begin{cases} F_\rho(\min S(\rho))\,, \text{ if } s_\rho \le \min S(\rho) = \kappa - |\delta|\,, \\ F_\rho(s_\rho)\,, \text{ if } \min S(\rho) \le s_\rho \le \max S(\rho) = \kappa + |\delta|\,, \\ F_\rho(\max S(\rho))\,, \text{ if } s_\rho \ge \max S(\rho) = \kappa + |\delta|\,, \end{cases}$$

$$= F_\rho\Big(\max\{\kappa-|\delta|\,,\min\{s_\rho\,,\kappa+|\delta|\}\}\Big)\,.$$

$$L(\sigma) = F_\sigma\Big(\max\{\kappa-|\delta|\,,\min\{s_\sigma\,,\kappa+|\delta|\}\}\Big)\,,$$

and

$$L = \max\left\{\begin{array}{l} F_\rho\Big(\max\{\kappa - |\delta|\,,\min\{\rho\,,\sqrt{2\beta}\,,\kappa + |\delta|\}\}\Big) \\ F_\sigma\Big(\max\{\kappa - |\delta|\,,\min\{\sigma\,,\sqrt{2\beta}\,,\kappa + |\delta|\}\}\Big) \end{array}\right\}.$$

At this point in our discussion, we have to introduce the abbreviations

$$\nu := \min\{\sqrt{2\beta}\,,\kappa + |\delta|\}\ ,\ \varphi(\rho) := F_\rho\Big(\max\{\kappa - |\delta|\,,\min\{\rho\,,\nu\}\}\Big).$$

With these abbreviations, $L = \max\{\varphi(\rho)\,,\varphi(\sigma)\}$.

Lemma 6.4. 1° *The function* φ *is not decreasing in* $(0\,,l(\kappa\,,a))$.
2° $\rho \le \sigma \iff |\delta| \ge 0.5\left(\sqrt{\kappa^2 + 2|a|} - \sqrt{\kappa^2 - 2|a|}\right)$.

Proof. 1° Denote $g(t) := 0.5t + \beta/t$. g is decreasing in $(0\,,\sqrt{2\beta})$ and increasing beyond, so that $g(t) \ge g\left(\sqrt{2\beta}\right) = \sqrt{2\beta}\,, \forall\, t \ge 0$. There are three possibilities:

$$(i)\ \rho \le \kappa - |\delta|\ ,\ (ii)\ \kappa - |\delta| < \rho < \nu\ ,\ (iii)\ \rho \ge \nu\,.$$

In the first case, $\varphi(\rho) = F_\rho(\kappa - |\delta|)$ depends on the relation between ρ, $\sqrt{2\beta}$, and $\kappa - |\delta|$. Namely, $\sqrt{2\beta} \le \rho \le \kappa - |\delta|$

$$\Longrightarrow \sqrt{2\beta} \le \rho = \max\{\rho\,,g(\rho)\} \le \kappa - |\delta|$$

$$\Longrightarrow F_\rho(\kappa - |\delta|) = 2(\kappa - |\delta|) - \sqrt{(\kappa - |\delta|)^2 + 2\beta} - \sqrt{(\kappa - |\delta|)^2 - 2\beta}$$

and $\rho \le \min\{\sqrt{2\beta}\,,\kappa - |\delta|\}$

$$\Longrightarrow \rho \le g(\rho) = \max\{\rho\,,g(\rho)\} \le \kappa - |\delta|$$

$$\Longrightarrow F_\rho(\kappa - |\delta|) = \kappa - |\delta| + \rho - \sqrt{(\kappa - |\delta|)^2 + 2\rho(\kappa - |\delta|) - \rho^2}$$

In the second case, $\rho < \sqrt{2\beta}$, so that $\varphi(\rho) = F_\rho(\rho) = \left(2 - \sqrt{2}\right)\rho$ is increasing in ρ. In the third case, $\nu \le \min\left\{\rho\,,\sqrt{2\beta}\right\}$, so that $\varphi(\rho) = F_\rho(\nu) = \left(2 - \sqrt{2}\right)\nu$. Thus,

$$\varphi(\rho) = \begin{cases} 2(\kappa - |\delta|) - \sqrt{(\kappa - |\delta|)^2 + 2\beta} - \sqrt{(\kappa - |\delta|)^2 - 2\beta}\,, \\ \qquad\qquad\qquad \text{if } \sqrt{2\beta} \le \rho \le \kappa - |\delta|\,, \\ \kappa - |\delta| + \rho - \sqrt{(\kappa - |\delta|)^2 + 2\rho(\kappa - |\delta|) - \rho^2}\,, \\ \qquad\qquad\qquad \text{if } \rho \le \min\left\{\sqrt{2\beta}\,,\kappa - |\delta|\right\}, \\ \left(2 - \sqrt{2}\right)\rho\,,\ \text{if } \kappa - |\delta| \le \rho \le \nu\,, \\ \left(2 - \sqrt{2}\right)\nu\,,\ \text{if } \rho \ge \nu\,. \end{cases} \tag{6.16}$$

We see that in any case φ is either increasing or constant with respect to ρ.

2° As δ is negative (see (6.14)),

$$\rho \le \sigma \Longleftrightarrow \kappa - |\delta| - \sqrt{\kappa^2 - 2|a|} \le \kappa + |\delta| - \sqrt{\kappa^2 + 2|a|}$$
$$\Longleftrightarrow |\delta| \ge \frac{\sqrt{\kappa^2 + 2|a|} - \sqrt{\kappa^2 - 2|a|}}{2}.$$

□

By the lemma,

$$L = \begin{cases} \varphi(\rho), \text{ if } |\delta| \le 0.5\left(\sqrt{\kappa^2+2|a|} - \sqrt{\kappa^2-2|a|}\right), \\ \varphi(\sigma), \text{ if } |\delta| \ge 0.5\left(\sqrt{\kappa^2+2|a|} - \sqrt{\kappa^2-2|a|}\right). \end{cases}$$

This value depends on δ, the coordinate of the position of x_1 within the segment (6.11). So, the optimal position is that which minimizes L as a function of $|\delta|$ subject to (6.14). Hence, our next task is the problem

$$\min_{u \le |\delta| \le v} \max\{\varphi(\rho), \varphi(\sigma)\}, \tag{6.17}$$

where (for brevity)

$$u := \sqrt{\kappa^2 + 2|a|} - \kappa, \quad v := \kappa - \sqrt{\kappa^2 - 2|a|}. \tag{6.18}$$

Clearly, this minimum

$$= \min \left\{ \begin{array}{l} \min_{|\delta|}\{\varphi(\rho) \mid u \le |\delta| \le v \;\;\&\;\; \varphi(\rho) \ge \varphi(\sigma)\} \\ \min_{|\delta|}\{\varphi(\sigma) \mid u \le |\delta| \le v \;\;\&\;\; \varphi(\rho) \le \varphi(\sigma)\} \end{array} \right\}.$$

By the lemma,

$$\varphi(\rho) \ge \varphi(\sigma) \Longleftrightarrow \rho \ge \sigma \Longleftrightarrow |\delta| \le \frac{u+v}{2}.$$

So, the minimum (6.17)

$$= \min\left\{ \min_{|\delta|}\left\{\varphi(\rho) \;\middle|\; u \le |\delta| \le \frac{u+v}{2}\right\}, \min_{|\delta|}\left\{\varphi(\sigma) \;\middle|\; \frac{u+v}{2} \le |\delta| \le v\right\}\right\}.$$

Besides, as follows from (6.14), $u \le |\delta| \le 0.5(u+v) \Longleftrightarrow 0.5(v-u) \le \rho \le v-u$ and $0.5(u+v) \le |\delta| \le v \Longleftrightarrow 0.5(v-u) \le \sigma \le v-u$. Therefore, the minimum

$$= \min\left\{ \min_{\rho}\left\{\varphi(\rho) \;\middle|\; \frac{v-u}{2} \le \rho \le v-u\right\}, \min_{\sigma}\left\{\varphi(\sigma) \;\middle|\; \frac{v-u}{2} \le \sigma \le v-u\right\}\right\}$$

$$= \min_{\rho}\left\{\varphi(\rho) \;\middle|\; \frac{v-u}{2} \le \rho \le v-u\right\} = \varphi\left(\frac{v-u}{2}\right),$$

because φ is not decreasing, by the lemma. As we have seen above, the maximum possible value of $\varphi(0.5(v-u))$ is $\left(1-0.5\sqrt{2}\right)(v-u)$. Correspondingly, the optimal positioning of x_1 reduces the solution's entropy at least by the factor $1-0.5\sqrt{2}=0.29\ldots$. Moreover, the existence condition $\left|f(x_0)\right| \le 0.5 f'(x_0)^2$ constitutes also a convergence condition of the optimal method

$$x_+ := x - \frac{sign\big(f(x)f'(x)\big)}{2}\left(\sqrt{f'(x)^2+2\left|f(x)\right|}-\sqrt{f'(x)^2-2\left|f(x)\right|}\right), \tag{6.19}$$

since all subsequent iterations x_k satisfy, by design, the similar condition $\left|f(x_k)\right| \le 0.5 f'(x_k)^2$ and so similarly reduce the solution's entropy. Thus we have obtained the convergence domain of the optimal method and its rate of convergence. It should be noted, however, that this method is not defined outside its convergence domain. Therefore, it can be used only after iterations generated by some other converging method enter its convergence domain.

It is interesting to compare the method (6.19) with Newton's method. To this end, note that, for any two real x and y, $sign(xy)=sign(x)\,sign(y)$ and $sign(x)\,|x|=x$, so that x_+-x in (6.19)

$$= -sign\big(f(x)\big)\,sign\big(f'(x)\big)\left|f'(x)\right|\frac{1}{2}\left(\sqrt{1+\frac{2\left|f(x)\right|}{f'(x)^2}}-\sqrt{1-\frac{2\left|f(x)\right|}{f'(x)^2}}\right)$$

$$= -sign\big(f(x)\big)\frac{f'(x)}{2\left|f(x)\right|}\left|f(x)\right|\left(\sqrt{1+\frac{2\left|f(x)\right|}{f'(x)^2}}-\sqrt{1-\frac{2\left|f(x)\right|}{f'(x)^2}}\right)$$

$$= -\frac{f(x)}{f'(x)}\cdot\frac{f'(x)^2}{2\left|f(x)\right|}\left(\sqrt{1+\frac{2\left|f(x)\right|}{f'(x)^2}}-\sqrt{1-\frac{2\left|f(x)\right|}{f'(x)^2}}\right).$$

The function $g(t):=\left(\sqrt{1+t}-\sqrt{1-t}\right)/t$ is increasing in $[0,1]$ from $g(0)=1$ to $g(1)=\sqrt{2}$. So, in the neighborhood of solution the method (6.19) is almost Newton's, while in proximity to the boundary of its convergence domain its steps are greater than Newton's.

6.4 Optimal methods for scalar equations (regular smoothness)

In this section, we are trying to extend the results of the previous section to a more general class of functions, ω-regularly smooth ones. So, here f is assumed to be ω-regularly smooth on $\mathbb{R}$ in the sense of Definition (2.2):

$$\omega^{-1}\Big(\min\{|f'(x)|,|f'(x')|\}+|f'(x')-f'(x)|\Big)-\omega^{-1}\Big(\min\{|f'(x)|,|f'(x')|\}\Big)\le|x'-x|. \tag{6.20}$$

The symbols a, b, κ and the functions w, p, q are as in Proposition 6.1:

$$a:=f(x_0),\ b:=f'(x_0),\ \kappa:=\omega^{-1}(|b|),$$

$$w(t):=\int_0^t\omega(\tau)\,d\tau,\ \Psi(\alpha,t)=\begin{cases} t\omega(\alpha)-w(\alpha)+w(\alpha-t), & \text{if } 0\le t\le\alpha, \\ \alpha\omega(\alpha)-2w(\alpha)+w(t), & \text{if } t\ge\alpha\ge 0. \end{cases}$$

$$p(\alpha,t):=t\,\omega(\alpha)+\Psi(\alpha,t),\ q(\alpha,t):=t\,\omega(\alpha)-\Psi(\alpha,t).$$

We begin with an x_0 satisfying $|a|\le w(\kappa)$. The existence of a zero of f either in the segment $\big[x_0-q_-^{-1}(\kappa,|a|),\,x_0-p^{-1}(\kappa,|a|)\big]$, if $ab>0$, or in $\big[x_0+p^{-1}(\kappa,|a|),\,x_0+q_-^{-1}(\kappa,|a|)\big]$, if $ab<0$, is guaranteed by Proposition 6.1, and we need to choose from the corresponding segment the next approximation x_1 to that zero. Consider first the case $ab>0$. In this case, we should look for a zero in the segment $\big[x_0-q_-^{-1}(\kappa,|a|),\,x_0-p^{-1}(\kappa,|a|)\big]$:

$$x_0-q_-^{-1}(\kappa,|a|)\le x_1\le x_0-p^{-1}(\kappa,|a|) \tag{6.21}$$

subject to the condition that the new existence segment

$$\big[x_1-q_-^{-1}(\kappa_1,|a_1|),\,x_1-p^{-1}(\kappa_1,|a_1|)\big] \tag{6.22}$$

(here $a_1:=f(x_1)$, $b_1:=f'(x_1)$, and $\kappa_1:=\omega^{-1}(|b_1|)$), if $a_1b_1>0$, or

$$\big[x_1+p^{-1}(\kappa_1,|a_1|),\,x_1+q_-^{-1}(\kappa_1,|a_1|)\big], \tag{6.23}$$

if $a_1b_1<0$, be defined and confined within the appropriate part of the old:

$$|a_1|\le w(\kappa_1), \tag{6.24}$$

and

$$x_1-q_-^{-1}(\kappa_1,|a_1|)\ge x_0-q_-^{-1}(\kappa,|a|)\ \ \&\ \ x_1-p^{-1}(\kappa_1,|a_1|)\le x_1, \tag{6.25}$$

if $a_1b_1>0$, or

$$x_1+p^{-1}(\kappa_1,|a_1|)\ge x_1\ \ \&\ \ x_1+q_-^{-1}(\kappa_1,|a_1|)\le x_0-p^{-1}(\kappa,|a|), \tag{6.26}$$

if $a_1b_1<0$. Besides, a_1 and b_1 are conditioned by the ω-regular smoothness of f. By (6.20) and because, for any real b and b_1, $|b_1-b|\ge\big||b_1|-b\big|$,

$$\begin{aligned}|x_1-x_0|&\ge\omega^{-1}(\min\{|b|,|b_1|\}+|b_1-b|)-\omega^{-1}(\min\{|b|,|b_1|\})\\&\ge\omega^{-1}(\min\{|b|,|b_1|\}+||b_1|-b|)-\omega^{-1}(\min\{|b|,|b_1|\}).\end{aligned}$$

If $b \geq |b_1|$, then this difference $= \omega^{-1}\big(|b_1| + b - |b_1|\big) - \omega^{-1}\big(|b_1|\big) = \kappa - \kappa_1$. Otherwise, there are three possibilities:

$$(i)\ 0 \leq b \leq |b_1|,\ (ii)\ -|b_1| \leq b < 0,\ (iii)\ b < -|b_1|.$$

In the first case, the difference $= \omega^{-1}\big(b + |b_1| - b\big) - \omega^{-1}(b) = \kappa_1 - \kappa$. In the second case, $|b| \leq |b_1|$, so that the difference

$$= \omega^{-1}\big(|b| + |b_1| + |b|\big) - \omega^{-1}\big(|b|\big) > \omega^{-1}\big(|b_1|\big) - \omega^{-1}\big(|b|\big) = \kappa_1 - \kappa\,.$$

In the third case, it

$$= \omega^{-1}\big(|b_1| + |b_1| + |b|\big) - \omega^{-1}\big(|b_1|\big) > \omega^{-1}\big(|b|\big) - \omega^{-1}\big(|b_1|\big) = \kappa - \kappa_1\,.$$

Thus, in any case,

$$|x_1 - x_\circ| \geq |\kappa_1 - \kappa|\,. \tag{6.27}$$

Moreover, by the Newton-Leibnitz theorem,

$$f(x_1) = f(x_\circ) + f\,'(x_\circ)(x_1 - x_\circ) + \int_{x_0}^{x_1} [f\,'(t) - f\,'(x_\circ)]dt$$

and

$$|f(x_1) - f(x_\circ) - f\,'(x_\circ)(x_1 - x_\circ)| \leq \int_{\min\{x_0,x_1\}}^{\max\{x_0,x_1\}} |f\,'(t) - f\,'(x_\circ)|dt\ ,$$

where, as in the proof of Theorem 5.3, $|f\,'(t) - f\,'(x_\circ)| \leq e\big(\kappa\,, |t - x_\circ|\big)$. So,

$$\big|a_1 - a - b(x_1 - x_\circ)\big| \leq \int_{\min\{x_0,x_1\}}^{\max\{x_0,x_1\}} e\big(\kappa\,, |t - x_\circ|\big)\,dt = \Psi\big(\kappa\,, |x_1 - x_\circ|\big)$$

and

$$a + b(x_1 - x_\circ) - \Psi\big(\kappa\,, |x_1 - x_\circ|\big) \leq a_1 \leq a + b(x_1 - x_\circ) + \Psi\big(\kappa\,, |x_1 - x_\circ|\big)\,. \tag{6.28}$$

Since a priori each position of the zero in the segments (6.22) or (6.23) is equally probable, we assume that its position is distributed there uniformly. Then the entropy of the zero's position is proportional to the length of the segment

$$l\big(\kappa_1\,, |a_1|\big) := q_-^{-1}\big(\kappa_1\,, |a_1|\big) - p^{-1}\big(\kappa_1\,, |a_1|\big)\,.$$

According to the worst case philosophy, we have to find x_1 in either the range

$$\big[x_\circ - q_-^{-1}\big(\kappa\,, |a|\big)\,,\ x_\circ - p^{-1}\big(\kappa\,, |a|\big)\big]\ ,$$

if $ab > 0$, or in $\big[x_\circ + p^{-1}\big(\kappa\,, |a|\big)\,,\ x_\circ + q_-^{-1}\big(\kappa\,, |a|\big)\big]$, if $ab < 0$, that minimizes the maximum value L of $l\big(\kappa_1\,, |a_1|\big)$ over all pairs $(\kappa_1\,, a_1)$ satisfying the constraints (6.24)-(6.28) or, symbolically, belonging to one of two sets

$$KA(\rho) := \left\{(\kappa_1\,, a_1)\ \middle|\ \begin{array}{c} |a_1| \leq w(\kappa_1)\ \ \&\ \ |\kappa_1 - \kappa| \leq |\delta|\ \ \&\ \ \alpha \leq a_1 \leq \beta \\ q_-^{-1}(\kappa_1\,, |a_1|) \leq \rho \end{array}\right\}, \tag{6.29}$$

$$KA(\sigma) := \left\{ (\kappa_1\,, a_1) \;\middle|\; \begin{array}{c} |a_1| \le w(\kappa_1) \;\;\&\;\; |\kappa_1 - \kappa| \le |\delta| \;\;\&\;\; \alpha \le a_1 \le \beta \\ q_-^{-1}(\kappa_1\,, |a_1|) \le \sigma \end{array} \right\}, \tag{6.30}$$

where (for brevity)

$$\delta := x_1 - x_0\,,\ \rho := \delta + q_-^{-1}\big(\kappa\,, |a|\big)\,,\ \sigma := -\delta - p^{-1}\big(\kappa\,, |a|\big)\,,$$

$$\alpha := a + b\delta - \Psi\big(\kappa\,, |\delta|\big)\,,\ \beta := a + b\delta + \Psi\big(\kappa\,, |\delta|\big)\,.$$

As seen from (6.21), $-q_-^{-1}(\kappa\,, |a|) \le \delta \le -p^{-1}(\kappa\,, |a|)$ or, equivalently,

$$p^{-1}(\kappa\,, |a|) \le |\delta| \le q_-^{-1}(\kappa\,, |a|)\,. \tag{6.31}$$

Lemma 6.5. $\alpha \le 0 \le \beta$.

Proof. Since δ is negative, $\alpha \le 0 \Longleftrightarrow a - b|\delta| - \Psi\big(\kappa\,, |\delta|\big) \le 0$. If $b > 0$ (and so $\alpha > 0$), then $b = |b| = \omega(\kappa)$ and

$$\alpha \le 0 \Longleftrightarrow |\delta|\omega(\kappa) + \Psi\big(\kappa\,, |\delta|\big) \ge |a| \Longleftrightarrow p\big(\kappa\,, |\delta|\big) \ge |a|$$

$$\Longleftrightarrow |\delta| \ge p^{-1}\big(\kappa\,, |a|\big)\,,$$

which is true by (6.31). Else, $b < 0\,,\ \alpha < 0\,,\ -b = |b| = \omega(\kappa)$, and

$$\alpha \le 0 \Longleftrightarrow -|a| + |\delta|\omega(\kappa) - \Psi\big(\kappa\,, |\delta|\big) \le 0 \Longleftrightarrow q\big(\kappa\,, |\delta|\big) \le |a| \Longleftrightarrow |\delta| \le q_-^{-1}\big(\kappa\,, |a|\big)\,,$$

which is also true by (6.31). Likewise, $\beta \ge 0 \Longleftrightarrow a - b|\delta| + \Psi\big(\kappa\,, |\delta|\big) \ge 0$. If $b > 0\,, a > 0\,,$ then $b = \omega(\kappa)$ and

$$\beta \ge 0 \Longleftrightarrow |\delta|\omega(\kappa) - \Psi\big(\kappa\,, |\delta|\big) \le |a| \Longleftrightarrow q\big(\kappa\,, |\delta|\big) \le |a| \Longleftrightarrow |\delta| \le q_-^{-1}\big(\kappa\,, |a|\big)\,.$$

Otherwise, $b < 0\,, \alpha < 0\,, -b = \omega(\kappa)$, and

$$\beta \ge 0 \Longleftrightarrow |\delta|\omega(\kappa) + \Psi\big(\kappa\,, |\delta|\big) \ge |a| \Longleftrightarrow p\big(\kappa\,, |\delta|\big) \ge |a| \Longleftrightarrow |\delta| \ge p^{-1}\big(\kappa\,, |a|\big)\,.$$

□

Clearly, $L = \max\{L(\rho)\,, L(\sigma)\}$, where $L(\rho) := \max\limits_{(\kappa_1, a_1) \in KA(\rho)} l\big(\kappa_1\,, |a_1|\big)$. While evaluating $L(\rho)$, we use the same abbreviated symbols for the variables as previously: $s := \kappa_1$ and $t := a_1$. In terms of s and t,

$$L(\rho) = \max_{(s,t) \in ST(\rho)} l(s\,, t)\,,$$

$$ST(\rho) := \left\{ (s\,, t) \;\middle|\; \begin{array}{c} |t| \le w(s) \;\;\&\;\; |s - \kappa| \le |\delta| \;\;\&\;\; \alpha \le t \le \beta \\ q_-^{-1}(s\,, |t|) \le \rho \end{array} \right\}.$$

By the lemma on sections (Lemma 1.4), $L(\rho) = \max_{s \in S(\rho)} \max_{t \in T(\rho,s)} l\big(s\,,|t|\big)$, where

$$T(\rho, s) := \big\{t \mid (s,t) \in ST(\rho)\big\}$$
$$= \begin{cases} \Big\{t \;\Big|\; |t| \le w(s) \;\;\&\;\; \alpha \le t \le \beta \;\;\&\;\; q_-^{-1}\big(s\,,|t|\big) \le \rho\Big\} \\ \qquad\qquad\qquad\qquad \text{if } |s-\kappa| \le |\delta|\,, \\ \varnothing\,, \text{ otherwise} \end{cases}$$

and $S(\rho) := \big\{s \mid T(\rho, s) \ne \varnothing\big\}$. Inasmuch as for $\rho \le s$

$$q_-^{-1}\big(s\,,|t|\big) \le \rho \iff |t| = q\big(s\,, q_-^{-1}\big(s\,,|t|\big)\big) \le q(s\,,\rho) = w(s) - w(s-\rho)\,,$$

we infer that

$$\begin{aligned} |s-\kappa| \le |\delta| \Longrightarrow T(\rho, s) &= \big\{t \mid |t| \le q(s\,,\rho) \;\;\&\;\; \alpha \le t \le \beta\big\} \\ &= \big\{t \mid \max\{\alpha\,, -q(s\,,\rho)\} \le t \le \min\{\beta\,, q(s\,,\rho)\}\big\} \end{aligned}$$

and so

$$\begin{aligned} T(\rho, s) \ne \varnothing &\iff |s-\kappa| \le |\delta| \;\;\&\;\; \alpha \le q(s\,,\rho) \;\;\&\;\; -q(s\,,\rho) \le \beta \\ &\iff |s-\kappa| \le |\delta| \;\;\&\;\; q(s\,,\rho) \ge \max\{\alpha\,, -\beta\} =: \gamma'\,. \end{aligned}$$

Thus, $S(\rho) = \big\{s \mid |s-\kappa| \le |\delta| \;\;\&\;\; q(s\,,\rho) \ge \gamma'\big\}$ and

$$L(\rho) = \max_{s \in S(\rho)} \max_t \Big\{l\big(s\,,|t|\big) \;\Big|\; \max\{\alpha\,, -q(s\,,\rho)\} \le t \le \min\{\beta\,, q(s\,,\rho)\}\Big\}. \tag{6.32}$$

The interior maximum

$$= \max \left\{ \begin{array}{c} \max_t \Big\{l\big(s\,,|t|\big) \;\Big|\; \max\{\alpha\,, -q(s\,,\rho)\} \le t \le 0\Big\} \\ \max_t \Big\{l(s\,,t) \;\Big|\; 0 \le t \le \min\{\beta\,, q(s\,,\rho)\}\Big\} \end{array} \right\}.$$

The first maximum in braces $= \max_t \Big\{l(s\,,t) \;\Big|\; 0 \le t \le \min\{-\alpha\,, q(s\,,\rho)\}\Big\}$. Hence, the interior maximum

$$\begin{aligned} &= \max \left\{ \begin{array}{c} \max_t \Big\{l(s\,,t) \;\Big|\; 0 \le t \le \min\{-\alpha\,, q(s\,,\rho)\}\Big\} \\ \max_t \Big\{l(s\,,t) \;\Big|\; 0 \le t \le \min\{\beta\,, q(s\,,\rho)\}\Big\} \end{array} \right\} \\ &= \max_t \Big\{l(s\,,t) \;\Big|\; 0 \le t \le \max\big\{\min\{-\alpha\,, q(s\,,\rho)\}\,,\, \min\{\beta\,, q(s\,,\rho)\}\big\}\Big\}. \end{aligned}$$

It is not difficult to show that the last maximum is in fact

$$\min\big\{\max\{-\alpha\,, \beta\}\,, q(s\,,\rho)\big\}.$$

Indeed, let γ denotes $\max\{-\alpha\,,\beta\}$. It ≥ 0 by Lemma 6.5. If $-\alpha \geq q(s\,,\rho)$, then $\min\{-\alpha\,,q(s\,,\rho)\} = q(s\,,\rho) \geq \min\{\beta\,,q(s\,,\rho)\}$ and so the maximum $= q(s\,,\rho)$. On the other hand, $q(s\,,\rho) \leq -\alpha \leq \gamma$, so that $\min\{\gamma\,,q(s\,,\rho)\} = q(s\,,\rho)$. If $-\alpha < q(s\,,\rho) \leq \beta$, then $\min\{-\alpha\,,q(s\,,\rho)\} = -\alpha$, $\min\{\beta\,,q(s\,,\rho)\} = q(s\,,\rho)$, and the maximum

$$= \max\{-\alpha\,,q(s\,,\rho)\} = q(s\,,\rho) \leq \beta \leq \gamma\,,$$

so that

$$\min\{\gamma\,,q(s\,,\rho)\} = q(s\,,\rho) = \max\{\min\{-\alpha\,,q(s\,,\rho)\}\,,\,\min\{\beta\,,q(s\,,\rho)\}\,.$$

Finally, if $-\alpha < \beta < q(s\,,\rho)$, then $\min\{-\alpha\,,q(s\,,\rho)\} = -\alpha$, $\min\{\beta\,,q(s\,,\rho)\} = \beta$, and the maximum $= \gamma = \min\{\gamma\,,q(s\,,\rho)\}$. So, the interior maximum in (6.32)

$$= \max_t\Big\{l(s\,,t)\;\Big|\;0 \leq t \leq \min\{\gamma\,,q(s\,,\rho)\}\Big\}\,.$$

Lemma 6.6. *The function $t \mapsto l(s\,,t)$ is increasing in the segment $[0\,,w(s)]$.*

Proof. It suffices to show that the derivative of the function

$$g(t) := q_-^{-1}(s\,,t) - p^{-1}(s\,,t)$$

is positive in $\big[0\,,w(s)\big]$. Recalling the definition of $q(\alpha\,,t) := t\omega(\alpha) - \Psi(\alpha\,,t)$, differentiate the identity $t = q\big(s\,,q_-^{-1}(s\,,t)\big)$ with respect to t. The result is

$$1 = \frac{\partial q}{\partial t}\big(s\,,q_-^{-1}(s\,,t)\big)\,\frac{\partial}{\partial t}q_-^{-1}(s\,,t)$$

and

$$\frac{\partial}{\partial t}q_-^{-1}(s\,,t) = \frac{1}{\dfrac{\partial q}{\partial t}\big(s\,,q_-^{-1}(s\,,t)\big)} = \frac{1}{\omega(s) - \dfrac{\partial \Psi}{\partial t}\big(s\,,q_-^{-1}(s\,,t)\big)}\,.$$

Because $q_-^{-1}(s\,,t) \leq s$, we have $\Psi(s\,,t) = t\omega(s) - w(s) + w(s-t)$, $\partial\Psi/\partial t(s\,,t) = \omega(s) - \omega(s-t)$, and $\partial\Psi/\partial t\big(s\,,q_-^{-1}(s\,,t)\big) = \omega(s) - \omega\big(s - q_-^{-1}(s\,,t)\big)$. It follows that

$$\frac{\partial}{\partial t}q_-^{-1}(s\,,t) = \frac{1}{\omega\big(s - q_-^{-1}(s\,,t)\big)} > \frac{1}{\omega(s)}\,.$$

Similarly,

$$\frac{\partial}{\partial t}p^{-1}(s\,,t) = \frac{1}{\dfrac{\partial p}{\partial t}\big(s\,,p^{-1}(s\,,t)\big)} = \frac{1}{2\omega(s) - \omega\big(s - p^{-1}(s\,,t)\big)} < \frac{1}{\omega(s)}\,.$$

Then $g'(t) = \dfrac{\partial}{\partial t}q_-^{-1}(s\,,t) - \dfrac{\partial}{\partial t}p^{-1}(s\,,t) > 0.$ □

By the lemma, the interior maximum $= l\Big(s\,,\min\{\gamma\,,q(s\,,\rho)\}\Big)$, where (for brevity) $\gamma := \max\{-\alpha\,,\beta\} \geq 0$ by Lemma 6.5. Therefore,

$$L(\rho) = \max_s \left\{ l\Big(s\,,\min\{\gamma\,,q(s\,,\rho)\}\Big) \;\Big|\; |s-\kappa| \leq |\delta| \;\;\&\;\; q(s\,,\rho) \geq \gamma'\right\}.$$

For $s \geq \rho$, the function $s \mapsto q(s\,,\rho) = w(s) - w(s-\rho)$ is increasing in $[\rho\,,\infty)$ from $w(\rho)$ to ∞. As $\gamma' \leq 0$, the constraint $q(s\,,\rho) \geq \gamma'$ is redundant and can be dropped. So,

$$L(\rho) = \max_s \left\{ l\Big(s\,,\min\{\gamma\,,q(s\,,\rho)\}\Big) \;\Big|\; |s-\kappa| \leq |\delta| \right\} = \max\{L_1(\rho)\,,\,L_2(\rho)\}\,,$$

where

$$\begin{aligned} L_1(\rho) &:= \max_s \left\{ l(s\,,\gamma) \;\big|\; |s-\kappa| \leq |\delta| \;\;\&\;\; q(s\,,\rho) \geq \gamma \right\}\,, \\ L_2(\rho) &:= \max_s \left\{ l(s\,,q(s\,,\rho)) \;\big|\; |s-\kappa| \leq |\delta| \;\;\&\;\; \gamma \geq q(s\,,\rho) \right\}\,. \end{aligned}$$

Because q is increasing with respect to the first argument, the equation $q(s\,,\rho) = \gamma$ for s is uniquely solvable for any $\gamma \geq q(\rho\,,\rho) = w(\rho)$. Denote the solution $G(\rho\,,\gamma)$:

$$q\big(G(\rho\,,\gamma)\,,\rho\big) = \gamma\,. \tag{6.33}$$

The function $\gamma \mapsto G(\rho\,,\gamma)$ is defined and increasing in $\big[w(\rho)\,,\infty\big)$. Indeed, as q is increasing with respect to the first argument,

$$\gamma < \gamma' \iff q\big(G(\rho\,,\gamma)\,,\rho\big) < q\big(G(\rho\,,\gamma')\,,\rho\big) \implies G(\rho\,,\gamma) < G(\rho\,,\gamma')\,.$$

By the same reasoning, G is increasing in ρ too. It follows that the constraint $q(s\,,\rho) \geq \gamma$ is equivalent to $s \geq G(\rho\,,\gamma)$. Hence,

$$\begin{aligned} L_1(\rho) &= \max_s \Big\{ l(s\,,\gamma)) \;\Big|\; |s-\kappa| \leq |\delta| \;\;\&\;\; s \geq G(\rho\,,\gamma) \Big\} \\ &= \max_s \Big\{ l(s\,,\gamma) \;\Big|\; \max\{\kappa - |\delta|\,, G(\rho\,,\gamma)\} \leq s \leq \kappa + |\delta| \Big\}. \end{aligned}$$

Lemma 6.7. *The function* $s \mapsto l(s,t)$ *is decreasing in* $\big[w^{-1}(t)\,,\infty\big)$.

Proof. As w is increasing in $[0\,,\infty)$, $s \geq w^{-1}(t) \iff q(s\,,s) = w(s) \geq t$ and so $s = q_-^{-1}\big(s\,,q(s\,,s)\big) \geq q_-^{-1}(s\,,t)$. It follows that

$$\Psi\big(s\,,q_-^{-1}(s\,,t)\big) = \omega(s)q_-^{-1}(s\,,t) - w(s) + w\big(s - q_-^{-1}(s\,,t)\big)\,.$$

For $t \in [0\,,s]$, $\Psi(s\,,t) = t\omega(s) - w(s) + w(s-t)$, so that

$$\frac{\partial \Psi}{\partial s}(s\,,t) = t\omega'(s) - \omega(s) + \omega(s-t)\,.$$

Then, recalling the definitions of $p(s,t) := t\omega(s) + \Psi(s,t)$ and $q(s,t) := t\omega(s) - \Psi(s,t)$, we obtain that

$$\frac{\partial p}{\partial s}(s,t) = 2t\,\omega'(s) - \omega(s) + \omega(s-t)\,, \quad \frac{\partial q}{\partial s}(s,t) = \omega(s) - \omega(s-t)\,.$$

In particular,

$$\frac{\partial p}{\partial s}(s, q_-^{-1}(s,t)) = 2\,\omega'(s)\,q_-^{-1}(s,t) - \omega(s) + \omega\big(s - q_-^{-1}(s,t)\big)$$

and

$$\frac{\partial q}{\partial s}(s, q_-^{-1}(s,t)) = \omega(s) - \omega\big(s - q_-^{-1}(s,t)\big)\,.$$

Differentiating the identity $t = q\big(s\,,\, q_-^{-1}(s,t)\big)$ with respect to s results in

$$0 = \frac{\partial q}{\partial s}\big(s\,,\, q_-^{-1}(s,t)\big) + \frac{\partial q}{\partial t}\big(s\,,\, q_-^{-1}(s,t)\big)\,\frac{\partial}{\partial s} q_-^{-1}(s,t)$$

and

$$\frac{\partial}{\partial s} q_-^{-1}(s,t) = -\frac{\frac{\partial q}{\partial s}\big(s\,,\, q_-^{-1}(s,t)\big)}{\frac{\partial q}{\partial t}\big(s\,,\, q_-^{-1}(s,t)\big)} = -\frac{\omega(s) - \omega\big(s - q_-^{-1}(s,t)\big)}{\omega\big(s - q_-^{-1}(s,t)\big)}$$

$$= 1 - \frac{\omega(s)}{\omega\big(s - q_-^{-1}(s,t)\big)}\,.$$

Similarly, $t = p\big(s\,,\, p^{-1}(s,t)\big)$

$$\Longrightarrow 0 = \frac{\partial p}{\partial s}\big(s\,,\, p^{-1}(s,t)\big) + \frac{\partial p}{\partial t}\big(s\,,\, p^{-1}(s,t)\big)\,\frac{\partial}{\partial s} p^{-1}(s,t)$$

$$\Longrightarrow \frac{\partial}{\partial s} p^{-1}(s,t) = -\frac{\frac{\partial p}{\partial s}\big(s\,,\, p^{-1}(s,t)\big)}{\frac{\partial p}{\partial t}\big(t\,,\, p^{-1}(s,t)\big)}$$

$$= -\frac{2\,\omega'(s)p^{-1}(s,t) - \omega(s) + \omega\big(s - p^{-1}(s,t)\big)}{2\,\omega(s) - \omega\big(s - p^{-1}(s,t)\big)}$$

$$= 1 - \frac{2\omega'(s)p^{-1}(s,t) + \omega(s)}{2\,\omega(s) - \omega\big(s - p^{-1}(s,t)\big)}\,.$$

So, the derivative of the function $g(s) := q_-^{-1}(s,t) - p^{-1}(s,t)$

$$
\begin{aligned}
g'(s) &= \frac{\partial}{\partial s} q_-^{-1}(s,t) - \frac{\partial}{\partial s} p^{-1}(s\,,t) \\
&= \frac{2\,\omega'(s) p^{-1}(s,t) + \omega(s)}{2\,\omega(s) - \omega\big(s - p^{-1}(s,t)\big)} - \frac{\omega(s)}{\omega\big(s - q_-^{-1}(s,t)\big)} \\
&< \frac{2\,\omega'(t) p^{-1}(s,t) + \omega(s)}{2\,\omega(s) - \omega\big(s - p^{-1}(s,t)\big)} - \frac{\omega(s)}{\omega\big(s - p^{-1}(s,t)\big)}.
\end{aligned}
$$

This difference is rewritten as the ratio whose denominator is clearly positive. So, the sign of the difference coincide with the sign of the nominator

$$
\begin{aligned}
&= \omega\big(s - p^{-1}(s,t)\big)\Big(2\,\omega'(s) p^{-1}(s,t) + \omega(s)\Big) - \\
&\qquad \omega(s)\Big(2\,\omega(t) - w\big(s - p^{-1}(s,t)\big)\Big)\omega\big(s - p^{-1}(s,t)\big)\Big(2\,\omega(s) - \omega\big(s - p^{-1}(s,t)\big)\Big).
\end{aligned}
$$

It

$$
\begin{aligned}
&= 2\Big(\omega(s)\omega\big(s - p^{-1}(s,t)\big) + \omega'(s) p^{-1}(s,t)\omega\big(s - p^{-1}(s,t)\big) - \omega(s)^2\Big) \\
&< 2\Big(\omega(s)\omega\big(s - p^{-1}(s,t)\big) + \omega'(s) p^{-1}(s,t)\omega(s) - \omega(s)^2\Big) \\
&= 2\,\omega(s)\Big(\omega\big(s - p^{-1}(s,t)\big) - \omega(s) + \omega'(s) p^{-1}(s,t)\Big) \le 0\,,
\end{aligned}
$$

thanks to the positivity and concavity of ω. □

By the lemma,

$$
L_1(\rho) = \begin{cases} l\big(\kappa - |\delta|\,,\gamma\big)\,, \text{ if } G(\rho\,,\gamma) \le \kappa - |\delta|\,, \\ l\big(G(\rho\,,\gamma)\,,\gamma\big)\,, \text{ if } \kappa - |\delta| \le G(\rho\,,\gamma) \le \kappa + |\delta|\,, \\ -\infty\,, \text{ if } G(\rho\,,\gamma) > \kappa + |\delta|\,. \end{cases} \tag{6.34}
$$

Likewise,

$$
\begin{aligned}
L_2(\rho) &= \max_s \Big\{ l\big(s\,, q(s\,,\rho)\big) \;\Big|\; |s - \kappa| \le |\delta| \;\;\&\;\; s \le G(\rho\,,\gamma) \Big\} \\
&= \max_s \Big\{ l\big(s\,, q(s\,,\rho)\big) \;\Big|\; \kappa - |\delta| \le s \le \min\big\{\kappa + |\delta|\,, G(\rho\,,\gamma)\big\} \Big\}.
\end{aligned}
$$

Lemma 6.8. *The function* $s \mapsto l\big(s\,, q(s\,,\rho)\big)$ *is decreasing in* $[\rho\,,\infty)$.

Proof. By the definition of q_-^{-1}, $q_-^{-1}\big(s\,, q(s\,,\rho)\big) = \rho$, so that

$$
g(s) := l\big(s\,, q(s\,,\rho)\big) = q_-^{-1}\big(s\,, q(s\,,\rho)\big) - p^{-1}\big(s\,, q(s\,,\rho)\big) = \rho - p^{-1}\big(s\,, q(s\,,\rho)\big)
$$

and $g'(s) = -\partial/\partial s\, p^{-1}(s\,, q(s\,, \rho))$. Differentiate the identity

$$p(s\,, p^{-1}(s\,, q(s\,, t))) = q(s\,, t)$$

with respect to s. The result is

$$\frac{\partial q}{\partial s}(s, t) = \frac{\partial p}{\partial s}(s\,,\, p^{-1}(s, q(s, t))) + \frac{\partial p}{\partial t}(s\,,\, p^{-1}(s, q(s, t)))\,\frac{\partial}{\partial s} p^{-1}(s, q(s, t))$$

and

$$\frac{\partial}{\partial s} p^{-1}(s, q(s, t)) = \frac{\dfrac{\partial q}{\partial s}(s, t) - \dfrac{\partial p}{\partial s}(s\,,\, p^{-1}(s, q(s, t)))}{\dfrac{\partial p}{\partial t}(s\,,\, p^{-1}(s, q(s, t)))}\,.$$

The denominator is positive, since $p(s, t)$ is increasing in t. As to the numerator, note that

$$q(s, t) < p(s, t) \Longrightarrow p^{-1}(s\,, q(s, t)) < p^{-1}(s\,, p(s, t)) = t \le s$$

and so $q(s\,, p^{-1}(s\,, q(s, t))) = w(s) - w(s - p^{-1}(s\,, q(s, t)))$, $\partial q/\partial s(s, t) = \omega(s) - \omega(s - t)$, and

$$\frac{\partial p}{\partial s}(s\,, p^{-1}(s\,, q(s\,, t))) = 2\omega'(s) p^{-1}(s\,, q(s\,, t)) - \omega(s) + \omega(s - p^{-1}(s\,, q(s\,, t)))\,.$$

So, the nominator

$$\begin{aligned} &= (\omega(s) - \omega(s - t)) - (2\omega'(s) p^{-1}(s, q(s, t)) - \omega(s) + \omega(s - p^{-1}(s, q(s, t)))) \\ &= 2\omega(s) - \omega(s - t) - 2\omega'(s) p^{-1}(s\,, q(s, t)) - \omega(s - p^{-1}(s\,, q(s, t)))\,, \end{aligned}$$

where $\omega(s - t) < \omega(s - p^{-1}(s, q(s, t)))$. Therefore, the numerator

$$> 2\Big(\omega(s) - \omega'(s) p^{-1}(s, q(s, t)) - \omega(s - p^{-1}(s, q(s, t)))\Big) \ge 0\,,$$

because of the concavity of ω. It follows that $\dfrac{\partial}{\partial s} p^{-1}(s, q(s, \rho)) > 0$ and $g'(s) < 0$. □

By the lemma,

$$L_2(\rho) = \begin{cases} l(\kappa - |\delta|\,, q(\kappa - |\delta|\,, \rho)\,, & \text{if } \kappa - |\delta| \le G(\rho\,, \gamma)\,, \\ -\infty\,, & \text{otherwise.} \end{cases}$$

Comparison of this result with (6.34) shows that

$$L(\rho) = \begin{cases} l(\kappa - |\delta|\,, \gamma)\,, & \text{if } G(\rho\,, \gamma) \le \kappa - |\delta|\,, \\ \max\{l(G(\rho\,, \gamma)\,, \gamma)\,, l(\kappa - |\delta|\,, q(\kappa - |\delta|\,, \rho))\}\,, & \\ & \text{if } \kappa - |\delta| \le G(\rho\,, \gamma) \le \kappa + |\delta|\,, \\ l(\kappa - |\delta|\,, q(\kappa - |\delta|\,, \rho))\,, & \text{if } G(\rho\,, \gamma) \ge \kappa + |\delta|\,. \end{cases}$$

In view of (6.33), $l(G(\rho,\gamma),\gamma) = l\Big(G(\rho,\gamma), q(G(\rho,\gamma),\rho)\Big)$, so that $\kappa - |\delta| \leq G(\rho,\gamma) \leq \kappa + |\delta| \Longrightarrow$

$$\begin{aligned} L(\rho) &= \max\Big\{ l(G(\rho,\gamma), q(G(\rho,\gamma),\rho)\,,\, l(\kappa - |\delta|, q(\kappa - |\delta|,\rho))\Big\} \\ &= l(\kappa - |\delta|, q(\kappa - |\delta|,\rho)) \end{aligned}$$

by Lemma 6.8. Therefore,

$$\begin{aligned} L(\rho) &= \begin{cases} l(\kappa - |\delta|,\gamma)\,, & \text{if } G(\rho,\gamma) \leq \kappa - |\delta|\,, \\ l(\kappa - |\delta|, q(\kappa - |\delta|,\rho))\,, & \text{if } G(\rho,\gamma) \geq \kappa - |\delta|\,. \end{cases} \\ &= \begin{cases} l(\kappa - |\delta|, q(G(\rho,\gamma),\rho))\,, & \text{if } q(G(\rho,\gamma),\rho) \leq q(\kappa - |\delta|,\rho)\,, \\ l(\kappa - |\delta|, q(\kappa - |\delta|,\rho))\,, & \text{if } q(G(\rho,\gamma),\rho) \geq q(\kappa - |\delta|,\rho)\,, \end{cases} \\ &= l\Big(\kappa - |\delta|, \min\{q(G(\rho,\gamma),\rho))\,,\, q(\kappa - |\delta|,\rho)\}\Big) \\ &= l\Big(\kappa - |\delta|, q(\min\{G(\rho,\gamma)\,,\,\kappa - |\delta|\},\rho)\Big)\,, \end{aligned}$$

$$L(\sigma) = l\Big(\kappa - |\delta|, q(\min\{G(\sigma,\gamma)\,,\,\kappa - |\delta|\},\sigma)\Big)\,,$$

and

$$L = \max\left\{ \begin{array}{c} l\Big(\kappa - |\delta|, q(\min\{G(\rho,\gamma)\,,\,\kappa - |\delta|\},\rho)\Big) \\ l\Big(\kappa - |\delta|, q(\min\{G(\sigma,\gamma)\,,\,\kappa - |\delta|\},\sigma)\Big) \end{array} \right\}.$$

By Lemma 6.6, l is increasing with respect to the second argument. So, this maximum

$$= l\Big(\kappa - |\delta|, \max\{q(\min\{G(\rho,\gamma)\,,\,\kappa - |\delta|\},\rho)\,,\, q(\min\{G(\sigma,\gamma)\,,\,\kappa - |\delta|\},\sigma)\}\Big).$$

Lemma 6.9. $\max\Big\{q(\min\{G(\rho,\gamma)\,,\,z\},\rho), q(\min\{G(\sigma,\gamma)\,,\,z\},\sigma)\Big\} =$

$$\min\Big\{q(z, \max\{\rho,\sigma\})\,,\gamma\Big\}.$$

Proof. Because of the symmetry $\rho \leftrightarrow \sigma$, it suffices to focus on the case $\rho \geq \sigma$. Then the claim becomes

$$\max\Big\{q(\min\{G(\rho,\gamma)\,,\,z\},\rho)\,,\, q(\min\{G(\sigma,\gamma)\,,\,z\},\sigma)\Big\} = \min\{q(z,\rho),\gamma\}\,.$$

As G is increasing with respect to the first argument, $\rho \geq \sigma \Longrightarrow G(\rho,\gamma) \geq G(\sigma,\gamma)$, so that there are three possibilities:

$$(i)\ z < G(\sigma,\gamma)\,,\ (ii)\ G(\sigma,\gamma) \leq z \leq G(\rho,\gamma)\,,\ (iii)\ z > G(\rho,\gamma)\,.$$

In the first case, $q(\min\{G(\rho,\gamma), z\},\rho) = q(z,\rho)$, $q(\min\{G(\sigma,\gamma), z\},\sigma) =$

$q(z\,,\rho)$, and so the maximum $=q(z\,,\rho)$. In the second case, $q\big(\min\{G(\rho,\gamma),z\},\rho\big) = q(z,\rho)$, $q\big(\min\{G(\sigma,\gamma)\,,\,z\},\sigma\big) = q\big(G(\sigma,\gamma)\,,\sigma\big) = \gamma$ (by (6.33)), and the maximum $= \max\big\{q(z\,,\rho)\,,\gamma\big\}$. As follows from the monotonicity of the function G,

$$G(\sigma\,,\gamma) \le z \le G(\rho\,,\gamma) \Longleftrightarrow q(z\,,\sigma) \le \gamma \le q(z\,,\rho)\,.$$

Therefore, $\max\big\{q(z\,,\rho)\,,\gamma\big\} = q(z\,,\rho)$. Hence, in the case (ii), the maximum $= q(z\,,\rho)$. Finally, (iii) implies $q\big(\min\{G(\rho,\gamma)\,,\,z\},\rho\big) = q\big(G(\rho,\gamma)\,,\rho\big) = \gamma$ and

$$q\big(\min\{G(\sigma,\gamma)\,,\,z\},\sigma\big) = q\big(G(\sigma,\gamma)\,,\sigma\big) = \gamma,$$

so that the maximum $= \gamma$. Thus, the maximum

$$= \begin{cases} q(z\,,\rho)\,, \text{ if } z \le G(\rho,\gamma)\,, \\ \gamma\,, \text{ if } z \ge G(\rho,\gamma)\,. \end{cases}$$

Thanks to monotonicity of G, $z \le G(\rho,\gamma) \Longleftrightarrow q(z\,,\rho) \le \gamma$. Therefore, the maximum

$$= \begin{cases} q(z\,,\rho)\,, \text{ if } q(z\,,\rho) \le \gamma\,, \\ \gamma\,, \text{ if } q(z\,,\rho) \ge \gamma\,. \end{cases} = \min\big\{q(z\,,\rho)\,,\gamma\big\}.$$

□

By the lemma,

$$L = l\Big(\kappa - |\delta|\,, \min\big\{q(\kappa - |\delta|\,, \max\{\rho\,,\sigma\})\,,\gamma\big\}\Big)\,,$$

where, in accordance with the definitions of ρ, σ, and γ,

$$\max\{\rho\,,\sigma\} = \begin{cases} q_-^{-1}(\kappa\,,|a|) - |\delta|\,, \text{ if } |\delta| \le 0.5\Big(p^{-1}(\kappa\,,|a|) + q_-^{-1}(\kappa\,,|a|)\Big)\,, \\ |\delta| - p^{-1}(\kappa\,,|a|)\,, \text{ if } |\delta| \ge 0.5\Big(p^{-1}(\kappa\,,|a|) + q_-^{-1}(\kappa\,,|a|)\Big)\,, \end{cases}$$

and

$$\gamma = \max\Big\{-a + b|\delta| + \Psi\left(\kappa\,,|\delta|\right)\,, a - b|\delta| + \Psi\left(\kappa\,,|\delta|\right)\Big\} = \big|a - b|\delta|\big| + \Psi\left(\kappa\,,|\delta|\right)\,.$$

As $|\delta| \in \big[p^{-1}(\kappa\,,|a|),\, q_-^{-1}(\kappa\,,|a|)\big]$ by (6.31) and $|a| \le w(\kappa) = q(\kappa\,,\kappa) \Longrightarrow q_-^{-1}\left(\kappa\,,|a|\right) \le \kappa$, we have $|\delta| \le \kappa$ and so

$$\Psi\left(\kappa\,,|\delta|\right) = |\delta|\omega(\kappa) - w(\kappa) + w\left(\kappa - |\delta|\right) = |b\delta| - w(\kappa) + w\left(\kappa - |\delta|\right)$$

and $\gamma = \big|a - b|\delta|\big| + |b\delta| - w(\kappa) + w\left(\kappa - |\delta|\right)$. Consequently,

$$L = l\Big(\kappa - |\delta|\,, \min\big\{q(\kappa - |\delta|\,, \max\{\rho\,,\sigma\})\,, \big|a - b|\delta|\big| + |b\delta| - w(\kappa) + w\left(\kappa - |\delta|\right)\big\}\Big)\,,$$

where $\rho = q_-^{-1}(\kappa, |a|) - |\delta|$ and $\sigma = |\delta| - p^{-1}(\kappa, |a|)$.

This result completes the first stage of our task (maximization). The second stage is minimization of L, as a function of $|\delta|$, over the range $[p^{-1}(\kappa, |a|), q_-^{-1}(\kappa, |a|)]$. While minimizing, we use the abbreviations

$$u := p^{-1}(\kappa, |a|)\,,\ v := q_-^{-1}(\kappa, |a|)\,,\ \gamma(\tau) := |a - b\tau| + |b|\tau - w(\kappa) + w(\kappa - \tau)\,,$$

$$\varphi(\tau) := \min\{q(\kappa - \tau, \max\{v - \tau, \tau - u\}), \gamma(\tau)\}\,.$$

With these abbreviations, the problem now is to find a minimizer for

$$m(a,b) := \min_{\tau}\Big\{l(\kappa - \tau, \varphi(\tau))\ \Big|\ u \le \tau \le v\Big\}. \tag{6.35}$$

In this connection, it is worth while to note that

Lemma 6.10.

$$\frac{u+v}{2} \le \left|\frac{a}{b}\right| \le v = \kappa - w^{-1}(w(\kappa) - |a|)\,.$$

Proof. By the definition of q,

$$q(\kappa, \kappa - w^{-1}(w(\kappa) - |a|)) = w(\kappa) - w(\kappa - (\kappa - w^{-1}(w(\kappa) - |a|))) = |a|$$

and so $\kappa - w^{-1}(w(\kappa) - |a|) = q_-^{-1}(\kappa, |a|) =: v$. Besides, by the convexity of w,

$$w(\kappa - |a/b|) \ge w(\kappa) + w'(\kappa)((\kappa - |a/b|) - \kappa) = w(\kappa) - \omega(\kappa)\left|\frac{a}{b}\right| = w(\kappa) - |a|\,,$$

whence $\kappa - w^{-1}(w(\kappa) - |a|) \ge |a/b|$. These remarks prove the second inequality. The first is equivalent to

$$g(|a|) := p^{-1}(\kappa, |a|) + q_-^{-1}(\kappa, |a|) - 2\left|\frac{a}{b}\right| \ge 0\,.$$

To prove it, differentiate g twice.

$$g'(t) = \frac{\partial}{\partial t}p^{-1}(\kappa, t) + \frac{\partial}{\partial t}q_-^{-1}(\kappa, t) - \frac{2}{|b|}\,,$$

$$g''(t) = \frac{\partial^2}{\partial t^2}p^{-1}(\kappa, t) + \frac{\partial^2}{\partial t^2}q_-^{-1}(\kappa, t)\,.$$

Differentiating the identity $t = p(\kappa, p^{-1}(\kappa, t))$, we get

$$\begin{aligned} 1 &= \frac{\partial p}{\partial}(\kappa, p^{-1}(\kappa, t)) \cdot \frac{\partial s}{\partial t}(\kappa, p^{-1}(\kappa, t)) + \frac{\partial p}{\partial t}(\kappa, p^{-1}(\kappa, t)) \cdot \frac{\partial}{\partial t}p^{-1}(\kappa, t) \\ &= (2\omega(\kappa) - \omega(\kappa - p^{-1}(\kappa, t)))\frac{\partial}{\partial t}p^{-1}(\kappa, t)\,, \end{aligned}$$

whence $\frac{\partial}{\partial t}p^{-1}(\kappa\,,t) = \big(2\omega(\kappa) - \omega(\kappa - p^{-1}(\kappa\,,t))\big)^{-1}$. Similarly, $t = q(\kappa\,,q_-^{-1}(\kappa\,,t)) \Longrightarrow$

$$1 = \frac{\partial q}{\partial s}\big(\kappa\,,q_-^{-1}(\kappa\,,t)\big)\cdot\frac{\partial s}{\partial t}\big(\kappa\,,q_-^{-1}(\kappa\,,t)\big) + \frac{\partial q}{\partial t}\big(s\,,q_-^{-1}(\kappa\,,t)\big)\cdot\frac{\partial}{\partial t}q_-^{-1}(\kappa\,,t)$$

$$= \omega\big(\kappa - q_-^{-1}(\kappa\,,t)\big)\frac{\partial}{\partial t}q_-^{-1}(\kappa\,,t)$$

and $\frac{\partial}{\partial t}q_-^{-1}(\kappa\,,t) = \big(\omega(\kappa) - \omega(\kappa - t)\big)^{-1}$. So,

$$g'(t) = \frac{1}{2\omega(\kappa) - \omega\big(\kappa - p^{-1}(\kappa\,,t)\big)} + \frac{1}{\omega\big(\kappa - q_-^{-1}(\kappa\,,t)\big)} - \frac{2}{|b|}$$

and

$$g''(t) = \frac{-\omega'\big(\kappa - p^{-1}(\kappa\,,t)\big)}{\big(2\omega(\kappa) - \omega\big(\kappa - p^{-1}(\kappa\,,t)\big)\big)^2}\left(-\frac{\partial}{\partial t}p^{-1}(\kappa\,,t)\right) + \frac{-\omega'\big(\kappa - q_-^{-1}(\kappa\,,t)\big)}{\omega\big(\kappa - q_-^{-1}(\kappa\,,t)\big)\big)^2}\left(-\frac{\partial}{\partial t}q_-^{-1}(\kappa\,,t)\right)$$

$$= \frac{\omega'\big(\kappa - p^{-1}(\kappa\,,t)\big)}{\big(2\omega(\kappa) - \omega\big(\kappa - p^{-1}(\kappa\,,t)\big)\big)^3} + \frac{\omega'\big(\kappa - q_-^{-1}(\kappa\,,t)\big)}{\omega\big(\kappa - q_-^{-1}(\kappa\,,t)\big)^3}\,.$$

As $u \le v$ and because of the concavity of ω, we have $\omega'(\kappa - v) \ge \omega'(\kappa - u)$, and so

$$g''(t) \ge \frac{\omega'(\kappa - v)}{\big(2\omega(\kappa) - \omega(\kappa - u)\big)^3} + \frac{\omega'(\kappa - v)}{\omega\big(\kappa - p^{-1}(\kappa\,,t)\big)\big)^3}$$

$$= \omega'(\kappa - v)\left(\frac{1}{\big(2\omega(\kappa) - \omega(\kappa - u)\big)^3} + \frac{1}{\omega(\kappa - v)^3}\right) \ge 0\,.$$

So, g' is not decreasing, $\min_{0\le t\le\kappa} g'(t) = g'(0) = 0$. It follows that $g'(t) \ge 0\,, \forall t \in [0\,,\kappa]$, g is not decreasing and $\min_{0\le t\le\kappa} g(t) = g(0) = 0$. Thus, $g(t) \ge 0$. □

To evaluate the minimum (6.35), we have to investigate the restriction of the objective $l\big(\kappa - \tau\,,\varphi(\tau)\big)$ to the segment $[u\,,v]$. As $v - \tau > \tau - u \Longleftrightarrow \tau < 0.5(u + v)$,

$$\varphi(\tau) = \begin{cases} \min\big\{q(\kappa - \tau\,,v - \tau)\,,\gamma(\tau)\big\}\,, & \text{if } u \le \tau \le 0.5(u + v)\,, \\ \min\big\{q(\kappa - \tau\,,\tau - u)\,,\gamma(\tau)\big\}\,, & \text{if } 0.5(u + v) \le \tau \le v\,. \end{cases}$$

By Lemma 6.10, $v-\tau \le \kappa-\tau$ and so $\Psi(\kappa-\tau\,,v-\tau)$

$$= (v-\tau)\,\omega(\kappa-\tau) - w(\kappa-\tau) + w\big((\kappa-\tau)-(v-\tau)\big)$$

$$= (v-\tau)\,\omega(\kappa-\tau) - w(\kappa-\tau) + w\big(\kappa-\big(\kappa-w^{-1}\big(w(\kappa)-|a|\big)\big)\big)$$

$$= (v-\tau)\,\omega(\kappa-\tau) - w(\kappa-\tau) + w(\kappa) - |a|\,.$$

Consequently,

$$q(\kappa-\tau\,,v-\tau) = (v-\tau)\,\omega(\kappa-\tau) - \Psi(\kappa-\tau\,,v-\tau) = w(\kappa-\tau) - w(\kappa) + |a|\,.$$

Besides, by Lemma 6.10, $u \le \tau \le 0.5(u+v) \Longrightarrow \gamma(\tau) = |a| - w(\kappa) + w(\kappa-\tau)$. So,

$$u \le \tau \le \frac{u+v}{2} \Longrightarrow l\big(\kappa-\tau\,,\varphi(\tau)\big) = l\big(\kappa-\tau\,,w(\kappa-\tau)-w(\kappa)+|a|\big)\,.$$

Lemma 6.11. *The function* $s \mapsto l\big(s\,,w(s)-w(\kappa)+|a|\big)$ *is increasing in its domain.*

Proof. Let for brevity $t(s) := w(s) - w(\kappa) + |a|$ and

$$g(s) := l\big(s\,,t(s)\big) = q_-^{-1}\big(s\,,t(s)\big) - p^{-1}\big(s\,,t(s)\big)\,.$$

With these abbreviations, $g'(s) = \frac{\partial}{\partial s}q_-^{-1}\big(s\,,t(s)\big) - \frac{\partial}{\partial s}p^{-1}\big(s\,,t(s)\big)$. Differentiating the identity $t(s) = q\big(s\,,q_-^{-1}\big(s\,,t(s)\big)\big)$, we obtain

$$\frac{dt}{ds} = \frac{\partial q}{\partial s}\big(s\,,q_-^{-1}\big(s\,,t(s)\big)\big) + \frac{\partial q}{\partial t}\big(s\,,q_-^{-1}\big(s\,,t(s)\big)\big)\cdot\frac{\partial}{\partial s}q_-^{-1}\big(s\,,t(s)\big)$$

and so

$$\frac{\partial}{\partial s}q_-^{-1}\big(s\,,t(s)\big) = \frac{\omega(s) - \dfrac{\partial q}{\partial s}\big(s\,,q_-^{-1}\big(s\,,t(s)\big)\big)}{\dfrac{\partial q}{\partial t}\big(s\,,q_-^{-1}\big(s\,,t(s)\big)\big)}\,.$$

Since the restriction of $t \mapsto q(s\,,t)$ to $[0\,,s]$ is $w(s)-w(s-t)$, the derivatives $\frac{\partial q}{\partial s}(s\,,t) = \omega(s)-\omega(s-t)$, $\frac{\partial q}{\partial t} = \omega(s-t)$, so that

$$\frac{\partial q}{\partial s}\big(s\,,q_-^{-1}\big(s\,,t(s)\big)\big) = \omega(s) - \omega\big(s - q_-^{-1}\big(s\,,t(s)\big)\big)$$

$$\frac{\partial q}{\partial t}\big(s\,,q_-^{-1}\big(s\,,t(s)\big)\big) = \omega\big(s - q_-^{-1}\big(s\,,t(s)\big)\big)\,,$$

and $\frac{\partial}{\partial s}q_-^{-1}\big(s\,,t(s)\big) = 1$. Similarly, $t(s) = p\big(s\,,p^{-1}\big(s\,,t(s)\big)\big)$

$$\Longrightarrow \frac{dt}{ds} = \frac{\partial p}{\partial s}\big(s\,,p^{-1}\big(s\,,t(s)\big)\big) + \frac{\partial p}{\partial t}\big(s\,,p^{-1}\big(s\,,t(s)\big)\big)\cdot\frac{\partial}{\partial s}p^{-1}\big(s\,,t(s)\big)$$

$$\Longrightarrow \frac{\partial}{\partial s}p^{-1}\big(s\,,t(s)\big) = \frac{\omega(s) - \dfrac{\partial p}{\partial s}\big(s\,,p^{-1}\big(s\,,t(s)\big)\big)}{\dfrac{\partial p}{\partial t}\big(s\,,p^{-1}\big(s\,,t(s)\big)\big)}$$

and $p(s,t) = 2t\omega(s) - w(s) + w(s-t)$ implies

$$\frac{\partial p}{\partial s}\big(s, p^{-1}(s, t(s))\big) = 2\omega'(s)p^{-1}\big(s, t(s)\big) - \omega(s) + \omega\big(s - p^{-1}(s, t(s))\big)$$

and

$$\frac{\partial p}{\partial t}\big(s, p^{-1}(s, t(s))\big) = 2\omega(s) - \omega\big(s - p^{-1}(s, t(s))\big).$$

Consequently,

$$\begin{aligned}\frac{\partial}{\partial s} p^{-1}\big(s, t(s)\big) &= \frac{2\omega(s) - \omega\big(s - p^{-1}(s, t(s))\big) - 2\omega'(s)p^{-1}\big(s, t(s)\big)}{2\omega(s) - \omega\big(s - p^{-1}(s, t)\big)} \\ &= 1 - 2\frac{\omega'(s)p^{-1}\big(s, t(s)\big)}{2\omega(s) - \omega\big(s - p^{-1}(s, t(s))\big)}.\end{aligned}$$

It follows that

$$g'(s) = \frac{2\Big(\omega'(s)p^{-1}\big(s, t(s)\big)\Big)}{\Big(2\omega(s) - \omega\big(s - p^{-1}(s, t(s))\big)\Big)} > 0,$$

so that g is increasing. □

By the lemma, the objective $l\big(\kappa - \tau, \varphi(\tau)\big)$ is decreasing in the first half of $[u, v]$, so that the minimizer in (6.35) is in the second half $\big[0.5(u+v), v\big]$. This segment is divided into two parts $\big[0.5(u+v), |a/b|\big]$ and $\big[|a/b|, v\big]$ by Lemma 6.10. This division determines the value of $\gamma(\tau)$:

$$\gamma(\tau) = \begin{cases} |a| - w(\kappa) + w(\kappa - \tau), & \text{if } 0 \le \tau \le \left|\dfrac{a}{b}\right|, \\ 2|b|\tau - |a| - w(\kappa) + w(\kappa - \tau), & \text{if } \left|\dfrac{a}{b}\right| \le \tau \le \kappa. \end{cases}$$

The value of $q(\kappa - \tau, \tau - u)$ depends on the position of $0.5(\kappa + u)$ within $\big[0.5(u+v), v\big]$. As it turns out,

Lemma 6.12. $0.5(\kappa + u) > |a/b|$.

Proof. Suppose that (contrary to the claim) $0.5(\kappa + u) \le |a/b|$. Then this inequality would hold also in the special case $\omega(t) = t$, when $0.5(\kappa + u) = 0.5\sqrt{b^2 + 2|a|}$, i.e., we would have the inequality $2|a/b| \ge \sqrt{b^2 + 2|a|}$. But it

$$\Longleftrightarrow \frac{2|a|}{b^2} \ge \sqrt{1 + \frac{2|a|}{b^2}} \Longleftrightarrow \left(\frac{2|a|}{b^2}\right)^2 - \frac{2|a|}{b^2} - 1 \ge 0 \Longleftrightarrow \frac{2|a|}{b^2} \ge \frac{1 + \sqrt{5}}{2} > 1,$$

in contradiction to the assumption $b^2 \ge 2|a|$. So, the supposition is wrong:

$$0.5(\kappa + u) > |a/b|.$$

□

By the lemma, $0.5(u+v) \le \tau \le |a/b|$

$$\Longrightarrow \begin{cases} q(\kappa-\tau\,,\tau-u) = w(\kappa-\tau) - w(\kappa+u-2\tau) \\ \gamma(\tau) = w(\kappa-\tau) - \big(w(\kappa) - |a|\big) \end{cases}$$

$$\Longrightarrow \varphi(\tau) = \min\big\{w(\kappa-\tau) - w(\kappa+u-2\tau)\,, w(\kappa-\tau) - \big(w(\kappa)-|a|\big)\big\}$$

$$\Longrightarrow \varphi(\tau) = \begin{cases} w(\kappa-\tau) - w(\kappa+u-2\tau)\,, \text{ if } \tau \le \dfrac{\kappa+u-w^{-1}\big(w(\kappa)-|a|\big)}{2}\,, \\ w(\kappa-\tau) - \big(w(\kappa)-|a|\big)\,, \text{ if } \tau \ge \dfrac{\kappa+u-w^{-1}\big(w(\kappa)-|a|\big)}{2}\,. \end{cases}$$

By Lemma 6.10, $\kappa + u - w^{-1}\big(w(\kappa) - |a|\big) = u + v$, so that

$$\frac{u+v}{2} \le \tau \le \left|\frac{a}{b}\right| \Longrightarrow \tau \ge \frac{\kappa + u - w^{-1}\big(w(\kappa) - |a|\big)}{2}$$
$$\Longrightarrow \varphi(\tau) = w(\kappa-\tau) - \big(w(\kappa) - |a|\big)$$
$$\Longrightarrow l\big(\kappa-\tau\,,\varphi(\tau)\big) = l\big(\kappa-\tau\,, w(\kappa-\tau) - w(\kappa) + |a|\big)\,,$$

which function is decreasing by Lemma 6.11. So, the minimizer is in $\big[|a/b|\,, v\big]$, where $\gamma(\tau) = 2|b|\tau - |a| - w(\kappa) + w(\kappa-\tau)$. As to $q(\kappa-\tau\,,\tau-u)$, there are two possibilities:

$$(i)\ \left|\frac{a}{b}\right| \le \frac{\kappa+u}{2} < v\,,\ (ii)\ \frac{\kappa+u}{2} \ge v\,. \tag{6.36}$$

In the first case, $q(\kappa-\tau\,,\tau-u)$

$$= \begin{cases} w(\kappa-\tau) - w(\kappa+u-2\tau)\,, \text{ if } \left|\dfrac{a}{b}\right| \le \tau \le \dfrac{\kappa+u}{2}\,, \\ (2\tau-\kappa-u)\omega(\kappa-\tau) + 2w(\kappa-\tau) - w(\tau-u)\,, \text{ if } \dfrac{\kappa+u}{2} \le \tau \le v\,, \end{cases}$$

and so

$$\varphi(\tau) = \begin{cases} \min\big\{w(\kappa-\tau) - w(\kappa+u-2\tau)\,, 2|b|\tau - |a| - w(\kappa) + w(\kappa-\tau)\big\}\,, \\ \qquad\qquad \text{if } \left|\dfrac{a}{b}\right| \le \tau \le \dfrac{\kappa+u}{2}\,, \\ \min\left\{\begin{matrix} (2\tau-\kappa-u)\omega(\kappa-\tau) + 2w(\kappa-\tau) - w(\tau-u) \\ 2|b|\tau - |a| - w(\kappa) + w(\kappa-\tau) \end{matrix}\right\}, \\ \qquad\qquad \text{if } \dfrac{\kappa+u}{2} \le \tau \le v\,. \end{cases} \tag{6.37}$$

The first minimum

$$= w(\kappa-\tau) - w(\kappa+u-2\tau) - w(\kappa) - |a| + \min\big\{2|b|\tau + w(\kappa+u-2\tau)\,, w(\kappa) + |a|\big\}\,.$$

The function $g(\tau) := 2|b|\tau + w(\kappa+u-2\tau)$ is increasing in $\big[|a/b|\,, 0.5(\kappa+u)\big]$ (look at its derivative) and so

$$\left|\frac{a}{b}\right| < \tau < \frac{\kappa+u}{2} \Longrightarrow 2|a| + w\Big(\kappa+u-2\left|\frac{a}{b}\right|\Big) = g\Big(\left|\frac{a}{b}\right|\Big) < g(\tau) < g\Big(\frac{\kappa+u}{2}\Big)$$
$$= |b|(\kappa+u)\,.$$

Lemma 6.13. $2|a| + w\big(\kappa+u-2|a/b|\big) \geq w(\kappa) + |a|$.

Proof. The claim is equivalent to $w\big(\kappa+u-2|a|/|b|\big) \geq w(\kappa) - |a|$. If the converse is true, then we would have the inequality $w\big(\kappa+u-2|a|/|b|\big) < w(\kappa) - |a|$ also in the special case, when $\omega(t) = t$, $\kappa = |b|$, $w(\kappa) = 0.5b^2$, and $u = \sqrt{b^2+2|a|} - |b|$, that is, the inequality $0.5\left(|b| + \sqrt{b^2+2|a|} - |b| - 2|a/b|\right)^2 < 0.5b^2 - |a|$. But it

$$\Longleftrightarrow \left(\sqrt{b^2+2|a|} - 2\left|\frac{a}{b}\right|\right)^2 < b^2 - 2|a|$$

$$\Longleftrightarrow b^2 + 2|a| - 4\left|\frac{a}{b}\right|\sqrt{b^2+2|a|} + 4\frac{a^2}{b^2} < b^2 - 2|a|$$

$$\Longleftrightarrow |a| + \frac{a^2}{b^2} > |a|\sqrt{1+\frac{2|a|}{b^2}}$$

$$\Longleftrightarrow 1 + \frac{|a|}{b^2} < \sqrt{1+2\frac{|a|}{b^2}} \Longleftrightarrow 1 + 2\frac{|a|}{b^2} + \frac{a^2}{b^4} < 1 + 2\frac{|a|}{b^2}\,,$$

which is impossible. Hence, the claim. □

By the lemma, the first minimum in (6.37) $= w(\kappa-\tau) - w(\kappa+u-2\tau)$. The second

$$= (2\tau-\kappa-u)\omega(\kappa-\tau) + 2w(\kappa-\tau) - w(\tau-u) + \min\{0\,, h(\tau) - w(\kappa) - |a|\}\,,$$

where

$$h(\tau) := 2|b|\tau - w(\kappa-\tau) + w(\tau-u) - (2\tau-\kappa-u)\omega(\kappa-\tau)\,.$$

The function h is increasing in $[u\,,\kappa]$ (and, in particular, in $\big[|a/b|\,, v\big]$), because its derivative

$$h'(\tau) = 2\omega(\kappa) - \omega(\kappa-\tau) + \omega(\tau-u) + (2\tau-\kappa-u)\,\omega'(\kappa-\tau)$$

is positive there. So, it is invertible. Denote

$$\chi := \chi(a\,,b) := h^{-1}\big(w(\kappa) + |a|\big)\,.$$

The second minimum in (6.37)

$$= (2\tau-\kappa-u)\omega(\kappa-\tau) + 2w(\kappa-\tau) - w(\tau-u) \Longleftrightarrow \tau > \chi\,.$$

Lemma 6.14. $\chi \le |a/b|$.

Proof. Suppose the converse is true: $\chi > |a/b|$ or, equivalently,

$$h\big(|a/b|\big) < w(\kappa) + |a|.$$

Then this inequality would hold also in the special case $\omega(t) = t$, when $\kappa = |b|$, $w(t) = 0.5t^2$, $u = \sqrt{b^2+2|a|} - |b|$,

$$h\left(\left|\frac{a}{b}\right|\right) = 2|a| - \frac{1}{2}\left(|b| - \left|\frac{a}{b}\right|\right)^2 + \frac{1}{2}\left(\left|\frac{a}{b}\right| - \sqrt{b^2+2|a|} + |b|\right)^2 - \left(2\left|\frac{a}{b}\right| - \sqrt{b^2+2|a|}\right)\left(|b| - \left|\frac{a}{b}\right|\right)$$

$$= b^2\left(2\frac{|a|}{b^2} - \frac{1}{2}\left(1 - \frac{|a|}{b^2}\right)^2 + \frac{1}{2}\left(\frac{|a|}{b^2} - \sqrt{1+2\frac{|a|}{b^2}} + 1\right)^2 - \left(2\frac{|a|}{b^2} - \sqrt{1+2\frac{|a|}{b^2}}\right)\left(1 - \frac{|a|}{b^2}\right)\right)$$

and $w(\kappa) + |a| = 0.5b^2 + |a| = 0.5b^2(1 + 2|a|/b^2)$, i.e., we would have the inequality

$$2\frac{|a|}{b^2} - \frac{1}{2}\left(1 - \frac{|a|}{b^2}\right)^2 + \frac{1}{2}\left(\frac{|a|}{b^2} - \sqrt{1+2\frac{|a|}{b^2}} + 1\right)^2 - \left(2\frac{|a|}{b^2} - \sqrt{1+2\frac{|a|}{b^2}}\right)\left(1 - \frac{|a|}{b^2}\right) < \frac{1}{2} + \frac{|a|}{b^2},$$

which is equivalent to $1 + |a|/b^2 < \sqrt{1+2|a|/b^2}$ or $1 + 2|a|/b^2 + a^2/b^4 < 1 + 2|a|/b^2$. But this is impossible. So, the supposition is wrong. Hence, the claim. □

By the lemma, $|a/b| \le \tau \le v$

$$\Longrightarrow \varphi(\tau) = \begin{cases} w(\kappa-\tau) - w(\kappa+u-2\tau), \text{ if } \left|\dfrac{a}{b}\right| \le \tau \le \dfrac{\kappa+u}{2}, \\ (2\tau-\kappa-u)\omega(\kappa-\tau) + 2w(\kappa-\tau) - w(\tau-u), \\ \qquad\qquad \text{if } \dfrac{\kappa+u}{2} \le \tau \le v, \end{cases}$$

$$\Longrightarrow l(\kappa-\tau, \varphi(\tau)) = \begin{cases} l(\kappa-\tau, w(\kappa-\tau) - w(\kappa+u-2\tau)), \\ \qquad\qquad \text{if } \left|\dfrac{a}{b}\right| \le \tau \le \dfrac{\kappa+u}{2}, \\ l(\kappa-\tau, (2\tau-\kappa-u)\,\omega(\kappa-\tau) + 2w(\kappa-\tau) - w(\tau-u)), \\ \qquad\qquad \text{if } \dfrac{\kappa+u}{2} \le \tau \le v, \end{cases}$$

Lemma 6.15. *The function* $s \mapsto l\big(s\,,w(s)-w(2s-\kappa+u)\big)$ *is decreasing in* $\big(0.5(\kappa-u)\,,\kappa-|a/b|\big)$.

Proof. Let (for brevity) $\alpha := \kappa - u$, $t(s) := w(s) - w(2s-\alpha)$, and

$$g(s) := l\big(s\,,t(s)\big) = q_-^{-1}\big(s\,,t(s)\big) - p^{-1}\big(s\,,t(s)\big).$$

Then $g'(s) = \frac{d}{ds}q_-^{-1}\big(s\,,t(s)\big) - \frac{d}{ds}p^{-1}\big(s\,,t(s)\big)$. It suffices to prove negativity of g' in $\big(0.5\alpha\,,\kappa-|a/b|\big)$, that is, the inequality $\frac{d}{ds}q_-^{-1}\big(s\,,t(s)\big) < \frac{d}{ds}p^{-1}\big(s\,,t(s)\big)$. Differentiating the identity $t(s) = q\big(s\,,q_-^{-1}\big(s\,,t(s)\big)\big)$, we obtain

$$t'(s) = \frac{\partial q}{\partial s}\Big(s\,,q_-^{-1}\big(s\,,t(s)\big)\Big) + \frac{\partial q}{\partial t}\Big(s\,,q_-^{-1}\big(s\,,t(s)\big)\Big)\cdot\frac{d}{ds}q_-^{-1}\big(s\,,t(s)\big)\,,$$

whence

$$\begin{aligned}\frac{d}{ds}q_-^{-1}\big(s\,,t(s)\big) &= \frac{t'(s) - \dfrac{\partial q}{\partial s}\Big(s\,,q_-^{-1}\big(s\,,t(s)\big)\Big)}{\dfrac{\partial q}{\partial t}\Big(s\,,q_-^{-1}\big(s\,,t(s)\big)\Big)}\\ &= \frac{\omega(s) - 2\omega(2s-\alpha) - \dfrac{\partial q}{\partial s}\Big(s\,,q_-^{-1}\big(s\,,t(s)\big)\Big)}{\dfrac{\partial q}{\partial t}\Big(s\,,q_-^{-1}\big(s\,,t(s)\big)\Big)}. \qquad (6.38)\end{aligned}$$

As

$$\begin{aligned} t\le s &\Longrightarrow q(s\,,t) = w(s) - w(s-t)\\ &\Longrightarrow \frac{\partial q}{\partial s}(s\,,t) = \omega(s)-\omega(s-t) \;\;\&\;\; \frac{\partial q}{\partial t}(s\,,t) = \omega(s-t)\\ &\Longrightarrow \begin{cases} \dfrac{\partial q}{\partial s}\Big(s\,,q_-^{-1}\big(s\,,t(s)\big)\Big) = \omega(s) - \omega\Big(s - q_-^{-1}\big(s\,,t(s)\big)\Big)\\ \dfrac{\partial q}{\partial t}\Big(s\,,q_-^{-1}\big(s\,,t(s)\big)\Big) = \omega\Big(s - q_-^{-1}\big(s\,,t(s)\big)\Big), \end{cases}\end{aligned}$$

(6.38) yields

$$\begin{aligned}\frac{d}{ds}q_-^{-1}\big(s\,,t(s)\big) &= \frac{\omega(s) - 2\omega(2s-\alpha) - \omega(s) + \omega\Big(s - q_-^{-1}\big(s\,,t(s)\big)\Big)}{\omega\Big(s - q_-^{-1}\big(s\,,t(s)\big)\Big)}\\ &= 1 - 2\frac{\omega(2s-\alpha)}{\omega\Big(s - q_-^{-1}\big(s\,,t(s)\big)\Big)}.\end{aligned}$$

Similarly, $t(s) = p\Big(s\,,p^{-1}(s\,,t(s))\Big)$

$$\Longrightarrow t'(s) = \frac{\partial p}{\partial s}\Big(s\,,p^{-1}(s\,,t(s))\Big) + \frac{\partial p}{\partial t}\Big(s\,,p^{-1}(s\,,t(s))\Big)\cdot\frac{d}{ds}p^{-1}(s\,,t(s))$$

$$\Longrightarrow \frac{d}{ds}p^{-1}(s\,,t(s)) = \frac{\omega(s) - 2\omega(2s-\alpha) - \frac{\partial p}{\partial s}\Big(s\,,p^{-1}(s\,,t(s))\Big)}{\frac{\partial p}{\partial t}\Big(s\,,p^{-1}(s\,,t(s))\Big)}.$$

As $p(s\,,t) = 2t\omega(s) - w(s) + w(s-t)$

$$\Longrightarrow \frac{\partial p}{\partial s}(s\,,t) = 2t\omega'(s) - \omega(s) + \omega(s-t) \;\;\&\;\; \frac{\partial p}{\partial t}(s\,,t) = 2\omega(s) - \omega(s-t)$$

$$\Longrightarrow \begin{cases} \frac{\partial p}{\partial s}\Big(s,p^{-1}(s,t(s))\Big) = 2\omega'(s)p^{-1}(s\,,t(s)) - \omega(s) + \omega\Big(s - p^{-1}(s\,,t(s))\Big) \\ \frac{\partial p}{\partial t}\Big(s\,,p^{-1}(s\,,t(s))\Big) = 2\omega(s) - \omega\Big(s - p^{-1}(s\,,t(s))\Big), \end{cases}$$

we get $\frac{d}{ds}p^{-1}(s\,,t(s))$

$$= \frac{\omega(s) - 2\omega(2s-\alpha) - 2\omega'(s)p^{-1}(s\,,t(s)) + \omega(s) - \omega\Big(s - p^{-1}(s\,,t(s))\Big)}{2\omega(s) - \omega\Big(s - p^{-1}(s\,,t(s)\Big)}$$

$$= \frac{2\omega(s) - 2\omega(2s-\alpha) - 2\omega'(s)p^{-1}(s\,,t(s)) - \omega\Big(s - p^{-1}(s\,,t(s))\Big)}{2\omega(s) - \omega\Big(s - p^{-1}(s\,,t(s)\Big)}$$

$$= 1 - 2\frac{\omega(2s-\alpha) + \omega'(s)p^{-1}(s\,,t(s))}{2\omega(s) - \omega\Big(s - p^{-1}(s\,,t(s)\Big)}.$$

It follows that $\frac{d}{ds}q_-^{-1}(s\,,t(s)) < \frac{d}{ds}p^{-1}(s\,,t(s))$

$$\Longleftrightarrow \frac{\omega(2s-\alpha)}{\omega\Big(s - q_-^{-1}(s\,,t(s))\Big)} > \frac{\omega(2s-\alpha) + \omega'(s)p^{-1}(s\,,t(s))}{2\omega(s) - \omega\Big(s - p^{-1}(s\,,t(s)\Big)}$$

$$\Longleftrightarrow \omega(2s-\alpha)\Big[2\omega(s) - \omega\Big(s - p^{-1}(s\,,t(s))\Big) - \omega\Big(s - q_-^{-1}(s\,,t(s))\Big)\Big] >$$

$$\omega'(s)p^{-1}(s\,,t(s))\omega\Big(s - q_-^{-1}(s\,,t(s))\Big). \tag{6.39}$$

The concavity of ω implies that

$$\omega\Big(s - p^{-1}(s\,,t(s)\Big) \leq \omega(s) + \omega'(s)\Big(s - p^{-1}(s\,,t(s)) - s\Big) = \omega(s) - \omega'(s)p^{-1}(s\,,t(s))$$

and $\omega\Big(s-q_-^{-1}(s\,,t(s))\Big) \le \omega(s)-\omega'(s)q_-^{-1}(s\,,t(s))$, so that the expression on the left in (6.39) $\ge \omega(2s-\alpha)\omega'(s)\Big(p^{-1}(s\,,t(s))+q_-^{-1}(s\,,t(s))\Big)$. To get the claim, it remains to see that

$$\omega(2s-\alpha)\Big(p^{-1}(s\,,t(s))+q_-^{-1}(s\,,t(s))\Big) > p^{-1}(s\,,t(s))\omega\Big(s-q_-^{-1}(s\,,t(s))\Big)\,.$$

As $p^{-1}(s\,,t(s)) < q_-^{-1}(s\,,t(s))$, the left side of this inequality $<$ $2p^{-1}(s\,,t(s))\omega(2s-\alpha)$. Therefore, the claim follows from the inequality $2\omega(2s-\alpha) > \omega\Big(s-q_-^{-1}(s\,,t(s))\Big)$. Suppose that this is not true: $2\omega(2s-\alpha) \le \omega\Big(s-q_-^{-1}(s\,,t(s))\Big)$. Then we would have the same also in the special case $\omega(t)=t$, when $w(s)=0.5s^2$, $t(s)=0.5s^2-0.5(2s-\alpha)^2$, and

$$q_-^{-1}(s\,,t(s)) = s-\sqrt{s^2-2t(s)} = s-\sqrt{(2s-\alpha)^2} = s-(2s-\alpha) = \alpha-s$$

(for $2s>\alpha$ by assumption). Thus, the supposition implies $4s-2\alpha \le s-(\alpha-s) = 2s-\alpha$ or $2s\le\alpha$, contrary to the assumption. So, it is wrong and the claim is proved. □

By the lemma, the objective $l\big(\kappa-\tau\,,\varphi(\tau)\big)$ is increasing in $\big[|a/b|\,,0.5(\kappa+u)\big]$.

Lemma 6.16. *The function* $s\mapsto l\big(s\,,(\kappa-u-2s)\,\omega(s)+2w(s)-w(\kappa-u-s)\big)$ *is increasing in* $\big[0.5(\kappa+u)\,,v\big]$.

Proof. Let for brevity $t(s) := (\kappa-u-2s)\,\omega(s)+2w(s)-w(\kappa-u-s)\big)$ and

$$g(s) := l\big(s\,,t(s)\big) = q_-^{-1}(s\,,t(s)) - p^{-1}(s\,,t(s))\,.$$

Then

$$\begin{aligned} t'(s) &= -2\omega(s)+(\kappa-u-2s)\,\omega'(s)+2\omega(s)+\omega(\kappa-u-s) \\ &= (\kappa-u-2s)\,\omega'(s)+\omega(\kappa-u-s) \end{aligned}$$

and $g'(s)=\frac{\partial}{\partial s}q_-^{-1}(s\,,t(s))-\frac{\partial}{\partial s}p^{-1}(s\,,t(s))$. As in the proof of Lemma 6.11,

$$\frac{\partial}{\partial s}q_-^{-1}(s\,,t(s)) = \frac{t'(s)-\omega(s)+\omega\big(s-q_-^{-1}(s\,,t(s))\big)}{\omega\big(s-q_-^{-1}(s\,,t(s))\big)} = 1-\frac{\omega(s)-t'(s)}{\omega\big(s-q_-^{-1}(s\,,t(s))\big)}$$

and $\frac{\partial}{\partial s}p^{-1}(s\,,t(s))$

$$\begin{aligned} &= \frac{t'(s)-2\omega'(s)p^{-1}(s\,,t(s))+\omega(s)-\omega\big(s-p^{-1}(s\,,t(s))\big)}{2\omega(s)-\omega\big(s-p^{-1}(s\,,t(s))\big)} \\ &= 1-\frac{\omega(s)-t'(s)+2\omega'(s)p^{-1}(s\,,t(s))}{2\omega(s)-\omega\big(s-p^{-1}(s\,,t(s))\big)}\,. \end{aligned}$$

So,

$$g'(s) = \frac{\omega(s) - t'(s) + 2p^{-1}(s, t(s))\omega'(s)}{2\omega(s) - \omega(s - p^{-1}(s, t(s)))} - \frac{\omega(s) - t'(s)}{\omega(s - q_-^{-1}(s, t(s)))}.$$

Let (for brevity)

$$h_1(s) := \omega(s) - t'(s) = \omega(s) - \omega(\kappa - u - s) - (\kappa - u - 2s)\,\omega'(s),$$
$$h_2(s) := 2\omega(s) - \omega(s - p^{-1}(s, t(s))) - \omega(s - q_-^{-1}(s, t(s))),$$

and rewrite the difference on the right as the ratio

$$\frac{2p^{-1}(s, t)\,\omega'(s)\,\omega(s - q_-^{-1}(s, t(s))) - h_1(s)h_2(s)}{\big(2\omega(s) - \omega(s - p^{-1}(s, t(s)))\big)\omega(s - q_-^{-1}(s, t(s)))}.$$

The denominator of this ratio is positive. Look at the numerator. Due to the monotonicity of ω, h_1 is increasing in the interval $(\kappa - |a/b|, \kappa - 0.5(\kappa + u))$ from $h_1(\kappa - |a/b|)$ to $h_1(0.5(\kappa - u)) = 0$. So, $h_1(s) < 0, \forall s \in (\kappa - |a/b|, \kappa - 0.5(\kappa + u))$. Inasmuch as the terms $2p^{-1}(s, t)\,\omega'(s)\,\omega(s - q_-^{-1}(s, t(s)))$ and h_2 are positive, it follows that the numerator is positive. Hence, $g'(s) > 0$ and g is increasing. □

By the lemma, if $|a/b| \le 0.5(\kappa + u) \le v$, then $l(\kappa - \tau, \varphi(\tau))$ is decreasing in $[0.5(\kappa + u), v]$. Thus, in this case, the objective in (6.35) is increasing in $[|a/b|, 0.5(\kappa + u)]$ and decreasing in $[0.5(\kappa + u), v]$, so that the minimum

$$m = \min\left\{l\left(\kappa - \left|\frac{a}{b}\right|, \varphi\left(\left|\frac{a}{b}\right|\right)\right),\ l(\kappa - v, \varphi(v))\right\}$$
$$= \min\left\{\begin{array}{c} l\left(\kappa - \left|\frac{a}{b}\right|, w\left(\kappa - \left|\frac{a}{b}\right|\right) - w\left(\kappa + u - 2\left|\frac{a}{b}\right|\right)\right) \\ l(\kappa - v, (2v - \kappa - u)\,\omega(\kappa - v) + 2w(\kappa - v) - w(v - u)) \end{array}\right\}.$$

In the special case $\omega(t) = t$, when $\kappa = |b|$, $w(t) = 0.5t^2$, $p^{-1}(s, t(s)) = \sqrt{s^2 + 2t(s)} - s$, and $q_-^{-1}(s, t(s)) = s - \sqrt{s^2 - 2t(s)}$,

$$l(s, t(s)) = 2s - \sqrt{s^2 + 2t(s)} - \sqrt{s^2 - 2t(s)}.$$

In particular, for $s := |b| - |a/b|$ and $t(s) := 0.5(|b| - |a/b|)^2 - 0.5\left(2|a/b| - \sqrt{b^2 + 2|a|}\right)^2$, $l(s, t(s))$

$$= 2\left(|b| - \left|\frac{a}{b}\right|\right) - \sqrt{\left(|b| - \left|\frac{a}{b}\right|\right)^2 + \left(\left(|b| - \left|\frac{a}{b}\right|\right)^2 - \left(2\left|\frac{a}{b}\right| - \sqrt{b^2 + 2|a|}\right)^2\right)} -$$
$$\sqrt{\left(|b| - \left|\frac{a}{b}\right|\right)^2 - \left(\left(|b| - \left|\frac{a}{b}\right|\right)^2 - \left(2\left|\frac{a}{b}\right| - \sqrt{b^2 + 2|a|}\right)^2\right)}.$$

The expressions under the radicals are

$$2\left(|b|-\left|\frac{a}{b}\right|\right)^2-\left(2\left|\frac{a}{b}\right|-\sqrt{b^2+2|a|}\right)^2=2b^2\left(1-\frac{|a|}{b^2}\right)^2-b^2\left(2\frac{|a|}{b^2}-\sqrt{1+2\frac{|a|}{b^2}}\right)^2$$

and

$$\left(2\left|\frac{a}{b}\right|-\sqrt{b^2+2|a|}\right)^2=b^2\left(2\frac{|a|}{b^2}-\sqrt{1+2\frac{|a|}{b^2}}\right)^2,$$

respectively. So, $l(s,t(s))$

$$= |b|\left[2\left(1-\frac{|a|}{b^2}\right)-\sqrt{2\left(1-\frac{|a|}{b^2}\right)^2-\left(2\frac{|a|}{b^2}-\sqrt{1+2\frac{|a|}{b^2}}\right)^2}-\right.$$

$$\left.\sqrt{\left(2\frac{|a|}{b^2}-\sqrt{1+2\frac{|a|}{b^2}}\right)^2}\right].$$

As $0\le t\le 1 \Longrightarrow \sqrt{1+t}>t$, the last radical $=\sqrt{1+2|a|/b^2}-2|a|/b^2$. Therefore,

$$l(s,t(s))$$

$$= |b|\left[2-2\frac{|a|}{b^2}-\sqrt{2\left(1-\frac{|a|}{b^2}\right)^2-\left(2\frac{|a|}{b^2}-\sqrt{1+2\frac{|a|}{b^2}}\right)^2}-\sqrt{1+2\frac{|a|}{b^2}}+2\frac{|a|}{b^2}\right]$$

$$= |b|\left[2-\sqrt{2\left(1-\frac{|a|}{b^2}\right)^2-\left(2\frac{|a|}{b^2}-\sqrt{1+2\frac{|a|}{b^2}}\right)^2}-\sqrt{1+2\frac{|a|}{b^2}}\right]$$

$$= |b|\left[2-\sqrt{1-6\frac{|a|}{b^2}-2\frac{a^2}{b^4}+4\frac{|a|}{b^2}\sqrt{1+2\frac{|a|}{b^2}}}-\sqrt{1+2\frac{|a|}{b^2}}\right]. \qquad (6.40)$$

In the same special case $\omega(t)=t$, $\kappa-v=\sqrt{b^2-2|a|}$,

$$2v-\kappa-u=2|b|-\sqrt{b^2+2|a|}-2\sqrt{b^2-2|a|},$$

$w(\kappa - v) = 0.5b^2 - |a|$, and $(2v - \kappa - u)\omega(\kappa - v) + 2w(\kappa - v) - w(v - u)$

$$= \left(2|b| - \sqrt{b^2 + 2|a|} - 2\sqrt{b^2 - 2|a|}\right)\sqrt{b^2 - 2|a|} + b^2 - 2|a| - \frac{1}{2}\left(2|b| - \sqrt{b^2 + 2|a|} - \sqrt{b^2 - 2|a|}\right)^2$$

$$= 2|b|\sqrt{b^2 - 2|a|} - 2b^2 + 4|a| - \sqrt{b^4 - 4a^2} + b^2 - 2|a| - \frac{1}{2}\Big(4b^2 + b^2 + 2|a| + b^2 - 2|a| - 4|b|\sqrt{b^2 + 2|a|} - 4|b|\sqrt{b^2 - 2|a|} + 2\sqrt{b^4 - 4a^2}\Big)$$

$$= 4|b|\sqrt{b^2 - 2|a|} - 4b^2 + 2|a| - 2\sqrt{b^4 - 4a^2} + 2|b|\sqrt{b^2 + 2|a|}$$

$$= 2b^2\left(2\sqrt{1 - 2\frac{|a|}{b^2}} - 2 + \frac{|a|}{b^2} - \sqrt{1 - 4\frac{a^2}{b^4}} + \sqrt{1 + 2\frac{|a|}{b^2}}\right).$$

So, for $s := \kappa - v = \sqrt{b^2 - 2|a|} = |b|\sqrt{1 - 2|a|/b^2}$ and

$$\begin{aligned} t(s) &:= (2v - \kappa - u)\omega(\kappa - v) + 2w(\kappa - v) - w(v - u) \\ &= 2b^2\left(2\sqrt{1 - 2\frac{|a|}{b^2}} - 2 + \frac{|a|}{b^2} - \sqrt{1 - 4\frac{a^2}{b^4}} + \sqrt{1 + 2\frac{|a|}{b^2}}\right), \end{aligned}$$

we have $l(s, t(s))$

$$= |b|\left(2\sqrt{1 - 2\frac{|a|}{b^2}} - \sqrt{1 - 2\frac{|a|}{b^2} + 4\left(2\sqrt{1 - 2\frac{|a|}{b^2}} - 2 + \frac{|a|}{b^2} - \sqrt{1 - 4\frac{a^2}{b^4}} + \sqrt{1 + 2\frac{|a|}{b^2}}\right)} - \sqrt{1 - \frac{2|a|}{b^2} - 4\left(2\sqrt{1 - 2\frac{|a|}{b^2}} - 2 + \frac{|a|}{b^2} - \sqrt{1 - 4\frac{a^2}{b^4}} + \sqrt{1 + 2\frac{|a|}{b^2}}\right)}\right).$$

Comparison of plots of the functions

$$l_1(t) := 2 - \sqrt{2(1 - t)^2 - \left(2t - \sqrt{1 + 2t}\right)^2} - \sqrt{1 + 2t}$$

and

$$l_2(t) := 2\sqrt{1-2t} - \sqrt{1-2t+4\left(2\sqrt{1-2t}-2+t-\sqrt{1-4t^2}+\sqrt{1+2t}\right)} - \sqrt{1-2t-4\left(2\sqrt{1-2t}-2+t-\sqrt{1-4t^2}+\sqrt{1+2t}\right)}$$

on the interval $(0\,,\tau_*)$, where $\tau_* = 0.903...$ is the zero of the function

$$\tau \mapsto 2-\sqrt{1+\tau}-2\sqrt{1-\tau}\,,$$

clearly shows that, if $0.5(\kappa+u) < v$ (that is, $2|a|/b^2 < \tau_*$), then $l_1(t) < l_2(t)$, $\forall\, t \in (0\,,\tau_*)$. This observation provokes

Hypothesis 6.17. *If* $|a/b| \le 0.5(\kappa+u) < v$ *(case (i) in* (6.36)*), then the minimum* (6.35)

$$m(a\,,b) = l\left(\kappa - \left|\frac{a}{b}\right|\,,\, w\left(\kappa-\left|\frac{a}{b}\right|\right) - w\left(\kappa+u-2\left|\frac{a}{b}\right|\right)\right).$$

I have no clue how to prove it in the general case of nonlinear ω.

When $0.5(\kappa+u) > v$ (case (ii) in (6.36)), $\varphi(\tau) = w(\kappa-\tau) - w(\kappa+u-2\tau)\,,\ \forall\,\tau \in \big[|a/b|\,,v\big]$, and so the objective $l\big(\kappa-\tau\,,w(\kappa-\tau)-w(\kappa+u-2\tau\big)$ is increasing there by Lemma 6.15. Then the minimum $m(a\,,b)$ is again

$$= l\left(\kappa - \left|\frac{a}{b}\right|\,,\, w\left(\kappa-\left|\frac{a}{b}\right|\right) - w\left(\kappa+u-2\left|\frac{a}{b}\right|\right)\right).$$

Thus, if Hypothesis 6.17 is true, then

$$ab > 0 \Longrightarrow \min_{|\delta|}\left\{L \;\middle|\; p^{-1}(\kappa\,,|a|) \le |\delta| \le q_-^{-1}(\kappa\,,|a|)\right\} = l\left(\kappa - \left|\frac{a}{b}\right|\,,\, w\left(\kappa-\left|\frac{a}{b}\right|\right) - w\left(\kappa+u-2\left|\frac{a}{b}\right|\right)\right). \tag{6.41}$$

We now address the case $ab < 0$. In this case, we are looking for a zero in the segment $\big[x_0 + p^{-1}(\kappa\,,|a|)\,, x_0 + q_-^{-1}(\kappa\,,|a|)\big]$ and so this is where x_1 should be positioned under the condition that the new existence segment be defined (that is, satisfy (6.24)) and remain in the right part of the old:

$$x_1 - q_-^{-1}(\kappa_1\,,|a_1|) \ge x_0 + p^{-1}(\kappa\,,|a|) \;\;\&\;\; x_1 - p^{-1}(\kappa_1\,,|a_1|) < x_1\,, \tag{6.42}$$

if $a_1 b_1 > 0$, or

$$x_1 + p^{-1}(\kappa_1\,,|a_1|) > x_1 \;\;\&\;\; x_1 + q_-^{-1}(\kappa_1\,,|a_1|) \le x_0 + q_-^{-1}(\kappa\,,|a|)\,, \tag{6.43}$$

if $a_1 b_1 < 0$. The restrictions on the possible values of a_1 and b_1 dictated by the ω-regular smoothness of f do not depend on the sign of ab and so remain the same as in (6.27) and (6.28). Therefore, in the case $ab < 0$, we have to maximize the length

$$l(\kappa_1, |a_1|) = q_-^{-1}(\kappa_1, |a_1|) - p^{-1}(\kappa_1, |a_1|)$$

of the new existence segment subject to the constraints (6.24), (6.27), (6.28), and either (6.42) or (6.43) and then minimize the maximum value of l over all $\delta := x_1 - x_0$ compatible with these constraints. The first (maximization) problem is to find

$$\max_{\kappa_1, a_1}\Big\{ l(\kappa_1, |a_1|) \;\Big|\; (\kappa_1, a_1) \in KA(\sigma_1) \cup KA(\rho_1) \Big\},$$

where

$$KA(\sigma_1) := \left\{ (\kappa_1, a_1) \;\middle|\; \begin{array}{c} |a_1| \le w(\kappa_1) \;\;\&\;\; |\kappa_1 - \kappa| \le |\delta| \;\;\&\;\; \alpha \le a_1 \le \beta \\ q_-^{-1}(\kappa_1, |a_1|) \le \sigma_1 \end{array} \right\},$$

$$KA(\rho_1) := \left\{ (\kappa_1, a_1) \;\middle|\; \begin{array}{c} |a_1| \le w(\kappa_1) \;\;\&\;\; |\kappa_1 - \kappa| \le |\delta| \;\;\&\;\; \alpha \le a_1 \le \beta \\ q_-^{-1}(\kappa_1, |a_1|) \le \rho_1 \end{array} \right\},$$

$$\sigma_1 := \delta - p^{-1}(\kappa, |a|) \;\;,\;\; \rho_1 := q_-^{-1}(\kappa, |a|) - \delta\,.$$

Comparing these definitions of σ_1 and ρ_1 with those of ρ and σ, we see that they differ only in the sign of δ, which plays no role either in the value of L (that depends only on the absolute value of δ) or in the value of its minimum over δ. Hence, in the case of $ab < 0$, the result of optimization of the pair (κ_1, a_1) is the same as for $ab > 0$ (see (6.41)). It follows that the optimal position of x_1 is that for which

$$l(\kappa_1, |a_1|) = l\left(\kappa - \left|\frac{a}{b}\right|, \, w\left(\kappa - \left|\frac{a}{b}\right|\right) - w\left(\kappa + u - 2\left|\frac{a}{b}\right|\right)\right).$$

This equation yields

$$x_1 = x_0 - sign(ab)\left|\frac{a}{b}\right| = x_0 - \frac{f(x_0)}{f'(x_0)}, \tag{6.44}$$

which is the familiar Newton's method for scalar equations. So, the great I. Newton invented the (entropy) optimal method, though the notion of entropy was conceived centuries later.

Let us sum up the discussion of this section:

Proposition 6.18. *The (entropy) optimal position of x_1 coincides with that given by Newton's method* (6.44).

6.5 Existence and uniqueness of solutions (Lipschitz continuity of dd)

Proposition 6.1 is inapplicable to nondifferentiable functions. In this section, we prove its analog under a weaker assumption: Lipschitz continuity of divided differences. In its statement below

$$a := f(x_0)\ ,\ b := \frac{f(x_0) - f(x_{-1})}{x_0 - x_{-1}}\ ,\ \kappa := c^{-1}|b|\ ,\ \gamma := |x_0 - x_{-1}|\,.$$

Proposition 6.19. *Let the divided difference* $\big(f(x_1) - f(x_2)\big)/(x_1 - x_2)$ *of a function* $f : \mathbb{R} \to \mathbb{R}$ *be Lipschitz continuous on* $\mathbb{R}$*:*

$$\left|\frac{f(x_1) - f(x_2)}{x_1 - x_2} - \frac{f(u_1) - f(u_2)}{u_1 - u_2}\right| \le c\big(|x_1 - u_1| + |x_2 - u_2|\big)\ , \forall\, x_1, x_2, u_1, u_2 \in \mathbb{R}. \tag{6.45}$$

1° *If* $ab > 0$ & $4|a| > c(\kappa - \gamma)^2$, *then* f *has no zeroes in the interval*

$$\left(x_0 + \frac{\kappa + \gamma - \sqrt{(\kappa + \gamma)^2 + 4c^{-1}|a|}}{2}\ ,\ x_0 + \frac{\kappa - \gamma + \sqrt{(\kappa - \gamma)^2 + 4c^{-1}|a|}}{2}\right).$$

2° *If* $ab > 0$ & $4|a| \le c(\kappa - \gamma)^2$, *then* f *has a zero in the segment*

$$\left[x_0 - \frac{\kappa - \gamma - \sqrt{(\kappa - \gamma)^2 - 4c^{-1}|a|}}{2}\ ,\ x_0 - \frac{\sqrt{(\kappa + \gamma)^2 + 4c^{-1}|a|} - \kappa - \gamma}{2}\right]$$

and this zero is the only one in the interval

$$\left(x_0 - \frac{\kappa - \gamma + \sqrt{(\kappa - \gamma)^2 - 4c^{-1}|a|}}{2}\ ,\ x_0 + \frac{\kappa - \gamma + \sqrt{(\kappa - \gamma)^2 + 4c^{-1}|a|}}{2}\right).$$

3° *If* $ab < 0$ & $4|a| > c(\kappa - \gamma)^2$, *then* f *has no zeroes in the interval*

$$\left(x_0 - \frac{\sqrt{(\kappa - \gamma)^2 + 4c^{-1}|a|} + \kappa - \gamma}{2}\ ,\ x_0 + \frac{\sqrt{(\kappa - \gamma)^2 + 4c^{-1}|a|} - \kappa - \gamma}{2}\right).$$

4° *If* $ab < 0$ & $4|a| \le c(\kappa - \gamma)^2$, *then* f *has a zero in the segment*

$$\left[x_0 + \frac{\sqrt{(\kappa + \gamma)^2 + 4c^{-1}|a|} - \kappa - \gamma}{2}\ ,\ x_0 + \frac{\kappa - \gamma - \sqrt{(\kappa - \gamma)^2 - 4c^{-1}|a|}}{2}\right]$$

and this zero is the only one in the interval

$$\left(x_0 - \frac{\sqrt{(\kappa - \gamma)^2 + 4c^{-1}|a|} + \kappa - \gamma}{2}\ ,\ x_0 + \frac{\sqrt{(\kappa - \gamma)^2 + 4c^{-1}|a|} - \kappa - \gamma}{2}\right).$$

Proof. Let x_* be a zero of f and Δ denote the difference $x_* - x_0$. Then

$$f(x_0) = \frac{f(x_0) - f(x_*)}{x_0 - x_*}(x_0 - x_*)$$

and

$$\begin{aligned} a + b\Delta &= \frac{f(x_0) - f(x_*)}{x_0 - x_*}(x_0 - x_*) + \frac{f(x_0) - f(x_{-1})}{x_0 - x_{-1}}(x_* - x_0) \\ &= \Delta\left(\frac{f(x_0) - f(x_{-1})}{x_0 - x_{-1}} - \frac{f(x_0) - f(x_*)}{x_0 - x_*}\right). \end{aligned}$$

Due to the Lipschitz continuity of divided difference,

$$\begin{aligned} &\left|\frac{f(x_0) - f(x_{-1})}{x_0 - x_{-1}} - \frac{f(x_0) - f(x_*)}{x_0 - x_*}\right| \\ &\le c|x_* - x_{-1}| \le c\big(|x_* - x_0| + |x_0 - x_{-1}|\big) = c\big(|\Delta| + \gamma\big). \end{aligned}$$

Hence, $\big||a|sign(ab) + |b|\Delta\big| = |a + b\Delta| \le c|\Delta|\big(|\Delta|| + \gamma\big)$ or, equivalently,

$$-|b|\Delta - c|\Delta|\big(|\Delta| + \gamma\big) \le |a|sign(ab) \le -|b|\Delta + c|\Delta|\big(|\Delta| + \gamma\big). \tag{6.46}$$

In particular,

$$\begin{aligned} &ab > 0 \Longrightarrow -|b|\Delta - c|\Delta|\big(|\Delta| + \gamma\big) \le |a| \le -|b|\Delta + c|\Delta|\big(|\Delta| + \gamma\big) \\ &\Longleftrightarrow \begin{cases} |a| \le c\Delta(\Delta + \gamma) - |b|\Delta\,, \text{ if } \Delta > 0\,, \\ |b\Delta| - c|\Delta|\big(|\Delta| + \gamma\big) \le |a| \le |b\Delta| + c|\Delta|\big(|\Delta| + \gamma\big)\,, \text{ if } \Delta > 0\,. \end{cases} \end{aligned}$$

The system $\Delta > 0$ & $|a| \le c\Delta(\Delta + \gamma) - |b|\Delta$ is solved to give

$$\Delta \ge \frac{\kappa - \gamma + \sqrt{(\kappa - \gamma)^2 + 4c^{-1}|a|}}{2},$$

i.e., there are no zeroes to the left of $x_0 + 0.5\left(\kappa - \gamma + \sqrt{(\kappa - \gamma)^2 + 4c^{-1}|a|}\right)$. Similarly,

$$\Delta < 0 \;\&\; |b\Delta| + c|\Delta|\big(|\Delta| + \gamma\big) - |a| \ge 0 \Longleftrightarrow \Delta \le \frac{\kappa + \gamma - \sqrt{(\kappa + \gamma)^2 + 4c^{-1}|a|}}{2}$$

$\Big($no zeroes to the right of $x_0 + 0.5\left(\kappa + \gamma - \sqrt{(\kappa + \gamma)^2 + 4c^{-1}|a|}\right)\Big)$. The inequality $|a| \ge |b\Delta| - c|\Delta|\big(|\Delta| + \gamma\big)$ is trivial, if $|a| > \max_{t \ge 0}\Big(|b|t - ct(t + \gamma)\Big) = 0.25c(\kappa - \gamma)^2$. In this case, it contains no information about the position of x_* within the interval

$$\left(x_0 + \frac{\kappa + \gamma - \sqrt{(\kappa + \gamma)^2 + 4c^{-1}|a|}}{2}\,,\; x_0 + \frac{\kappa - \gamma + \sqrt{(\kappa - \gamma)^2 + 4c^{-1}|a|}}{2}\right).$$

So, all that can be said in this situation is that x_* is not in this interval. Alternatively, if $|a| \le 0.25c(\kappa-\gamma)^2$, then $\Delta < 0$ & $|a| \ge |b\Delta| - c|\Delta|(|\Delta|+\gamma)$

$$\iff \Delta \le -\frac{\kappa-\gamma+\sqrt{(\kappa-\gamma)^2-4c^{-1}|a|}}{2} \bigvee$$

$$-\frac{\kappa-\gamma-\sqrt{(\kappa-\gamma)^2-4c^{-1}|a|}}{2} \le \Delta < 0$$

$$\iff x_* \le x_0 - \frac{\kappa-\gamma+\sqrt{(\kappa-\gamma)^2-4c^{-1}|a|}}{2} \bigvee$$

$$x_0 - \frac{\kappa-\gamma-\sqrt{(\kappa-\gamma)^2-4c^{-1}|a|}}{2} \le x_* < x_0 .$$

It follows that there is a zero in the segment

$$\left[x_0 - \frac{\kappa-\gamma-\sqrt{(\kappa-\gamma)^2-4c^{-1}|a|}}{2}\,,\; x_0 - \frac{\sqrt{(\kappa+\gamma)^2+4c^{-1}|a|}-\kappa-\gamma}{2}\right]$$

and no other zeroes in

$$\left(x_0 - \frac{\kappa-\gamma+\sqrt{(\kappa-\gamma)^2-4c^{-1}|a|}}{2}\,,\; x_0 + \frac{\kappa-\gamma+\sqrt{(\kappa-\gamma)^2+4c^{-1}|a|}}{2}\right).$$

The case $ab<0$ is analyzed analogously. According to (6.46),

$$ab<0 \Longrightarrow |b|\Delta - c|\Delta|\big(|\Delta|+\gamma\big) \le |a| \le |b|\Delta + c|\Delta|\big(|\Delta|+\gamma\big)$$
$$\iff \begin{cases} |a| \le -|b\Delta| + c|\Delta|\big(|\Delta|+\gamma\big)\,, & \text{if } \Delta<0\,,\\ |b|\Delta - c\Delta(\Delta+\gamma) \le |a| \le |b|\Delta + c\Delta(\Delta+\gamma)\,, & \text{if } \Delta>0\,. \end{cases}$$

The system $\Delta<0$ & $c|\Delta|\big(|\Delta|+\gamma\big) - |b\Delta| \ge |a|$ implies

$$\Delta \le -\frac{\sqrt{(\kappa-\gamma)^2+4c^{-1}|a|}+\kappa-\gamma}{2} \iff x_* \le x_0 - \frac{\sqrt{(\kappa-\gamma)^2+4c^{-1}|a|}+\kappa-\gamma}{2},$$

that is, there are no zeroes to the right of $x_0 - 0.5\Big(\sqrt{(\kappa-\gamma)^2+4c^{-1}|a|}+\kappa-\gamma\Big)$. Likewise,

$$\Delta>0 \;\&\; |a| \le |b|\Delta + c\Delta(\Delta+\gamma) \iff \Delta \ge \frac{\sqrt{(\kappa+\gamma)^2+4c^{-1}|a|}-\kappa-\gamma}{2}$$
$$\iff x_* \ge x_0 + \frac{\sqrt{(\kappa+\gamma)^2+4c^{-1}|a|}-\kappa-\gamma}{2}$$

(no zeroes to the left of $x_0 + 0.5\left(\sqrt{(\kappa+\gamma)^2+4c^{-1}|a|}-\kappa-\gamma\right)$). Finally,

$\Delta > 0 \;\; \& \;\; |b|\Delta - c\Delta(\Delta + \gamma) \leq |a| \iff$

$$\begin{cases} \Delta>0\,, \text{ if } (\kappa-\gamma)^2 < 4c^{-1}|a|\,, \\ 0<\Delta\leq \dfrac{\kappa-\gamma-\sqrt{(\kappa+\gamma)^2-4c^{-1}|a|}}{2} \bigvee \Delta \geq \dfrac{\kappa-\gamma+\sqrt{(\kappa+\gamma)^2-4c^{-1}|a|}}{2}\,, \\ \qquad\qquad\qquad\qquad \text{if } (\kappa-\gamma)^2 \geq 4c^{-1}|a|\,. \end{cases}$$

Thus, $ab < 0 \;\; \& \;\; (\kappa - \gamma)^2 < 4c^{-1}|a|$ implies that there are no zeroes in the interval

$$\left(x_0 - \frac{\sqrt{(\kappa-\gamma)^2 + 4c^{-1}|a|} + \kappa - \gamma}{2} \, , \, x_0 + \frac{\sqrt{(\kappa-\gamma)^2 + 4c^{-1}|a|} - \kappa - \gamma}{2} \right), \tag{6.47}$$

while $ab < 0 \;\; \& \;\; (\kappa - \gamma)^2 \geq 4c^{-1}|a|$ guarantees the existence of a zero in the segment

$$\left[x_0 + \frac{\sqrt{(\kappa+\gamma)^2 + 4c^{-1}|a|} - \kappa - \gamma}{2} \, , \, x_0 + \frac{\kappa - \gamma - \sqrt{(\kappa-\gamma)^2 - 4c^{-1}|a|}}{2} \right]$$

and its uniqueness in the interval (6.47). □

6.6 Research projects

An iterative method for solving an operator equation bears a striking resemblance to a game between a method's designer and the operator: each move x_n of the designer prompts the operator's move $\mathbf{f}(x_n)$, which in turn prompts the designer to react with x_{n+1},. As far as I know, iterative methods have not been analyzed from the game-theoretic point of view. This point of view may produce interesting solutions to the problem of optimal strategy when dealing with certain classes of operators.

The result of Section 6.4 can be used for a refinement of Theorem 5.3. It guarantees the existence of a zero of the operator $\mathbf{f}$ in the set $D(x_0)$ (see (5.12)) bounded by two two spheres centered at x_0, but gives no clue about which part of that set deserves special attention. The refinement I have in mind should provide such a clue. The idea is this. Let's emit a ray from x_0. It intersects $D(x_0)$ by a segment. The restriction of the operator to this segment is a scalar function, to which the result of Section 6.4 applies. So, we can compute its zero. The union of all such zeroes contains a zero of $\mathbf{f}$. A satisfactory description of this union would be a very desirable addition to Theorem 5.3. Such a description could make possible a further attempt to narrow the set containing a solution.

Secant-type methods (6.1) seem to me a sufficiently interesting subject to be investigated in a multi-dimensional setting.

In this book, we have not mentioned so far *stability of iterative methods.* This subject, however, continues to draw the attention of numerous specialists in computational mathematics. Regarding the methods (6.1), the notion of stability can be defined as follows. Let both the operator $\mathbf{f}$ and the mapping F be continuous. Suppose that a starter x_0 causes the sequence $x_{n+1} := F\big(x_n\,,\mathbf{f}(x_n)\big)$ to converge to x_∞ (necessarily a fixed point of the operator $x \mapsto F\big(x\,,\mathbf{f}(x)\big)\big)$. A method (6.1) is *stable* if, for all $\varepsilon > 0$, there exists a $\delta_\varepsilon > 0$ such that

$$\|\hat{x}_0 - x_0\| < \delta_\varepsilon \Longrightarrow \hat{x}_{n+1} := F\big(\hat{x}_n\,,\mathbf{f}(\hat{x}_n)\big) \to x_\infty\,.$$

Having accepted this definition (of course, it is not the only one possible), it is natural to ask: what F results in a stable method?

Chapter 7

Majorant generators and their convergence domains

7.1 Motivation

Carrying out convergence analyses in the preceding chapters, we invariably faced the need to find the necessary and sufficient condition for convergence of the sequence generated by a majorant generator (a difference equation) of the form

$$u_+ := F(u,v) \ , \ v_+ := G(u,v) \ , u \in \mathbb{R} \ , v \in \mathbb{R}^m \ , \tag{7.1}$$

resulting from the use of Kantorovich's majorization technique (Propositions 2.11, 3.14, 4.5, 5.5). These conditions have been found as an inequality of the type $u_0 \le f_\infty(v_0)$, with the function f_∞ being the limit of the sequence f_n defined recursively. At the same time, it has been shown that this limit solves the system of a functional equation and an end condition. The use of the definition of f_∞ for its actual computing proves to be impractical because of slow convergence of the sequence f_n, especially when u_0 is close to its upper limit. This observation warrants a closer look at the system

$$x\big(G\big(x(v)\,,v\big)\big) = F\big(x(v)\,,v\big) \ \ \& \ \ x(0) = u_\infty \tag{7.2}$$

induced by the generator (7.1) and solved by the function f_∞. Except for some simplified generators as in (2.40), (3.36), (5.7), (5.21), the induced systems of the type (7.2) can be solved only numerically using some iterative procedure. In the present chapter, we try to develop such a procedure. But first we have to get a constructive description of the convergence domain of the generator (7.1). In other words, we have to answer the question: precisely which starters $(u_0\,,v_0)$ cause the sequence (u_n, v_n), generated by the generator from $(u_0\,,v_0)$, to converge.

7.2 Convergence domain of the generator (7.1)

Regarding the generator (7.1), we assume that

(a) the functions F and G are defined and continuous on a subset UV of

$$\mathbb{R}_+^{m+1} := \{(u\,,v) \in \mathbb{R} \times \mathbb{R}^m \mid u \geq 0 \ \ \& \ \ v \geq 0\}$$

containing a segment $[0\,,u^\circ]$, $u^\circ \leq \infty$, of the u-axis:

$$(u\,,0) \in UV\,,\ \forall\, u \in [0\,,u^\circ)\,.$$

(b) F is increasing and G is not decreasing in each of their arguments,

(c) $F(u,0) = u\ ,\ \ \forall\, u \in [0\,u^\circ)$,

(d) $G(u,0) = 0\ ,\ \ \forall\, u \in [0\,u^\circ)$.

The last assumption ensures that the interval $[0\,,u^\circ)$ is filled up by fixed points of the generator (7.1). A look at the specific generators (2.21), (3.18), (4.9), (5.7), (5.16), we have already dealt with, shows that they meet these conditions. Given UV, we define the sets

$$U(v) := \{u \in \mathbb{R} \mid (u\,,v) \in UV\}\ ,\ \ V := \{v \in \mathbb{R}^m \mid U(v) \neq \emptyset\}\,.$$

The next theorem provides a necessary and sufficient condition for a starter to belong to the convergence domain of the generator.

Theorem 7.1. 1° *If the functions F and G in (7.1) satisfy the conditions (a)–(d) above, then*

$$u_\infty < \infty \Longleftrightarrow u_0 \leq f_\infty(v_0)\,,$$

where $f_0(v)$ is the (unique) solution for u of the equation $F(u,v) = u_\infty$, and $f_{n+1}(v)$ is the (unique) solution for u of the equation

$$f_n\big(G(u\,,v)\big) = F(u\,,v)\,.$$

2° *The function f_∞ is the only continuous solution of the system (a functional equation with the end condition)*

$$x\big(G\big(x(v)\,,v\big)\big) = F\big(x(v)\,,v\big) \ \ \& \ \ x(0) = u_\infty\,. \tag{7.3}$$

Proof. If the sequence $(u_n\,,v_n)$ is unbounded, then $(u_0\,,v_0)$ is surely not in the convergence domain. Therefore, boundedness of $(u_n\,,v_n)$ is necessary for the starter to be in the convergence domain and we can take it for granted. By (7.1), $u_{n+1} \leq u_s \Longleftrightarrow F(u_n\,,v_n) \leq u_\infty$. As F is increasing in $u \in [0\,,u^\circ)$ (assumption (b)), the equality $F(u\,,v_n) = u_\infty$ is uniquely solvable for u. Let $f_0(v_n)$ be the solution:

$$F\big(f_0(v)\,,v\big) = u_\infty\ ,\ \ \forall\, v \in V\,. \tag{7.4}$$

Then the inequality $F(u_n, v_n) < u_\infty$ is equivalent to $u_n < f_0(v_n)$. The function f_0 is decreasing in V. Indeed, for example $0 \le v_1 < v_1' \Longrightarrow$

$$\begin{aligned} F\big(f_0(v_1, v_2, \dots, v_m), v_1, v_2, \dots, v_m\big) &= u_\infty \\ &= F\big(f_0(v_1', v_2, \dots, v_m), v_1', v_2, \dots, v_m\big) \\ &> F\big(f_0(v_1', v_2, \dots, v_m), v_1, v_2, \dots, v_m\big) \end{aligned}$$

$$\Longrightarrow f_0(v_1, v_2, \dots, v_m) > f_0(v_1', v_2, \dots, v_m)$$

(because F is increasing in each argument). Moreover, f_0 is continuous. Otherwise it would have a discontinuity at a point $v = (v_1, v_2, \dots, v_m)$, which (because of the monotonicity of f_0) must be a jump. Namely, $v_1' \nearrow v_1 \Longrightarrow f_0(v_1', v_2, \dots, v_m) \searrow u'$ and $v_1' \searrow v_1 \Longrightarrow f_0(v_1', v_2, \dots, v_m) \nearrow u'' < u'$. Then the continuity and monotonicity of F imply

$$v_1' \nearrow v_1 \Longrightarrow F\big(f_0(v_1', v_2, \dots, v_m), v\big) \searrow F(u', v)$$

and

$$v_1' \searrow v_1 \Longrightarrow F\big(f_0(v_1', v_2, \dots, v_m), v\big) \nearrow F(u'', v) < F(u', v).$$

We see that a jump of f_0 implies a jump of F, contrary to its continuity. Besides, by (d), (c) and $F(u_\infty, 0) = u_\infty$, while (7.4) implies $F\big(f_0(0), 0\big) = u_\infty$. So, $f_0(0) = u_\infty$.

Suppose that, for some $k \ge 0$, $u_{n+1} < u_\infty \Longleftrightarrow u_{n-k} < f_k(v_{n-k})$, where f_k is continuously decreasing on V and $f_k(0) = u_\infty$. Using (7.1), rewrite the last inequality as

$$\Phi_k(u_{n-k-1}, v_{n-k-1}) > 0, \tag{7.5}$$

where

$$\Phi_k(u, v) := f_k\big(G(u, v)\big) - F(u, v). \tag{7.6}$$

As F is continuously increasing in u, G is not decreasing (assumptions (a) and (b)), and f_k is continuously decreasing (by the induction hypothesis), the function Φ_k is continuously decreasing in all arguments. In particular, when u scans the segment $[0, u_\infty]$, Φ_k is continuously decreasing from

$$\Phi_k(0, v_{n-k-1}) \ge \Phi_k(u_{n-k-1}, v_{n-k-1}) > 0$$

(see (7.5)) to

$$\begin{aligned} \Phi_k(u_\infty, v_{n-k-1}) &= f_k\big(G(u_\infty, v_{n-k-1})\big) - F(u_\infty, v_{n-k-1}) \\ &< f_k\big(G(u_\infty, 0)\big) - F(u_\infty, 0) = f_k(0) - u_\infty = 0, \end{aligned}$$

by (d), (c), and the induction hypothesis. So, the equation $\Phi_k(u, v_{n-k-1}) = 0$ is uniquely solvable for u in $[0, u_\infty]$. Denote the solution $f_{k+1}(v_{n-k-1})$:

$$\Phi_k\big(f_{k+1}(v), v\big) = 0, \quad \forall v \in V. \tag{7.7}$$

Because Φ_k is decreasing with respect to the first argument, comparison with (7.5) shows that

$$u_{n-k-1} < f_{k+1}(v_{n-k-1})\,.$$

The function f_{k+1} is decreasing on $V: 0 < v_1 < v_1' \Longrightarrow$

$$\Phi_k\big(f_{k+1}(v_1, v_2, \dots, v_m)\,, v_1, v_2, \dots, v_m\big) = 0 =$$

$$\Phi_k\big(f_{k+1}(v_1', v_2, \dots, v_m)\,, v_1', v_2, \dots, v_m\big)$$

$$< \Phi_k\big(f_{k+1}(v_1', v_2, \dots, v_m)\,, v_1, v_2, \dots, v_m\big)$$

$$\Longrightarrow f_{k+1}(v_1, v_2, \dots, v_m) > f_{k+1}(v_1', v_2, \dots, v_m)$$

(for Φ_k is decreasing in all arguments). It is also continuous, for its jump implies (similar to f_0) a jump of the continuous Φ_k. Besides, in view of (d) and (c),

$$\Phi_k(u_\infty\,, 0) = f_k\big(G(u_s\,, 0)\big) - F(u_\infty\,, 0) = f_k(0) - u_\infty = 0$$

(by the induction hypothesis) and so (7.7) implies $f_{k+1}(0) = u_\infty$, since the equation $\Phi_k(u\,, 0) = 0$ for u can have only one solution. Thus, the equivalency

$$u_{n+1} < u_\infty \Longleftrightarrow u_{n-k} < f_k(v_{n-k})$$

implies

$$u_{n+1} < u_\infty \Longleftrightarrow u_{n-k-1} < f_{k+1}(v_{n-k-1})\,.$$

By induction, $u_{n+1} < u_\infty \Longleftrightarrow u_0 < f_n(v_0)$ and $\underset{n}{\&}\, u_n < u_\infty \Longleftrightarrow u_0 \leq \inf\limits_n f_n(v_0)$. The sequence f_n is not increasing pointwise:

$$\underset{n}{\&}\, f_{n+1}(v) \leq f_n(v)\ ,\ \forall\, v \in V\,. \tag{7.8}$$

This is verified inductively. First, we have to show that $f_1(v) \leq f_0(v)$ or, equivalently, that $\Phi_0\big(f_1(v)\,, v\big) = 0 \geq \Phi_0\big(f_0(v)\,, v\big)$, because Φ_0 is decreasing with respect to the first argument. According to (7.6),

$$\Phi_0\big(f_0(v)\,, v\big) = f_0\big(G\big(f_0(v)\,, v\big)\big) - F\big(f_0(v)\,, v\big) = f_0\big(G\big(f_0(v)\,, v\big)\big) - u_\infty\,,$$

for $F\big(f_0(v)\,, v\big) = u_\infty$ by (7.4). So,

$$v \geq 0 \Longrightarrow f_0\big(G\big(f_0(v)\,, v\big)\big) \leq f_0\big(G\big(f_0(v)\,, 0\big)\big) = f_0(0)$$

$$\Longrightarrow \Phi_0\big(f_0(v)\,, v\big) \leq f_0(0) - u_\infty = 0\,.$$

Suppose now that $f_n(v) \leq f_{n-1}(v)$, $\forall\, v \in V$, for some $n \geq 1$. Then

$$\Phi_{n-1}\big(f_n(v)\,, v\big) = 0 = \Phi_n\big(f_{n+1}(v)\,, v\big)$$

$$= f_n\big(G\big(f_{n+1}(v)\,, v\big)\big) - F\big(f_{n+1}(v)\,, v\big)$$

$$\leq f_{n-1}\big(G\big(f_{n+1}(v)\,, v\big)\big) - F\big(f_{n+1}(v)\,, v\big)$$

$$= \Phi_{n-1}\big(f_{n+1}(v)\,, v\big)\,.$$

Thus, $\Phi_{n-1}\big(f_n(v)\,,v\big) \le \Phi_{n-1}\big(f_{n+1}(v)\,,v\big)$ and so $f_{n+1}(v) \le f_n(v)$, for Φ_{n-1} is decreasing with respect to the first argument. By induction, the claim (7.8) is proved. It follows that $\inf\limits_n f_n = f_\infty$ and $\mathop{\&}\limits_n u_n \le u_\infty \iff u_0 \le f_\infty(v_0)$. Taking limits in (7.7) and in $f_n(0) = u_\infty$ yields

$$\Phi_\infty(f_\infty(v)\,,v) = 0 = f_\infty(0) - u_\infty\,,$$

that is, f_∞ is a solution of the system (7.3). Moreover, it is nonincreasing and continuous. Monotonicity is inherited from the monotonicity of f_n:

$$v_1' \le v_1 \Longrightarrow f_n(v_1', v_2, \ldots, v_m) \ge f_n(v_1, v_2, \ldots, v_m)\,,$$

whence (by forcing n to infinity) $f_\infty(v_1', v_2, \ldots, v_m) \ge f_\infty(v_1, v_2, \ldots, v_m)$. In turn, monotonicity ensures that each discontinuity of f_∞ is a jump. Suppose that f_∞ has a jump at a point $v \in V$:

$$\begin{aligned}
v_1' \nearrow v_1 &\Longrightarrow f_\infty(v_1', v_2, \ldots, v_m) \searrow u' \ge f_\infty(v_1, v_2, \ldots, v_m) = f_\infty(v),\\
v_1' \searrow v_1 &\Longrightarrow f_\infty(v_1', v_2, \ldots, v_m) \nearrow u'' \le f_\infty(v_1, v_2, \ldots, v_m) = f_\infty(v)\,,
\end{aligned}$$

and $u'' < u'$ with $f_\infty(v)$ somewhere in between. If $u'' \le f_\infty(v) < u'$, then

$$\begin{aligned}
v_1' \nearrow v_1 &\Longrightarrow f_n(v_1', v_2, \ldots, v_m) \searrow f_n(v) \;\;\&\;\; f_\infty(v_1', v_2, \ldots, v_m) \searrow u'\\
&\Longrightarrow f_n(v) \ge u' \Longrightarrow f_n(v) - f_\infty(v) \ge u' - f_\infty(v) > 0\,,
\end{aligned}$$

contrary to $f_n(v) \searrow f_\infty(v)$. If $f_\infty(v) = u' < f_0(v)$, then

$$\begin{aligned}
v' \nearrow v &\Longrightarrow f_n(v') \searrow f_n(v) \;\;\&\;\; f_\infty(v') \searrow u'\\
&\Longrightarrow f_n(v) \ge u' = f_\infty(v) \le f_0(v)\\
&\Longrightarrow f_n(v) - f_\infty(v) \ge u' - f_0(v) > 0\,,
\end{aligned}$$

the same contradiction. If $f_\infty(v) = u' = f_0(v)$, then $\mathop{\&}\limits_n f_n(v) = u'$, while

$$v' \searrow v \Longrightarrow f_\infty(v') \nearrow u'' \Longrightarrow f_\infty(v') \le u''.$$

Therefore, $v' \searrow v \Longrightarrow f_n(v') - f_\infty(v') \to u' - u''$ and so $f_n(v') - f_\infty(v') > (u' - u'')/2$ for all v' sufficiently close to v. On the other hand,

$$f_\infty(v') = \lim_{n\to\infty} f_n(v') \Longrightarrow f_n(v') - f_\infty(v') < (u' - u'')/2$$

for all n sufficiently large. Again, the jump hypothesis has led us to a contradiction. Thus, f_∞ is continuous.

To see that there is no other continuous solution, let x be a solution and consider the generator $\mathbf{g} : (p\,,q) \mapsto (p_+\,,q_+)$ defined as follows:

$$p_+ := F(p, q)\;,\; q_+ := G\big(x(q)\,,q\big)\,.$$

Then $(p\,,q) = (u\,,v)$ & $u = x(v)$ implies $p_+ := F(p\,,q) = F(u\,,v) = u_+$, $q_+ := G\big(x(q)\,,q\big) = G\big(x(v)\,,v\big) = G(u\,,v) = v_+$, and

$$u_+ := F(u\,,v) = F\big(x(v)\,,v\big) = x\big(G\big(x(v)\,,v\big)\big) = x\big(G(u\,,v)\big) = x(v_+)\,.$$

It follows (by induction) that $(p_0\,,q_0) = (u_0\,,v_0) \Longrightarrow \underset{n}{\&}(p_n\,,q_n) = (u_n\,,v_n)$, i.e., the generator $\mathbf{g}$ coincides with (7.1). Consequently, $x = f_\infty$. □

It is instructive to illustrate the theorem by the examples of generators from previous chapters, for which we already know the convergence domain. One such example is the generator (2.40):

$$\beta_+ := \beta\left(1+\sqrt{r}\right)\ ,\ r_+ := (r+s)^2\ ,\ s_+ := s(1+r+s)^2(0.5s+r).$$

For it, $m = 2$, $F(\beta, r, s) = \beta\,(1+\sqrt{r})$, and

$$G(\beta, r, s) = \Big((r+s)^2\,, s(1+r+s)^2(0.5s+r)\Big)$$

are clearly continuous and increasing functions on $\mathbb{R}^3_+$, $F(\beta, 0\,, 0) = \beta$ and $G(\beta, 0\,, 0) = (0\,, 0)$, that is, these F and G agree with the conditions (a)–(c). The system (7.3) for this generator becomes

$$x\big((r+s)^2\,,\ s(1+r+s)^2(0.5s+r)\big) = x(r, s)\left(1+\sqrt{r}\right)\quad \&\quad x(0\,,0) = \beta_\infty,$$

(cf. (2.41)). As stated by Proposition 2.12, this system is solved by the function

$$(r, s) \mapsto \beta_\infty \frac{\sqrt{(1-r)^2 - 2s}}{1+\sqrt{r}}\ ,\ r \ge 0 \le s \le 0.5(1-r)^2.$$

By the theorem, this is f_∞. Then the sequence $(\beta_n\,, r_n\,, s_n)$ converges if and only if

$$0 \le \beta_0 \le \frac{\beta_\infty}{1+\sqrt{r_0}}\sqrt{(1-r)^2 - 2s_0}\,.$$

So, the convergence domain $Q(\mathbf{g})$

$$= \left\{(\beta, r, s)\,\middle|\, r \ge 0 \le s \le 0.5(1-r)^2\ \&\ \beta \le \sup_{\beta_\infty > 0} \frac{\beta_\infty}{1+\sqrt{r_0}}\sqrt{(1-r)^2 - 2s_0}\right\}$$

$$= \left\{(\beta, \gamma, s)\ \middle|\ r \ge 0 \le s \le 0.5(1-r)^2\right\},$$

which is what is stated by Proposition 2.12.

Next, we apply the theorem to the generator (3.32), where (without loss of generality) $c = 1$:

$$\beta_+ := \beta(1+\gamma)\ ,\ \gamma_+ := \gamma^2 + \beta_+\delta\left(1+\gamma+\gamma^2\right)\ ,\ \delta_+ := \delta\gamma(1+\gamma)\,.$$

For it, $m=2$, $F(\beta,\gamma,\delta)=\beta+\beta\gamma$, and

$$G(\beta,\gamma,\delta)=\Big(\gamma^2+\beta\delta(1+\gamma)\left(1+\gamma+\gamma^2\right)\ ,\ \delta\gamma(1+\gamma)\Big)$$

are continuous increasing functions on $\mathbb{R}^3_+$, $F(\beta\,,0\,,0)=\beta$, and $G(\beta\,,0\,,0)=(0\,,0)$, so that assumptions (a)–(c) are satisfied. The system (7.3) takes the form of the system (3.33):

$$x\Big(\gamma^2+x(\gamma\,,\delta)\delta(1+\gamma)\left(1+\gamma+\gamma^2\right)\ ,\ \delta\gamma(1+\gamma)\Big)=x(\gamma\,,\delta)(1+\gamma)\ \ \&\ \ x(0\,,0)=\beta_\infty\,.$$

By the theorem, its solution (3.34) is f_∞:

$$f_\infty(\gamma,\delta)=\frac{(1-\gamma)^2}{\sqrt{4\gamma\delta^2+\beta_\infty^{-2}(1-\gamma)^2}+\delta(1+\gamma)}\,.$$

Then the convergence domain of the generator (3.32) consists of all nonnegative triples $(1,\gamma,\delta)$ with

$$\begin{aligned}1\le\sup_{\beta_\infty>0}\frac{(1-\gamma)^2}{\sqrt{4\gamma\delta^2+\beta_\infty^{-2}(1-\gamma)^2}+\delta(1+\gamma)}&=\frac{(1-\gamma)^2}{\delta\left(2\sqrt{\gamma}+1+\gamma\right)}\\&=\frac{(1-\gamma)^2}{\delta\left(1+\sqrt{\gamma}\right)^2}=\frac{\left(1-\sqrt{\gamma}\right)^2}{\delta}\,.\end{aligned}$$

This is the same conclusion concerning the generator (3.32) we arrived at in Section 3.3.

Consider also the generator (5.21)

$$t_+:=t+\delta\ ,\ \delta_+:=c\delta(2t+\delta+\gamma)\,,$$

where c and γ are constants. The corresponding functions $F(t,\delta)=t+\delta$ and $G(t,\delta)=c\delta(2t+\delta+\gamma)$ satisfy assumptions (a)–(c). The system (7.3) for this generator becomes

$$x\big(c\delta(2x(\delta)+\delta+\gamma)\big)=x(\delta)+\delta\ \ \&\ \ x(0)=t_\infty\,.$$

As stated by Lemma 5.7, it is solved by the function

$$\begin{aligned}f(\delta)&:=\frac{1}{2}\left(c^{-1}-\gamma-\sqrt{\left(c^{-1}-\gamma\right)^2-4c^{-1}(\delta_0-\delta)}\right)\\&=\frac{1}{2}\left(c^{-1}-\gamma-\sqrt{\left(c^{-1}-\gamma\right)^2+4c^{-1}\delta-4t_\infty\left(c^{-1}-\gamma\right)+4t_\infty^2}\right).\end{aligned}$$

So, the attraction basin of the fixed point $(t_\infty,0)$ of the generator is

$$\left\{(t\,,\delta)\ \middle|\ \delta\ge0\le t=\frac{1}{2}\left(c^{-1}-\gamma-\sqrt{\left(c^{-1}-\gamma\right)^2+4c^{-1}\delta-4t_\infty\left(c^{-1}-\gamma\right)+4t_\infty^2}\right)\right\}$$

and the convergence domain

$$\left\{(t,\delta)\,\middle|\,\delta\geq 0\leq t\leq \sup_{t_\infty>0}\frac{1}{2}\left(c^{-1}-\gamma-\sqrt{(c^{-1}-\gamma)^2+4c^{-1}\delta-4t_\infty(c^{-1}-\gamma)+4t_\infty^2}\right)\right\}$$

$$=\left\{(t,\delta)\,\middle|\,\delta\geq 0\leq t\leq \frac{1}{2}\left(c^{-1}-\gamma-2\sqrt{c^{-1}\delta}\right)\right\}$$

$$=\left\{(t,\delta)\,\middle|\,0\leq\delta\leq\left(\frac{c^{-1}-\gamma}{2}\right)^2\ \&\ 0\leq t\leq\frac{1}{2}\left(c^{-1}-\gamma-2\sqrt{c^{-1}\delta}\right)\right\}.$$

Again, we have recovered the result of Lemma 5.7.

7.3 Computation of the convergence domain

Given the sequence $(u_n\,,v_n)$, generated by the generator (7.1) from the starter $(u_0\,,v_0)$, we can define on the sequence v_n the function f (it depends on the parameter (u_0,v_0)) by setting $f(v_n):=u_n$. By (7.1),

$$\underset{n}{\&}\, f\big(G(f(v_n),v_n)\big)=F\big(f(v_n),v_n\big)\,,$$

i.e., on the sequence v_n , f satisfies the functional equation (7.2). According to Theorem 7.1, that equation has the unique continuous solution f_∞. So, the function f is in fact a restriction of f_∞ to the sequence v_n. It follows that $\underset{n}{\&}\, f_\infty(v_n)=u_n$. Thus, for all $n=0,1,\ldots$, f_∞ verifies the system

$$x\big(G(x(v)\,,v)\big)=F\big(x(v)\,,v\big)\ \ \&\ \ x(v_n)=u_n\ \ \&\ \ x(v_{n+1})=u_{n+1}\,.$$

This hints at the possibility to find f_∞ on the (m-dimensional) segment $[v_n\,,v_{n+1}]$ by application of an iterative method, capable of dealing with non-differentiable operators, to the operator equation

$$\mathbf{f}(x)(v):=x\big(G(x(v)\,,v)\big)-F\big(x(v)\,,v\big)=0\,, \tag{7.9}$$

starting from any $x_0(v)$ satisfying the initial and the end conditions $x(v_n)=u_n$, $x(v_{n+1})=u_{n+1}$. If $n>0$, the natural candidate for x_0 seems to be the solution of (7.9) on the previous segment $[v_{n-1}\,,v_n]$, since it satisfies these conditions automatically:

$$\begin{aligned}x_0(v_n)=u_n\Longrightarrow x(v_{n+1})&=x\big(G(u_n\,,v_n)\big)=x\big(G(x(v_n)\,,v_n)\big)=F\big(x(v_n)\,,v_n\big)\\&=F(u_n\,,v_n)=u_{n+1}\,.\end{aligned}$$

The case $n=0$, when the initial and the end conditions are not automatic, calls for a special effort to have them satisfied. For example, one can take for

x_0 an affine function $x_0(v) := u_0 + a^T(v - v_0)$ with the parameter a verifying $a^T(v_1 - v_0) = u_1 - u_0$. The process of piecewise approximating the solution of the system (7.2) terminates when the segment $[v_n\,, v_{n+1}]$ becomes too small.

The above discussion justifies the following generic algorithm for computation of f_∞.

1: Choose $u' > 0$ and $v_0 \neq 0$.
2: Running the generator (7.1), find k with
    $u_0 = 2^{-k}u' \Longrightarrow u_n \to u_\infty \leq u'$.
3: For all $n = 0, 1, \ldots$
    3.1: Compute (u_{n+1}, v_{n+1}) according to (7.1).
    3.2: Apply the chosen iterative method to the system

$$\mathbf{f}(x)(v) := x\big(G(x(v), v)\big) - x(v) - F\big(x(v), v\big) = 0 \;\;\&\;\; x(v_{n+1}) = u_{n+1}$$

starting from an affine function $v \mapsto u_n + a^T(v - v_n)$ satisfying

$$u_n + a^T(v_{n+1} - v_n) = u_{n+1}\,. \tag{7.10}$$

Denote the solution x_{n+1}.
    3.3: Compute $\|v_{n+1} - v_n\|$.
    3.4: If $\|v_{n+1} - v_n\| < \varepsilon$, terminate.
    3.5: Replace n by $n + 1$ and go to 3.1.
4: $x(v) := x_n(v)\ ,\ \forall\, v \in [v_n, v_{n+1}]$.

7.4 Research projects

Theorem 7.1 assumes that the functions F and G in (7.1) satisfy the conditions (a)–(c). There are however generators, which violate these conditions, most often condition (c). Such are, for example, the generators (3.36) and (4.24). So, the general model of a generator produced by a Kantorovich-type convergence analysis, which is the subject of Theorem 7.1, is not sufficiently general and should be extended to generators without property (c).

Functional equations related to majorant generators rarely admit analytical solutions as the equations (2.41), (3.38), (5.22). The most general approach to their solution is numerical through some iterative procedure. This is true not only for equations related to majorant generators, but also to functional equations in general. Numerical solution of functional equations looks to me like a blind spot in numerical analysis. However, promising approaches to the problem can be suggested. One is outlined below. It seems applicable to all operators acting on the Banach space of continuous functions, as, for example, Nemytsky's operator $\mathbf{N}(x)(t) := f(t, x(t))$ (if f is a continuous function) and

Uryson's integral operator

$$\mathbf{U}(x)(t) := \int_0^1 f\big(s, t, x(s)\big)\, ds\,.$$

Consider for example the system (7.3),

$$\mathbf{f}(x)(t) := x\big(G\big(x(t)\,, t\big)\big) - F\big(x(t)\,, t\big) = 0 \quad \& \quad x(0) = u_\infty\,, \tag{7.11}$$

where (for simplicity) $m = 1$. It is uniquely solvable by Theorem 7.1. Choose for the initial approximation to its solution an affine function $l_0(t) := bt + u_\infty$ satisfying the end condition $l_0(0) = u_\infty$, taking into account specifics of the functions F and G. A logical way to choose l_0 is to take

$$b := \arg\min_c \big\|\mathbf{f}(ct + u_\infty)\big\| = \arg\min_c \big\|cG\big(ct + u_\infty)\,, t\big) + u_\infty - F\big(ct + u_\infty)\,, t\big)\big\|\,.$$

Sometimes the value of b can be prescribed in advance, as in case of the system (4.25), when (4.25) $\Longrightarrow x\big(x(0)\big) = 0 \Longrightarrow x(0) = 0$ and $x(1) = a$, so that it is natural to choose for x_b the function $t \mapsto at$.

Let a mesh T_n :

$$0 = t_{n,0} < t_{n,1} < t_{n,2} < \ldots < t_{n,\,n-1} < t_{n,\,n} = 1$$

and an approximation l_n be given. Denote

$$f_{n,\,i} := \mathbf{f}(l_n)(t_{n,\,i})\ ,\ i = 0, 1, \ldots, n,$$

and define fl to be piecewise affine function determined by the table

t	0	$t_{n,\,1}$	$t_{n,\,2}$	$\ldots$	$t_{n,\,n-1}$	1
$fl(t)$	$f_{n,\,0}$	$f_{n,\,1}$	$f_{n,\,2}$	$\ldots$	$f_{n,\,n-1}$	$f_{n,\,n}$

that is,

$$fl(t) := f_{n,\,i-1} + \frac{f_{n,\,i} - f_{n,\,i-1}}{t_{n,\,i} - t_{n,\,i-1}}(t - t_{n,\,i})\ ,\ \forall\, t \in \big[t_{n,\,i-1}\,, t_{n,\,i}\big]\ ,\ i = 1, \ldots, n.$$

The operator $\mathbf{f}$ takes fl to $\mathbf{f}(fl) \in \mathbb{C}[0\,, 1]$, which induces the norm

$$\big\|\mathbf{f}(fl)\big\| = \max_{0 \le t \le 1} \big|\mathbf{f}(fl)(t)\big| = \max_{1 \le i \le n}\ \max_t \Big\{\big|\mathbf{f}(fl)(t)\big|\ \Big|\ t_{n,\,i-1} \le t \le t_{n,\,i}\Big\}\,. \tag{7.12}$$

This norm can be used for measuring the badness $bd(l_n)$ of the approximation l_n:

$$bd(l_n) := \big\|\mathbf{f}(fl)\big\|\,.$$

Adopting this measure, we should try to reduce it. Naturally, our attention is attracted to the segment $[t_{n,\,j-1}\,,t_{n,\,j}]$ with the greatest interior maximum in (7.12), that is,

$$j := \arg\max_{1\le i\le n}\max_t\Big\{\big|\mathbf{f}(fl)(t)\big|\ \Big|\ t_{n,\,i-1}\le t\le t_{n,\,i}\Big\}.$$

In order to reduce the badness $bd(l_n)$, we insert a new knot, say τ, into this segment, selecting it in accordance with the purpose. The addition of the knot τ to $[t_{n,\,j-1}\,,t_{n,\,j}]$ change fl on it to

$$fl_\tau(t) := \begin{cases} f_{n,\,j-1} + \dfrac{\mathbf{f}(fl)(\tau) - f_{n,\,j-1}}{\tau - t_{n,\,j-1}}(t - t_{n,\,j-1})\,, & \text{if } t_{n,\,j-1}\le t\le\tau\,,\\[2ex] \mathbf{f}(fl)(\tau) + \dfrac{f_{n,\,j} - \mathbf{f}(fl)(\tau)}{t_{n,\,j} - \tau}(t-\tau)\,, & \text{if } \tau\le t\le t_{n,\,j}\,. \end{cases}$$

Correspondingly, $\big\|\mathbf{f}(fl)\big\| = \max\limits_t\Big\{\big|\mathbf{f}(fl)(t)\big|\ \Big|\ t_{n,\,j-1}\le t\le t_{n,\,j}\Big\}$ in (7.12) is replaced by $\max\limits_t\Big\{\big|\mathbf{f}(fl_\tau)(t)\big|\ \Big|\ t_{n,\,i-1}\le t\le t_{n,\,i}\Big\}$, which is

$$\ge \min_\tau\Big\{\max_t\Big\{\big|\mathbf{f}(fl_\tau)(t)\big|\ \Big|\ t_{n,\,j-1}\le t\le t_{n,\,j}\Big\}\ \Big|\ t_{n,\,j-1}\le\tau\le t_{n,\,j}\Big\}. \tag{7.13}$$

The minimizing τ (let us call it τ_n) is added to T_n creating the next mesh T_{n+1}:

$$0 = t_{n+1,\,0} < t_{n+1,\,1} < t_{n+1,\,2} < \ldots < t_{n+1,\,n} < t_{n+1,\,n+1} = 1\,,$$

where

$$t_{n+1,\,i} := \begin{cases} t_{n,\,i}\,, & \text{if } 0\le i<j\,,\\ \tau_n\,, & \text{if } i=j\,,\\ t_{n,\,i+1}\,, & \text{if } j<i\le n+1\,, \end{cases}$$

The next approximation l_{n+1} coincides with l_n on the segments $[t_{n,\,i-1}\,,t_{n,\,i}]$, $i\ne j$, but on $[t_{n,\,j-1}\,,t_{n,\,j}]$ $l_{n+1} = fl_{\tau_n}$:

$$l_{n+1}(t) := \begin{cases} l_n(t)\,, & \text{if } t\in[t_{n,\,i-1}\,,t_{n,\,i}]\ \&\ i\ne j\,,\\ fl_{\tau_n}(t)\,, & \text{if } t\in[t_{n,\,j-1}\,,t_{n,\,j}]\,. \end{cases}$$

The badness of l_{n+1}

$$bd(l_{n+1}) = \max\left\{\begin{array}{c} \max\limits_{i\ne j}\max\limits_t\Big\{\big|\mathbf{f}(l_{n+1})(t)\big|\ \Big|\ t_{n,\,i-1}\le t\le t_{n,\,i}\Big\}\\[2ex] \max\limits_t\Big\{\big|\mathbf{f}(l_{n+1})(t)\big|\ \Big|\ t_{n,\,j-1}\le t\le t_{n,\,j}\Big\} \end{array}\right\}.$$

As $l_{n+1}(t) = l_n(t)$ for $t\in[t_{n,\,i-1}\,,t_{n,\,i}]$, $i\ne j$, the first maximum in braces $\le \big\|\mathbf{f}(fl)\big\|$, by the definition of j. Besides, $t\in[t_{n,\,j-1}\,,t_{n,\,j}] \Longrightarrow l_{n+1}(t) = fl_{\tau_n}(t)$

and so the second maximum $= \max_t \{\mathbf{f}(fl_{\tau_n})(t) \mid t_{n,j-1} \le t \le t_{n,j}\} \le \|\mathbf{f}(fl)\|$ by (7.13). It follows that $bd(l_{n+1}) \le \|\mathbf{f}(fl)\| = bd(l_n)$. Thus, this iterative method constructs the sequence of approximations l_n with monotonically decreasing (and so converging) badnesses $bd(l_n)$. The question that remains is that: is $\lim bd(l_n) = \inf_x \left\{ \|\mathbf{f}(x)\| \;\middle|\; x \in \mathbb{C}[0,1] \right\}$?
I leave it to you, the reader.

Another possible approach to functional equations of the type (7.3) takes into account the majorant generator that has induced the equation in question. In the example (7.3), it is the generator (7.1). Running this generator from the starter (u_0, v_0) creates the infinite table

u	u_0	u_1	...	u_n	...
v	v_0	v_1	...	v_n	...

where u_i are the values of the solution x of the system (7.3) corresponding to the values v_i of its argument. To approximate the solution x on the box $[v_0, v_1]$, we can apply an iterative method (like Ulm's or Broyden's) to the operator equation $\mathbf{f}(x) = 0$ starting from

$$x_0(v) := u_0 + \frac{\mathbf{f}(x)(v_1) - \mathbf{f}(x)(v_0)}{u_1 - u_0}(u - u_0)$$

and terminating when the norm of the function $v \to \mathbf{f}(x_n)(v)$ on the box $[v_0, v_1]$ becomes small. The approximation obtained is then taken for the starter of a similar process of approximation of the solution x on the box $[v_1, v_2]$ and so on.

In Section 1.6, we have shown that solving a functional equation induced by a majorant generator is equivalent to finding an invariant of that generator. It was noticed there also that each invariant itself is a solution of a functional equation. Moreover, this equation is a linear one. This observation sharpens our interest in a particular class of functional equations, linear equations. Such equations may have properties that facilitate their iterative solution. So, developing the theory of linear functional equations is a worthy research direction.

References

[1] M. Aigner. *Discrete Mathematics.* American Mathematical Society, Providence, R I, 2007.

[2] I.K. Argyros. On a mesh-independence principle for operator equations and the secant method, *Acta Math. Hung.*, **60** (1992), 7–19.

[3] J.P. Aubin, I. Ekeland. *Applied Nonlinear Analysis.* John Wiley & Sons, NY, 1984.

[4] C.G. Broyden. A class of methods for solving simultaneous equations. *Math. Comp.*, **19**(1965), 577–593.

[5] C.G. Broyden. The convergence of a class of double rank minimization algorithms: 2. The new algorithm. *J. Inst. Math. Appl.*, **6**(1970), 222–231.

[6] S. Chandrasekhar. *Radiative Transfer.* Dover, NY, 1960.

[7] J.E. Dennis, R. Schnabel. *Numerical Methods for Unconstrained Optimization and Nonlinear Equations.* Prentice-Hall Inc., Englewood Cliffs, NJ, 1983.

[8] Y. Eidelman, V. Milman, A. Tsolomitis. *Functional Analysis. An Introduction.* American Mathematical Society, Providence, RI, 2004.

[9] F. Facchinei, J.-S. Pang. *Finite-Dimensional Variational Inequalities and Complemetarity Problems.* Springer-Verlag, NY, 2003.

[10] R. Fletcher. A new approach to variable metric algorithms. *Computer J.*, **13**(1970), 317–322.

[11] A. Galperin, Z. Waksman. On pseudoinverses of operator products. *Linear Algebra and Its Applications*, **33**(1980), 123–131.

[12] A. Galperin, Z. Waksman. Newton's method under a weak smoothness assumption. *J. Comput. Appl. Math.*, **35**(1991), 207–215.

[13] A. Galperin, Z. Waksman. Regular smoothness and Newton's method. *Numer. Func. Anal. Optimiz.*, **15**(1994), 813–858.

[14] A. Galperin, Z. Waksman. Newton-type methods under regular smoothness. *Numer. Func. Anal. Optimiz.*, **17**(1996), 259–291.

[15] A. Galperin, Z. Waksman. Ulm's method under regular smoothness. *Numer. Funct. Anal. Optimiz.*, **19**(1998), 285–307.

[16] A. Galperin. Kantorovich's majorization and functional equations. *Numer. Funct. Anal. Optimiz.*, **24**(2003), 783–811.

[17] A. Galperin. On convergence domains of Newton's and modified Newton methods. *Numer. Funct. Anal. Optimiz.*, **26**(2005), 1–21.

[18] A. Galperin. Secant method with regularly continuous divided differences. *J. Comput. Appl. Math.*, **193**(2) (2006), 574–595.

[19] A. Galperin. On a class of systems of difference equations and their invariants. *J. Differ. Equations Appl.*, 13(2007), 357–381.

[20] A. Galperin. Ulm's method without derivatives. *Nonlinear Analysis.* **71**(2009), 2094–2113.

[21] A. Galperin. Optimal iterative methods for nonlinear equations. *Numer. Func. Anal. Optimiz.*, **30**(5-6)(2009), 499–522.

[22] A. Galperin. Optimal secant-type methods for operator equations. *Numer. Func. Anal. Optimiz.*, **32**(2)(2011), 177–188.

[23] A. Galperin. Optimal secant-updates of rank 1. *J. Comput. Appl. Math.*, **253**(2013), 66–79.

[24] A. Galperin. Broyden's method for operators with regularly continuous divided differences. *Journal of Korean Math. Soc.*, **52**(2015), 43–65.

[25] D. Goldfarb. A family of variable metric methods derived by variational means. *Math.Comput.* **23**(1970), 23–26.

[26] P.T. Harker, J.-S. Pang. Finite-dimensional variational inequality and nonlinear complementarity Problems: a survey of theory, algorithms and applications. *Mathematical Programming*, **48**(1990), 161–220.

[27] M.A. Hernández, M.J. Rubio. The secant method and divided differences Hölder continuous. *Appl. Math. Comput.*, **124**(2001), 139–149.

[28] M.A. Hernández, M.J. Rubio. Semilocal convergence of the secant method under mild convergence conditions of differentiability, *Int. J. Comp. Math.*, **44**(2002), 277–285.

[29] M.A. Hernández, M.J. Rubio. The secant method for nondifferential operators. *Appl.Math. Lett.*, **15**(2002), 395–399.

[30] M.A. Hernández, M.J. Rubio. A uniparametric family of iterative processes for solving nondifferentiable equations. *J. Math. Anal. Appl.*, **275**(2002), 821–834.

[31] Hernández, M.A.; Rubio, M.J.; Ezquerro, J.A. Secant-like methods for solving integral equations of the Hammerstein type. *J. Comput. Appl. Math.*, **115**(2000), 245–254.

[32] L.V. Kantorovich, G.P. Akilov. *Functional Analysis.* Pergamon Press, Elmsford, 1982.

[33] C.T. Kelley. *Iterative Methods for Linear and Nonlinear Equations.* SIAM, Philadelphia, 1995.

[34] H.Kneser. Sur un theoreme fondamental de la theorie des jeux. *C.R.Acad. Sci.*, **234**(1952), 25.

[35] M.R.S. Kulenović, O. Merino. *Discrete Dynamical Systems and Difference Equations with Mathematica.* Chapman & Hall/CRC, Boca Raton, FL, 2002.

[36] P. Lancaster, M. Tismenetsky. *The Theory of Matrices*, 2nd ed. Academic Press, London, 1985.

[37] G.G. Magaril-Ill'yaev, V.M. Tikhomirov. *Convex Analysis: Theory and Applications.* American Mathematical Society, Providence, RI, 2003.

[38] O.Mangasarian. Equivalence of the complementarity problem to a system of nonlinear equations. *SIAM J. on Applied Math.*, **31**(1976), 89–92.

[39] J.K. Moser. A new technique for construction of solutions of nonlinear differential equations. *Proc. Nat. Acad. Sci. USA*, **47**(1961), 1824–1831.

[40] J.K. Moser. A rapidly convergent iteration method and nonlinear differential equations. *Annali Scuola Norm. Sup. Pisa*, **20**(1966), 265–315, 499–535.

[41] Z. Nashed. Inner, outer, and generalized inverses in Banach and Hilbert spaces. *Numer. Funct. Anal. Optimiz.*, **9**(1987), 261–325.

[42] S.M. Nikol'sky. Fourier series with a given continuity modulus (Russian), *Doklady Acad. Nauk SSSR*, **52**(1946), 191–194.

[43] A.M. Ostrovsky. *Solutions of Equations and Systems of Equations.* Academic Press, NY, 1960.

[44] R. Penrose. A generalized inverse for matrices. *Proc. Cambridge Philos. Soc.*, **51**(1955), 406–413.

[45] R. Penrose. On best approximate solutions of linear matrix equations. *Proc. Cambridge Philos. Soc.*, **52**(1956), 17–19.

[46] G. Pimbley. Positive solutions of a quadratic integral equation. *Arch. Rational Mech. Anal.*, **24**(1967), 107–127.

[47] F.A. Potra. Sharp error bounds for a class of Newton-like methods, *Libertas Mathematica*, **5**(1985), 71–84.

[48] F.A. Potra, V. Ptak. *Nondiscrete Induction and Iterative Processes*. Pitman, London, 1983.

[49] L. Qi. On superlinear convergence of quasi-Newton methods for nonsmooth equations. *Operations Research Letters*, **20**(1997), 223–228.

[50] R.C. Robinson. *An Introduction to Dynamical Systems, Continuous and Discrete*. 2nd edition. American Mathematical Society, Providence, RI, 2012.

[51] R.T. Rockafellar. *Convex Analysis*. Princeton University Press, Princeton, 1967.

[52] M. Schechter. *Principles of Functional Analysis*, 2nd ed.. American Mathematical Society, Providence, RI, 2002.

[53] D.F. Shanno. Conditioning of quasi-Newton methods for function minimization. *Math.Comput.*, **24**(1970), 647–650.

[54] C. Shennon. A mathematical theory of communication. *Bell Systems Technical Journal*, **27**(1948), 379–423.

[55] M. Sion. On general minimax theorems. *Pac. J. Math.*, **8** (1958), 171–176.

[56] G. Teschl. *Ordinary Differential Equations and Dynamical Systems*. American Mathematical Society, Providence, RI, 2012.

[57] A.F. Timan. *Theory of Approximation of Functions of a Real Variable*. Pergamon Press, Oxford, 1963.

[58] S. Ulm. On iterative methods with successive aproximation of the inverse operator (Russian). *Izv. Akad. Nauk Est. SSR*, **16**(1967), 403–411.

[59] A.M. Yaglom, I.M. Yaglom. *Probability and Information*. Kluwer Academic Publishers, 1983.

Index